Lecture Notes in Computer Science 6100

Commenced Publication in 1973
Founding and Former Series Editors:
Gerhard Goos, Juris Hartmanis, and Jan van Leeuwen

Editorial Board

David Hutchison
  Lancaster University, UK
Takeo Kanade
  Carnegie Mellon University, Pittsburgh, PA, USA
Josef Kittler
  University of Surrey, Guildford, UK
Jon M. Kleinberg
  Cornell University, Ithaca, NY, USA
Alfred Kobsa
  University of California, Irvine, CA, USA
Friedemann Mattern
  ETH Zurich, Switzerland
John C. Mitchell
  Stanford University, CA, USA
Moni Naor
  Weizmann Institute of Science, Rehovot, Israel
Oscar Nierstrasz
  University of Bern, Switzerland
C. Pandu Rangan
  Indian Institute of Technology, Madras, India
Bernhard Steffen
  TU Dortmund University, Germany
Madhu Sudan
  Microsoft Research, Cambridge, MA, USA
Demetri Terzopoulos
  University of California, Los Angeles, CA, USA
Doug Tygar
  University of California, Berkeley, CA, USA
Gerhard Weikum
  Max Planck Institute for Informatics, Saarbruecken, Germany

Holger Giese Gabor Karsai Edward Lee
Bernhard Rumpe Bernhard Schätz (Eds.)

Model-Based Engineering of Embedded Real-Time Systems

International Dagstuhl Workshop
Dagstuhl Castle, Germany, November 4-9, 2007
Revised Selected Papers

 Springer

Volume Editors

Holger Giese
Hasso-Plattner-Institute
for Software Systems Engineering
Potsdam, Germany
E-mail: holger.giese@hpi.uni-potsdam.de

Gabor Karsai
Vanderbilt University
Nashville, TN, USA
E-mail: gabor.karsai@vanderbilt.edu

Edward Lee
University of California at Berkeley
Berkeley, USA
E-mail: eal@eecs.berkeley.edu

Bernhard Rumpe
RWTH Aachen University
Aachen, Germany
E-mail: rumpe@se-rwth.de

Bernhard Schätz
fortiss GmbH
Garching, Germany
E-mail: schaetz@fortiss.org

Library of Congress Control Number: 2010935675

CR Subject Classification (1998): D.2.9, D.2, D.3.3, C.3-4, F.3

LNCS Sublibrary: SL 2 – Programming and Software Engineering

ISSN   0302-9743
ISBN-10   3-642-16276-2 Springer Berlin Heidelberg New York
ISBN-13   978-3-642-16276-3 Springer Berlin Heidelberg New York

This work is subject to copyright. All rights are reserved, whether the whole or part of the material is concerned, specifically the rights of translation, reprinting, re-use of illustrations, recitation, broadcasting, reproduction on microfilms or in any other way, and storage in data banks. Duplication of this publication or parts thereof is permitted only under the provisions of the German Copyright Law of September 9, 1965, in its current version, and permission for use must always be obtained from Springer. Violations are liable to prosecution under the German Copyright Law.

springer.com

© Springer-Verlag Berlin Heidelberg 2010
Printed in Germany

Typesetting: Camera-ready by author, data conversion by Scientific Publishing Services, Chennai, India
Printed on acid-free paper   06/3180

Preface

The topic of "Model-Based Engineering of Real-Time Embedded Systems" brings together a challenging problem domain (real-time embedded systems) and a solution domain (model-based engineering). It is also at the forefront of integrated software and systems engineering, as software in this problem domain is an essential tool for system implementation and integration. Today, real-time embedded software plays a crucial role in most advanced technical systems such as airplanes, mobile phones, and cars, and has become the main driver and facilitator for innovation. Development, evolution, verification, configuration, and maintenance of embedded and distributed software nowadays are often serious challenges as drastic increases in complexity can be observed in practice.

Model-based engineering in general, and model-based software development in particular, advocates the notion of using models throughout the development and life-cycle of an engineered system. Model-based software engineering reinforces this notion by promoting models not only as the tool of abstraction, but also as the tool for verification, implementation, testing, and maintenance. The application of such model-based engineering techniques to embedded real-time systems appears to be a good candidate to tackle some of the problems arising in the problem domain.

Model-based development strategies and model-driven automatic code generation are becoming established technologies on the functional level. However, they are mainly applied within a limited scope only. The use of analogous modeling strategies on the system, technical, and configuration levels remains challenging, especially with the increasing shift to networks of systems, tight coupling between the control-engineering oriented and reactive parts of a system, and the growing number of variants introduced by product lines. Specific domain constraints such as real-time requirements, resource limitations, and hardware-specific dependencies often impede the acceptance of standard high-level modeling techniques and their application. Much effort in industry and academia therefore goes into the adaptation and improvement of object-oriented and component-based methods and model-based engineering that promise to facilitate the development, deployment, and reuse of software components embedded in real-time environments. The model-based development approach for embedded systems and their software proposes application-specific modeling techniques using domain specific concepts (e.g., time-triggered execution or synchronous data flow) to abstract "away" the details of the implementation, such as interrupts or method calls. Furthermore, analytical techniques (e.g., the verification of the completeness of function deployment and consistency of dynamic interface descriptions) and generative techniques (e.g., automatic schedule generation, default behavior generation) can then be applied to the resulting more abstract models to enable the efficient development of high-quality software.

Our Dagstuhl seminar brought together researchers and practitioners from the field of model-based engineering of embedded real-time systems. The topics covered included: frameworks and methods, validation, model-based integration technology, formal modeling of semantics, fault management, concurrency models and models of computation, requirements modeling, formal derivation of designs from requirements, test modeling and model-based test generation, quality assurance, design management, abstractions and extensions, and development techniques and problems of application domains. The broad spectrum of presentations clearly illustrate the prevalence of model-based techniques in the embedded systems area, as well as progress in the field.

This volume is a collection of long and short papers that survey the state of the art in model-based development of real-time embedded systems. It is composed of longer chapters that cover broad areas and short papers that discuss specific tools. The chapters are organized into sections as follows:

- **Foundations:** The chapters in this section survey general models of reactive systems, techniques, and approaches for model-based integration, and modeling and simulation of real-time applications.
- **Language Engineering:** The chapters here review metamodeling as a fundamental tool, the methods for specifying the semantics of models for dynamic behavior, and the requirements for modeling languages for real-time embedded systems.
- **Domain-Specific Issues:** Relevant issues of real-time embedded systems are discussed in this section, including the use of model-based techniques for safety-critical software, and analysis and development approaches to dependable systems.
- **Life-Cycle Issues:** These chapters discuss requirements modeling techniques for embedded systems, and the technology for model evolution and management

The short papers provide a state-of-the-art survey of existing tools that are being used in the model-based engineering of embedded real-time systems.

Finally, we would like to thank all authors and contributors to the project without whom such a large and complex project could not have been completed. It has been made possible by several researchers who supported the organizers and kept things going. In particular, we have to thank Claas Pinkernell, Markus Look, and Sven Bürger for their assistance in compiling this book. Thanks also goes to the Dagstuhl organization staff members who always make our meetings there a unique event.

Holger Giese
Gabor Karsai
Edward Lee
Bernhard Rumpe
Bernhard Schätz

Table of Contents

Part I: Foundation
---

1   Models of Reactive Systems: Communication, Concurrency, and Causality 3
Bernhard Schätz, Holger Giese
    1.1   Models and Abstraction 3
            1.1.1   Approach 4
            1.1.2   Overview 5
            1.1.3   Terminology 5
    1.2   Communication 6
    1.3   Concurrency 8
    1.4   Causality 9
    1.5   Models and Aspects 11
    1.6   Methodical Combination 12
    1.7   Conclusion and Summary 17

2   Model-Based Integration 17
Holger Giese, Stefan Neumann, Oliver Niggemann, Bernhard Schätz
    2.1   Introduction 17
    2.2   Integration 19
            2.2.1   Terminology 19
            2.2.2   Classification of Integration Problems 21
            2.2.3   Fundamental Integration Techniques 22
    2.3   State-of-the-Art Approach 29
            2.3.1   Function Development 31
            2.3.2   Function Integration 33
            2.3.3   Discussion 35
    2.4   Advanced Model-Based Solutions 36
            2.4.1   AUTOSAR 36
            2.4.2   ECHATRONIC UML 43
            2.4.3   Other Approaches 46
    2.5   Summary 48

Part II: Language Engineering
---

3   Metamodelling: State of the Art and Research Challenges 57
Jonathan Sprinkle, Bernhard Rumpe, Hans Vangheluwe, Gabor Karsai
    3.1   Metamodelling: State of the Art 57

		3.1.1	Concepts in Metamodelling	57
		3.1.2	Meta Object Facility (MOF)	61
		3.1.3	Essential MOF (EMOF)	61
		3.1.4	Eclipse Modelling Framework (EMF)	63
		3.1.5	Metamodelling of Languages	64
		3.1.6	Textual Metamodelling	65
		3.1.7	Concrete and Abstract Syntax...............	66
		3.1.8	Type System	67
		3.1.9	Merging of Metamodels.....................	70
	3.2	Metamodelling: Research Challenges		71
		3.2.1	Semantic Attachment	72
		3.2.2	Inference between Metamodels	72
		3.2.3	Evolution of Models Driven by Metamodel Evolution	73
	3.3	Conclusions...		73
4	Semantics of UML Models for Dynamic Behavior: A Survey of Different Approaches			77

Mass Soldal Lund, Atle Refsdal, Ketil Stølen

	4.1	Introduction		77
	4.2	Characterization of Scope, Main Notions, and Criteria for Evaluation		79
	4.3	Main Categories of Semantics		81
	4.4	Sequence Diagrams and Similar Notations		83
		4.4.1	Denotational Semantics.....................	85
		4.4.2	Denotational Semantics with Time	86
		4.4.3	Denotational Semantics with Probabilities	87
		4.4.4	Operational Semantics......................	87
		4.4.5	Operational Semantics with Time............	91
		4.4.6	Operational Semantics with Probabilities	91
	4.5	State Machines and Similar Notations.................		91
		4.5.1	Denotational Semantics.....................	92
		4.5.2	Denotational Semantics with Time	92
		4.5.3	Denotational Semantics with Probabilities	93
		4.5.4	Operational Semantics......................	93
		4.5.5	Operational Semantics with Time............	94
		4.5.6	Operational Semantics with Probabilities	95
	4.6	Evaluation and Comparison		95
	4.7	Summary and Conclusions............................		98

Part III: Modeling

5 Modeling and Simulation of TDL Applications 107
Stefan Resmerita, Patricia Derler, Wolfgang Pree, Andreas Naderlinger
- 5.1 Introduction 107
- 5.2 The Timing Definition Language 109
  - 5.2.1 TDL Description 109
  - 5.2.2 TDL Extensions for Control Applications 114
- 5.3 Simulation of TDL Models 117
  - 5.3.1 TDL Simulation in Simulink 117
  - 5.3.2 Using Ptolemy II 120
- 5.4 Related Work 125
- 5.5 Conclusions 126

6 Modeling Languages for Real-Time and Embedded Systems: Requirements and Standards-Based Solutions 129
Sébastien Gérard, Huascar Espinoza, François Terrier, Bran Selic
- 6.1 Introduction 129
- 6.2 Two Main Architectural Styles for Dealing with Abstraction 132
- 6.3 Modeling Needs for Real-Time and Embedded Systems Design 133
  - 6.3.1 Layering and Needs for RTES 133
  - 6.3.2 Slicing and Needs for RTES 134
- 6.4 MARTE, a Standard Real-Time and Embedded Modeling Language 136
  - 6.4.1 UML Profiling Capabilities 137
  - 6.4.2 MARTE Basics 139
  - 6.4.3 Architecture and Some Details of MARTE 140
  - 6.4.4 An Extract of the MARTE Specification 143
  - 6.4.5 Typical MARTE Usage Scenarios 145
- 6.5 Related Work 149
- 6.6 Conclusions and Perspectives 151

7 Requirements Modeling for Embedded Realtime Systems ... 155
Ingolf Krüger, Claudiu Farcas, Emilia Farcas, Massimiliano Menarini
- 7.1 Introduction and Overview 155
  - 7.1.1 What's in a Requirement? 156
  - 7.1.2 Why Requirements Engineering for ERS Is Hard 158
  - 7.1.3 Summary and Outline 166
- 7.2 Requirements Specifications and Modeling for ERS 167
  - 7.2.1 Requirements Models 167
  - 7.2.2 Programming Models 172

	7.3	Requirements Engineering Approaches: Processes and Practices	174
		7.3.1 Requirements Development and Management...	174
	7.4	Example: Failure Management in Automotive Software	179
		7.4.1 Central Locking System (CLS)	180
		7.4.2 Modeling the CLS Requirements	181
		7.4.3 Discussion	190
	7.5	Summary and Outlook	191

8 UML for Software Safety and Certification: Model-Based Development of Safety-Critical Software-Intensive Systems 201

Michaela Huhn, Hardi Hungar

	8.1	Introduction	201
	8.2	Development of Certifiable Software	203
	8.3	Safety-Related Extensions of UML....................	207
		8.3.1 The UML Profile for Developing Airworthiness-Compliant (RTCA DO-178B) Safety-Critical Software......................	208
		8.3.2 *rtUML* and the OMEGA-RT Profile...........	210
		8.3.3 Restricting UML for Specification and Programming in a Certification Context	211
		8.3.4 The UML Profile for Modeling and Analysis of Real-Time Embedded Systems (MARTE)....	213
		8.3.5 The Railway Control System Domain Profile (RCSD)	215
	8.4	Using UML in Certification-Oriented Processes.........	216
		8.4.1 Questions to Be Addressed by a Certification-Oriented Process	216
		8.4.2 Purpose and Scope of the Proposed Process	216
		8.4.3 Terms and Definitions	218
		8.4.4 Phases and Sub-processes	219
		8.4.5 The Use of UML in the Process	220
		8.4.6 Realization	221
	8.5	Verification and Validation Techniques	222
		8.5.1 General Remarks on Verification and Validation Techniques in Model-Based Development of Certifiable Software...........	222
		8.5.2 Testing	225
		8.5.3 (Formal) Verification	228
		8.5.4 Tool Support................................	229
	8.6	Conclusion	233

Part IV: Model Analysis

9 Model Evolution and Management 241
Tihamer Levendovszky, Bernhard Rumpe, Bernhard Schätz, Jonathan Sprinkle
- 9.1 Why Models Evolve and Need to Be Managed? 241
  - 9.1.1 Introduction 241
  - 9.1.2 Model Management 242
  - 9.1.3 Model Evolution 243
  - 9.1.4 Chapter Outline 243
- 9.2 Model Management 243
  - 9.2.1 Model Quality and Modeling Standards 244
  - 9.2.2 Model Transformation 249
  - 9.2.3 Model Versioning and Model Merging 252
- 9.3 Evolution 253
  - 9.3.1 Evolutionary Model Development 253
  - 9.3.2 Automating Evolutionary Transformations 255
  - 9.3.3 Semantics of Evolution 257
- 9.4 Modelling Language Evolution 259
  - 9.4.1 Syntactic Model Evolution 259
  - 9.4.2 Semantic Model Evolution 260
  - 9.4.3 Techniques for Automated Model Evolution ... 261
  - 9.4.4 Step-By-Step Model Evolution 262

10 Model-Based Analysis and Development of Dependable Systems 271
Christian Buckl, Alois Knoll, Ina Schieferdecker, Justyna Zander
- 10.1 Introduction 271
- 10.2 An Overview on Dependability 272
- 10.3 A Generic Model of Fault-Tolerant Systems 275
  - 10.3.1 System Operation without Faults 275
  - 10.3.2 Faults 277
  - 10.3.3 Fault-Tolerance Mechanism 277
  - 10.3.4 Summary: Modeling of Dependable Systems 279
- 10.4 Reliability and Safety Analysis 279
  - 10.4.1 The FMECA Method 280
  - 10.4.2 The Fault Tree Analysis Method 281
  - 10.4.3 Markov Analysis 282
  - 10.4.4 Testing and Model-Based Testing 283
  - 10.4.5 Summary: Reliability and Safety Analysis 284
- 10.5 Languages and Tool Support 284
  - 10.5.1 Models 285

		10.5.2	Implementations	288
		10.5.3	Summary: Language and Tool Support	289
	10.6		Conclusion and Research Challenges	289

Part V: Approaches

11 The EAST-ADL Architecture Description Language for Automotive Embedded Software 297
Philippe Cuenot, Patrick Frey, Rolf Johansson, Henrik Lönn, Yiannis Papadopoulos, Mark-Oliver Reiser, Anders Sandberg, David Servat, Ramin Tavakoli Kolagari, Martin Törngren, Matthias Weber

	11.1	Introduction	297
	11.2	Modeling and Analysis Capabilities of the EAST-ADL2	299
	11.3	A Small Case Study	301
		11.3.1 Vehicle Features: Vehicle Level	301
		11.3.2 Abstract Functional Description: Analysis Level	301
		11.3.3 Concrete Functional Description: Design Level	302
		11.3.4 Software Architecture: Implementation Level	304
	11.4	Related Work, Conclusions and Further Work	304

12 Fujaba4Eclipse Real-Time Tool Suite 309
Claudia Priesterjahn, Matthias Tichy, Stefan Henkler, Martin Hirsch, Wilhelm Schäfer

	12.1	Introduction	309
	12.2	Features	310
	12.3	Case Study: RailCab	313
	12.4	Conclusions and Future Work	314

13 AutoFocus 3 - A Scientific Tool Prototype for Model-Based Development of Component-Based, Reactive, Distributed Systems 317
Florian Hölzl, Martin Feilkas

	13.1	Introduction	317
	13.2	Capabilities of AUTOFOCUS 3	318
		13.2.1 Logical Architecture	318
		13.2.2 Technical Architecture	320
	13.3	Conclusion	321

14 MATE - A Model Analysis and Transformation Environment for MATLAB Simulink 323
Elodie Legros, Wilhelm Schäfer, Andy Schürr, Ingo Stürmer
- 14.1 Introduction 323
- 14.2 Approach 324
- 14.3 Application 326
- 14.4 Conclusion 328

15 Benefits of System Simulation for Automotive Applications 329
Oliver Niggemann, Anne Geburzi, Joachim Stroop
- 15.1 System Models 329
  - 15.1.1 State of the Art and AUTOSAR 330
- 15.2 System Simulation 332
- 15.3 Applications of System Simulation 334
  - 15.3.1 Specification Verification 334
  - 15.3.2 Software Component Tests 334
  - 15.3.3 ECU Tests 335
  - 15.3.4 Virtual Integration 335
- 15.4 Summary 335

16 Development of Tool Extensions with MOFLON 337
Ingo Weisemöller, Felix Klar, Andy Schürr
- 16.1 Introduction 337
- 16.2 History and Overview of Features 338
  - 16.2.1 MOF Editor and Code Generation for MOF Models 338
  - 16.2.2 Additional Frontends 339
  - 16.2.3 Model Transformations 339
  - 16.2.4 Triple Graph Grammar Editor 339
- 16.3 Usage Scenarios 340
  - 16.3.1 Tool Adapters 340
  - 16.3.2 Model Analysis and Repair 341
  - 16.3.3 Integration Framework 341
- 16.4 Conclusions and Future Work 342

17 Towards Model-Based Engineering of Self-configuring Embedded Systems 345
DeJiu Chen, Martin Törngren, Magnus Persson, Lei Feng, Tahir Naseer Qureshi
- 18.1 Introduction 345
- 18.2 Capabilities 346
- 18.3 Case Study 350
  - 18.3.1 Architecture Modelling with UML 350
  - 18.3.2 Verification and Validation through Analysis ... 351
  - 18.3.3 Run-Time Models 352
- 18.4 Conclusions and Future Work 352

18 Representation of Automotive Software Description Means in ASCET 355
Ulrich Freund
- 18.1 Introduction 355
- 18.2 Overview of Design Means for Automotive Software Design 356
  - 18.2.1 Description Means for Control Engineering..... 356
  - 18.2.2 Description Means for Software Engineering.... 356
- 18.3 Integration of the Design Approaches in ASCET 357
  - 18.3.1 Classes 358
  - 18.3.2 Modules 358
  - 18.3.3 Model-Types 359
  - 18.3.4 Tasks 359
  - 18.3.5 Implementations: Integer Arithmetic and Memory Section 359
  - 18.3.6 Codegeneration Approach................... 360
- 18.4 Conclusion 360

19 Papyrus: A UML2 Tool for Domain-Specific Language Modeling 361
Sébastien Gérard, Cédric Dumoulin, Patrick Tessier, Bran Selic
- 19.1 Introduction 361
- 19.2 Capabilities... 362
  - 19.2.1 Overview 363
  - 19.2.2 Global Architecture and Design Tenets 363
  - 19.2.3 UML2 Graphical Modeling Capabilities........ 364
  - 19.2.4 Building DSL Tools Profiling the UML2 366
- 19.3 Case Study 366
- 19.4 Conclusions and Future Work 367

20 The Model-Integrated Computing Tool Suite 369
Janos Sztipanovits, Gabor Karsai, Sandeep Neema, Ted Bapty
- 20.1 Introduction 369
- 20.2 Components of the MIC Tool Suite 370
  - 20.2.1 The Generic Modeling Environment (GME).... 370
  - 20.2.2 Transforming the Models: UDM and GReAT ... 371
  - 20.2.3 Integrating Design Tools: The Open Tool Integration Framework 372
  - 20.2.4 Design Space Exploration.................... 372
- 20.3 Application Example: Vehicle Control Platform 373
- 20.4 Conclusion 374

21 Application of Quality Standards to Multiple Artifacts with a Universal Compliance Solution 377
Tibor Farkas, Torsten Klein, Harald Röbig
    21.1   Introduction 377
    21.2   Idea: Meta-modeling for Constraint Definition.......... 378
    21.3   Approach: Universal Compliance Achievement 379
    21.4   Case Studies: Compliance with Modeling Standards..... 381
    21.5   Conclusion 382

Author Index 385

Part I

Foundation

1 Models of Reactive Systems
Communication, Concurrency, and Causality

Bernhard Schätz[1] and Holger Giese[2]

[1] fortiss GmbH, München, Germany
schaetz@fortiss.org
[2] Hasso Plattner Institute at the University of Potsdam, Germany
holger.giese@hpi.uni-potsdam.de

Abstract. In this chapter, *communication*, *concurrency*, and *causality* are introduced as basic aspects of reactive systems together with different levels of abstraction for each aspect, giving prominent examples of specific models as specifically useful combinations. By relating models along different dimension, we show how to set up development processes allowing not only to support step-wise adding of implementation details, but also to treat different aspects of a system in isolation and to combine the results, leading to a fork-and-join approach.

1.1 Models and Abstraction

Abstraction is a key attribute of modeling: by removing unnecessary and unwanted details, the complexity of the modeled object is reduced, allowing to effectively engineer systems by successively building more detailed models of the system under development. Therefore, a model-based development process is essentially shaped by the aspects removed or retained by the choice of models used in the development process. To build useful stacks of models, compatible and complementary layers of abstractions are necessary.

When considering models for embedded – or more general reactive – software systems, three central aspects can be distinguished: *communication*, *concurrency*, and *causality*. For each aspect, layers of abstraction can be defined, removing more details from the modeled (software) system; constructing specific combinations of abstractions for each aspect leads to models for specific purposes. Examples for the layers of *models of communication* are implicit variable based communication between co-routines, synchronous handshake between concurrent processes, and buffered message communication between concurrent processes. Examples for layers of *models of concurrency* are models using fairness sets to characterize behavior, behavior characterized by continuous stream-processing functions, and behaviors characterized by sets of finite traces. Examples for layers of *models of causality* are dense-time models, synchronous models, and linear models.

Each aspect heavily influences the usability of the development process: Communication influences its *functional* composability, concurrency its *operational* composability, and causality its *temporal* composability.

Table 1.1. Aspects of Reactive Systems

Development Step	Modeling Aspect	Abstraction
Composition: Properties of components still hold after combing them to a system	**Communication:** How are behaviors of components combined to describe behavior of a system?	Interference between systems
Modularization: Properties of components still hold after hiding internal structure'	**Concurrency:** How are observations combined to describe a behavior?	Scheduling of processes
Refinement: Properties of components actions still hold after refining actions'	**Causality:** How are actions combined to describe an observation?	Timing of actions

1.1.1 Approach

As shown in Table 1.1, the above introduced central aspects are related to principles of system description: **Communication** is related to the *composition* of system out of components and the interaction between these components. **Concurrency** is related to the *abstraction* from the internal structure of a system and the scheduling of internal activities. **Causality** is related to *refining* the interaction of a system and the dependencies between elements of an observation.

For each of these aspects, models can be grouped into few different basic classes, characterizing elementary categories of these aspects. Each category describes a different level of abstraction from concrete implementations as found in models of reactive systems. The corresponding classes are ordered concerning their capability to support these aspects. For each aspect only three essential level of abstraction are distinguished, avoiding an over-sophisticated classification of models while identifying the major variations.

The levels of abstractions are ordered concerning their reduction of (unnecessary) details. Here, we especially focus on abstractions that are relevant concerning the development method, with more abstract classes removing 'pathological cases' expressible within the more concrete classes. Thus, the abstraction with respect to *communication* removes synchronization problems like buffer limitations; abstraction with respect to *concurrency* removes scheduling problems like weak fairness; and abstraction with respect to *causality* removes timing problems like unclocked executions. Using more abstract models allows to concentrate on the essential design issues by ignoring problems, which can be caused by choosing the wrong or avoided by choosing the right implementation. Therefore, these levels of abstraction support a *methodical exploration of the design space*, by starting a more abstract level of system design and stepwise adding additional design details, finally arriving at a implementation in form of a concrete model.

Of course, when considering fine-grained mathematical classifications more complex orders are needed, as, e.g., [1] shows for behavioral models. Since, however, in this chapter the focus is put on the methodical principles behind these formalisms, these aspects are restricted to the most basic classes, concentrating on the aspect of combining them.

1.1.2 Overview

After providing some basic terminology in Subsection 1.1.3, in the following Sections the three core aspects of reactive systems are considered, which allow to classify models of reactive systems:

- Section 1.2 introduces the characteristics of *communication and interaction*
- Section 1.3 the characteristics of *concurrency and scheduling*
- Section 1.4 of *causality and timing*.

Since these classifications only cover a specific point of view for reactive systems, Section 1.5 shows examples of classification of specific models in all three dimensions.

As this classification allows to relate different models in a common scheme, it also provides a classification of different layers of abstractions of a system under development. Section 1.6 therefore discusses how these different abstractions can be combined and arranged to support a step-wise approach to design-space exploration.

Section 1.7 finally recapitulates the essential elements of models of reactive systems and relates them to the development of integrated and embedded systems.

1.1.3 Terminology

To compare the models of computation in the following, a short informal list of general concepts is introduced, needed to explain the aspects of communication, concurrency, and causality:

Interface: An interface separates a system from its environment, allowing to hide internal information and thus modularizing a system.
State: A state of a system assigns values to the variables of the system.
Component: A component is a unit of a system capsuling a state and supplying an interface.[1]
Observation: An observation describes the sequence of states observable at the interface of a system through the execution of a system, thus allowing to describe behavior abstracted from aspects hidden inside the interface.
Behavior: The behavior defines the collection of all possible observations of a system, thus allowing to define the construction of complex behavior by combining the respective observations.

Furthermore, besides providing those basic concepts, a model must support mechanisms for its modular construction:

---
[1] Note that a component itself can also be seen as a system, interpreting the remaining components as part of the environment of that component.

Composition: By means of composition, sub-parts of a system are combined into the overal systems, thus allowing to decompose a complex system into simpler parts.
Abstraction: Abstraction restricts an interface of a system, allowing to render parts of a system invisible to its environment.

Finally, it must be possible to compare content models of different products:

Refinement: By means of refinement, the behaviors of systems can be compared and related, making, e.g., one behavior a generalization of the other.

1.2 Communication

The aspect of communication deals with the different ways information can be exchanged between the interfaces of components, allowing a component to influence and be influenced by the state of another component. Via communication, the complex behavior of a system can be modularized – or *functionally decomposed* – into the simpler behaviors of the components of that system.

Therefore, as mentioned above, the corresponding methodical aspect of communication is the issue of *functional compositionality*, i.e., the capability to deduce the observations about the exchanged information of a composed system from the observations of its components. Since communication deals with the functional view of a system, issues like communication events or synchronization of executions play an important role here.

In interactive, or more general, reactive systems, the possibility to interact with its environment is an essential characteristic of a system. Observations capture these interactions. In denotational models, (e.g., traces [2], failures [3], or stream processing functions [4]), observations are defined by the sequences of interface events of these components; in algebraic ([5]) or operational models (e.g., CCS [6], I/O-Automata [7]) observations are defined in terms of events enabled in the current state and triggering transitions to further states.

In this dimension, we consider the following three levels of abstraction:

Implicit communication: This form of communication corresponds to implicit exchange of information, e.g., by using variables shared between the system and the environment. Since no explicit communication mechanism is used and both system and environment can read and write a variable, the environment can change the (shared) variables unnoticed by the component. Therefore, compositionality is dependent on the well-behavedness of the environment, e.g., by changing the variables only at well-defined times. Concerning modeling, this form is used, e.g., in approaches like co-procedures [8] or UNITY [9]; concerning implementations, this form is used in paradigms like co-rountines and thread-programming.
Explicit event based synchronous communication: This class of models offers an explicit communication mechanism enforcing a synchronization between the environment and the system, thus avoiding an unnoticed change of

information. However, the synchronization imposed is undirected, since there is no designated sender and receiver: Synchronization between the communicating partners takes place by the agreeing on a communication they are all ready to accept. Therefore, when embedding a system into its environment, in this model the possibility of blocking has to be considered.

Concerning modeling, this from of communication is found, e.g., in TCSP [3] or CCS [6], supporting only immediate synchronization between sender and received (i.e., zero-buffer communication); concerning implementations, this form is used, e.g., in remote procedure calls (RPC) or the client-server communication found, e.g., in AUTOSAR [10].

Explicit message based asynchronous communication: In this class of models there is an explicit communication mechanism with a clear distinction between sender and receiver concerning the synchronization of the information exchange. The communicating partners are decoupled, i.e., while the receiver has to wait until a message is available to be read, the sender can write a message without delay. Therefore, when embedding a system into its environment, no (output) blocking has to be considered. System and environment are always ready to accept a message.

Examples for this model are semantics for asynchronous circuits like [2], reactive modules [11], or stream processing functions like [4]; concerning implementation, this form is used, e.g., in (unbounded) buffered communication as used in SDL [12] or in queued communication with buffer-overwrite as used in OSEK-COM [13].

Note that increasing modularity is related to increasing abstraction from restrictions concerning compositionality: from compositionality with respect to freedom of interference (implicit), via compositionality with respect to deadlock (synchronous), to unrestricted compositionality (asynchronous). Therefore, with respect to functional compositionality, these classes provide layers of abstraction with implicit communication as the least abstract class of models and the explicit message based asynchronous communication as the most abstract class of models.

This can be methodically exploited in a design exploration process. Thus, high-level functional design often starts with loosely coupled components using asynchronous communication, e.g., with unbounded buffers, as model of computation. In a later development step, additionally, the size of buffers is limited, e.g., based on bounds of the execution order or relative speed of the components of the system. In a subsequent design step, these buffers are implemented via access routines to shared dedicated input or output variables, which can be easily implemented using a shared memory if implemented on a shared (single-/multi-core) processor or a shared communication medium if implemented on a multi-processor system.

1.3 Concurrency

The aspect of concurrency deals with the different ways the behavior of a system – described in form of its executions – can be obtained from the behaviors of its components. The executions of the system is obtained by scheduling the executions of the components of this system, thus composing sequence of states of the components to sequences of states of the system. Via concurrency, the complex behavior of a system can be abstracted out of – or *operationally composed* from – the behaviors of the components of that system.

Therefore, as mentioned before, the corresponding methodical aspect of concurrency is the issue of *operational compositionality*; i.e., the capability to deduce observations about the order of computation of a composed system from the observations about the components of this system. Since concurrency is related to the operational view of the system, issues like explicit parallelism via true concurrency of events as well as fairness of observations play an essential role here.

In parallel, or more generally, reactive systems, the possibility to perform (potentially) infinite runs is an important characteristic of a system. In denotational models (e.g., traces [2], failures [3], or stream processing functions [4]), observations are directly defined as the core elements of the models; in algebraic (e.g., [5]) or operational models (e.g., CCS [6], I/O-Automata [7]) observations are defined in terms of states and possible transitions from these states. Since these formalisms describe (potentially) nonterminating systems, infinite observations are included in the behavior of a system.

Concerning the issues of concurrency, we obtain the following range:

General Fairness: This class of models makes no assumptions about the scheduling of actions to form observations about the behavior of a system. Since no restrictions are imposed on the scheduling of these concurrent components in a composed system, these models allow to describe all forms of 'sound'/'strongly fair' as well as 'pathological'/unfair' executions, independently of the form of concurrency used.

Examples for this class of concurrency are TLA [14], state-machine approaches using fairness constructs like I/O automata [7], or general trace-based models like [5]. Concerning implementations, this class of concurrency is most often found in parameterized operating systems leaving the scheduling strategy to the designer.

Weak Fairness: This class of models supports the treatment of concurrent systems, ensuring the fair combination of executions of the (concurrent) components to the executions of the composed systems. Thus, for a single component, its infinite observations are directly deduced from fair and unfair extension of finite observations, leading to an admissible or continuous behavior for sequential executions. By supporting an explicit distinction between a sequential or parallel execution, this class provides weak fairness for concurrent components.

Concerning modeling, examples for this form of concurrency are models with 'true concurrency' in form of infinite observations like continuous stream processing functions [4] or synchronous languages like [15]. Concerning implementations, this form of concurrency is used in operating systems providing a weak fair scheduling strategy like the task queues used in OSEK/VDX [16].

Finite Observations: This class of models does not deal with the distinction of different forms of parallelism at all, ignoring the issue of fairness altogether. Therefore these models have no need for or do not allow the description of infinite behavior; they do not include a distinction between arbitrary sequentialization and parallel execution.

Concerning modeling, this form of concurrency is found in models based on finite traces like TCSP [3] or receptive processes [17] with the possibility of divergence, operational models with classical structural rules like CCS [6], or in approaches based on finite state machines like AutoFocus [18]. Concerning implementations, this form of concurrency is used in thread-models with unspecified fair scheduling strategies like the Java Virtual Machine.

The methodical aspect of this dimension is the increasing abstraction from scheduling details: from arbitray forms of scheduling allowing different forms of fairness (general fair), via a fair scheduling of parallel systems (weak fair), to scheduling treating parallel systems analogously to sequential ones (finite). Therefore, with respect to operational composability, these classes provide layers of abstraction with general fairness as the least abstract class of models and the finite models as the most abstract class of models.

This can be methodically exploited in a design exploration process. Thus, high-level functional design often starts with computations ignoring aspects of concurrency during black-box-specification, e.g., using finite models of computation. In a later development step, additionally, a glass-box-design is added, e.g., assuming a concurrent execution of the components of the system. In a subsequent design step, these components are deployed to processes with explicit scheduling assumptions, which can be easily implemented using a network of concurrently running distributed processing units, each one sequentially scheduling of these components.

1.4 Causality

The aspect of causality deals with the different possibilities of describing the behavior of a system through combining the actions performed by a system to form executions. Via *causality*, the complex behavior of a system can be refined – or *temporally decomposed* – into the behaviors of its constituents.

Thus, the corresponding methodical aspect of causality is the issue of temporal compositionality, i.e., the capability to deduce observations about the ordering of actions of a composed system from the observations of its components. As causality is related to the temporal view of a system, here the timing of actions and their relative ordering play an essential role.

In timed, and more general, reactive systems, the capability to describe the temporal dependencies between the interactions performed by a system is a core property of a modeling approach. In denotational models (e.g., traces [2], synchronous traces [19], or dense traces [20]) causality of actions is directly reflected in the ordering relations imposed on the observations forming the core elements of the models. In algebraic (e.g., [5]) or operational models (e.g., [21], [22]), causality is reflected in the ordering of events imposed by the transitions. Since those relations describe the unfolding of interactions in executions, those models always contain some (explicit or implicit) aspect of time.

Concerning causality we obtain the following range of this dimension:

Dense Ordering: This class of models imposes no restriction of the ordering of interaction events in the construction of observations: Between each two interactions there may be an unlimited number of causally dependent interaction events, with those two interactions forming a lower and upper bound of this chain of events. Thus, this class allows to model continuous time observations by assigning interactions to arbitrary temporal instances. Furthermore, in this class of models, there may be circular dependencies between interaction events, allowing to describe a mutual cause-and-effect relation between events.

Examples for this class of causality are dense traces as used in [20] or continuous-time streams [4]. Concerning implementations, this class is most suitable for the description of non-digital hardware as found, e.g., in hybrid systems.

Clocked Ordering: This class of models restricts the occurrence of events to fixed time instances, leading to models for discrete time or clocked systems: Between two interactions there is only a limited number of causality dependent observations, with these two interactions forming the least and greatest element of this chain of dependent events. Since, however, no further restriction is imposed, causality ordering may still form circular dependencies, allowing *perfect synchrony* between the causing and the effected event. Concerning modeling, this form of causality is used in synchronous traces [19] or some Statecharts variants [23]. Concerning implementations, this class is most suitable for clocked circuits or – generally by avoiding circular dependencies[2] – synchronous languages like Esterel [15].

Linear Ordering: This class of models imposes a linear ordering of events on the causality relation, by allowing only a finite number of dependent interactions between two events, and furthermore excluding causal loops. Concerning modeling, examples for this form of causality are trace-based (e.g., asynchronous circuits [2]) or state-based history semantics (TLA [14]). Concerning implementations, this approach is used, e.g., in time-triggered approaches.

The methodical aspect of this dimension is the increasing abstraction from timing aspects: From metric models explicitly dealing with real time (dense ordering), via sequentialization abstracting from the passing of time between events (clocked ordering), to restricted causality additionally separating cause and effect (linear ordering). Therefore, with respect to temporal composability, these

---

[2] The restricted class of synchronous languages without causal loops actually therefore uses a linear ordering.

classes provide layers of abstraction with dense ordering as the least abstract class of models and the linear models as the most abstract class of models.

Like in the case of communication and concurrency, this can be methodically exploited in a design exploration process. To that end, in high-level functional design often functional models without aspects of timing are used, e.g., using linear models of computation. In a later development step, additionally, timing aspects are added, enforcing a timed interaction between system and environment, e.g., by using a clocked execution scheme with logical time. In a sub-sequent design step, this model is enriched to a real-time model by including aspects like execution times, which can be effectively implemented on (digital) hardware.

1.5 Models and Aspects

The different aspects of computational models of reactive systems – introduced in Sections 1.2, 1.3, and 1.4 – correspond to a specific point of view when considering the characteristics of a model. Obviously, therefore, by classifying a model for reactive systems for each of these aspects, an overall classification of the model can be obtained.

To clarify the construction of Table 1.2, we use Esterel, Focus, and TCSP for illustration.

Esterel [15] uses signals to describe system states. An execution of a system consists of a sequence of clocked computing rounds, all signals computed perfectly synchronous, leading to a clocked model. Furthermore, communication is "message asynchronous", as the sender does not have to wait for the receiver. As Esterel uses infinite runs and without any explicit fairness conditions, it belongs to the class of weak fair systems.

Classical FOCUS [4] defines each component as a stream processing function. Such a stream implies a linear order on messages, so we classify the causality modeling as "strict sequentialization". Message delivery is not influenced by the receiver and therefore the communication mode is asynchronous. As Focus includes infinite histories of interaction to describe the behavior of a system, it is classified as a weak fair system.

In *TCSP* message exchange is realized by general synchronizing events; thus it is classified as using synchronous event-based communication. Since additionally TCSP considers only finite runs, it is classified as finite. Finally, since all interactions take place in a sequentialized fashion, it is classified as a linear model.

Using the same scheme of classification, we classify formalisms as shown in Table 1.2: TCSP [3], CCS [6], I/O-Automata [7], Asynchronous Traces [2] Esterel [15], Focus [4], TLA [14], Co-routines [8], fair process algebra [24] and Receptive Processes [17]. We also consider timed variants of some formalisms: Timed CSP [25], Timed Focus[4] and Timed TLA [26].

When combining aspects to form a specific modeling formalism, the question arises whether these combinations lead to reasonable models. While basically, from a technical point of view, arbitrary classifications can be chosen, Table 1.2 shows that from a methodical point of view, the selection depends on the levels

Table 1.2. Classification according to Communication, Concurrency, and Causality

Concur-rency	Communication	Causality		
		Dense	Clocked	Linear
General	Implicit	Timed TLA		TLA
	Synchronous			
	Asynchronous		I/O-Automata	
Weak	Implicit			UNITY
	Synchronous	Timed CSP/CCS	Fair Process Alg.	
	Asynchronous	Timed Focus	Esterel, Focus	Traces
Finite	Implicit			Co-routines
	Synchronous			TCSP/CCS
	Asynchronous			Receptive Processes

of abstraction needed in the application. Many formalisms provide an overall medium level or high level of abstraction, trying to balance low complexity of modeling with high simplicity of implementation.

The classification of Table 1.2 also shows that there is a justification for a wide range of different models, since specific combinations of aspects lead to models specially suited for application domains, from low-level models of unclocked, analogue circuits to high-level models of untimed dataflow functions. This, in turn, leads to the question of combining models of different levels of abstraction, discussed in the following section.

1.6 Methodical Combination

The different classes of models provide tailor-made views of a system under development. Thus, a asynchronous, weak, and clocked model may be most suited to describe a data-flow module of a system. On the other hand, a synchronous, finite, and linear model may be most suited to describe the event-handling of a system. Therefore, models can be used to describe the *same system at different levels of abstraction* as well as *different parts of the system at different levels of abstraction*. The first case corresponds to the use of homogeneous models, arranged in a stepwise development process. The second case corresponds to a combination of heterogeneous models.

In the first, *homogeneous* case, different classes of models are arranged in a development approach to support views of a system at different levels of abstraction. Here, the classification can be used to simplify the development process by providing *layers of abstractions* for a system under development. On the one hand, a more abstract model allows to focus on central aspects of a system, like the data flow between components, while ignoring other aspects, like limited buffers between communicating components, live-locks of concurrent processes, or respective timing of events. On the other hand, a more concrete model supports a simpler implementation on the target platform, since these limiting factors like buffer size, scheduling, or timing are explicitly considered in the model.

Here, the ordering within the dimensions can help to find appropriate models and corresponding refinement steps stating which properties must be considered explicitly (e.g., buffer sizes when moving from asynchronous to synchronous communication).

Therefore, to provide a *methodical development process supporting step-wise design space exploration*, these models mut be arranged in layers enabling smooth transitions between the layers by adding additional details without invalidating established properties of the system under development. Here, the classification scheme of Section 1.5 can help by providing levels of abstractions of classes of models, as well as systematic transitions between them. By integrating abstractions in a useful development process, leading from high-level abstract models to low-level implementations, abstractions can be taken away (e.g., adding buffers, adding access schedules) in a stepwise fashion, ensuring the properties on the lower level that have already been established on the higher level.

This approach is, e.g., used in the Metropolis design methodology [27]. Here, during subsequent design steps, additional implementation details concerning synchronization of shared resources, scheduling policies, and timing constraints are added to the system under development.

In the second, *heterogeneous* case, the system under development is constructed by combining components of the system formalized by different classes of models. Here, to obtain an integrated description of the system, these models must be interpreted in a common framework. Here, too, the classification scheme can help to provide such a common framework.

Since the scheme of Section 1.5 allows to classify models according to the three different dimensions, and each dimension provides a simple (linear) order, a partial order for all classes of models in this scheme is provided For two arbitrary classes of models, a common least abstract class can be identified by using the least abstract class in each dimension, thus providing a common framework allowing to capture the model of each part of the system.

The latter approach is, e.g., chosen in the Ptolemy II framework ([28], [29]). Ptolemy II is a Java based, network integrated framework, where multiple models of concurrency, communication, and causality can be mixed hierarchically. To support this form of combination, e.g., to provide a joint simulation of all models, the framework uses the joint least abstract model of computation. Obviously, due to the rich set of models provided by the framework, this is the implementation model with unrestricted communication, concurrency, and causality. Each computational model provided by Ptolemy —— and using concepts from high levels of abstraction is therefore interpreted in terms of this base model, allowing their immediate integration.

1.7 Conclusion and Summary

The classification of models for reactive systems according to the level of abstraction concerning communication, concurrency, and causality allows to relate different models along these dimensions. By means of this relation, the differences and commonalities between models can be identified to select a common concretization, thus supporting approaches from joint simulation of models to a guided design-space exploration.

Obviously, the introduced classification scheme only considers aspects of functionality, operation, and timing. In the development of embedded systems, often additional aspects play an important role, especially typical 'non-functional' properties like required footprint (e.g., in terms of FPGA elements or ASIC gates), consumed energy, or cost of production. Since current development processes generally use a two-stage approach, separating the 'functional' from the 'non-functional' models, here the focus was put on level of abstractions for the first aspect. However, by providing a more tight integration of both aspects, even more flexibility can be added to the exploration of the design space.

References

[1] van Glabbeek, R.J.H.: Comparative concurrency semantics and refinement of actions. Technical Report 109, Centrum voor Wiskunden en Informatica, CWI Tracts (1996)

[2] Dill, D.L.: Trace Theory for Automatic Hierarchical Verification of Speed Independent Circuits. In: ACM Distinguished Dissertations. The MIT Press, Cambridge (1989)

[3] Hoare, C.A.R.: Communicating Sequential Processes. Prentice-Hall International, Englewood Cliffs (1985)

[4] Broy, M., Stølen, K.: Specification and Development of Interactive Systems: FOCUS on Streams, Interfaces, and Refinement. Texts and Monographs in Computer Science. Springer, Heidelberg (2001)

[5] Bergstra, J.A., Ponse, A., Smolka, S.A.: Handbook of Process Algebra. Elsevier, Amsterdam (2001)

[6] Milner, R.: Communication and Concurrency. Series in Computer Science. Prentice-Hall, Englewood Cliffs (1989)

[7] Lynch, N., Tuttle, M.: An Introduction to Input/Output Automata. CWI Quarterly 2(3), 219–246 (1989)

[8] Owicki, S., Gries, D.: An Axiomatic Proof Technique for Parallel Programs. Acta Informatica 14 (1976)

[9] Chandy, K.M., Misra, J.: Parallel Program Design - A Foundation, 2nd edn. Addison-Wesley, Reading (May 1989)

[10] AUTOSAR GbR: Autosar Specification of RTE Software, Version 1.0.1 (2006)

[11] Alur, R., Henzinger, T.A.: Reactive modules. Formal Methods in System Design: An International Journal 15(1), 7–48 (1999)

[12] CCITT: Functional Specification and Description Language (SDL) Criteria for Using Formal Description Techniques, FDTs (1989)

[13] OSEK/VDX Group: OSEK/VDX-COM 2.2 Communication Specification (2000)

[14] Lamport, L.: Verification and Specification of Concurrent Programs. In: de Bakker, J.W., de Roever, W.-P., Rozenberg, G. (eds.) REX 1993. LNCS, vol. 803, pp. 347–374. Springer, Heidelberg (1994)
[15] Berry, G.: The Esterel v5 Language Primer. Technical report, INRIA (July 2000), http://www-sop.inria.fr/meije/esterel/esterel-eng.html (accessed August 19, 2002)
[16] OSEK/VDX Group: OSEK/VDX-COM 2.2 Communication Specification (2005)
[17] Josephs, M.B.: Receptive process theory. Acta Informatica 29(1), 17–31 (1992)
[18] Schätz, B.: Mastering the Complexity of Embedded Systems - The AutoFocus Approach. In: Kordon, F., Lemoine, M. (eds.) Formal Techniques for Embedded Distributed Systems: From Requirements to Detailed Design. Kluwer, Dordrecht (2004)
[19] Berry, G.: Synchronous Languages for Reactive Systems: Styles, Semantics, Implementations. In: Symposium on Principles of Programming Languages, ACM SIGPLAN-SIGACT (1993)
[20] Henzinger, T.A.: Masaccio: A Formal Model for Embedded Components. In: Watanabe, O., Hagiya, M., Ito, T., van Leeuwen, J., Mosses, P.D. (eds.) TCS 2000. LNCS, vol. 1872, pp. 549–563. Springer, Heidelberg (2000)
[21] Alur, R., Dill, D.L.: Automata for modeling real-time systems. In: Paterson, M. (ed.) ICALP 1990. LNCS, vol. 443, pp. 322–335. Springer, Heidelberg (1990)
[22] Maler, O., Manna, Z., Pnueli, A.: From timed to hybrid systems. In: Huizing, C., de Bakker, J.W., Rozenberg, G., de Roever, W.-P. (eds.) REX 1991. LNCS, vol. 600. Springer, Heidelberg (1992)
[23] von der Beeck, M.: Comparison of Statecharts Variants. In: FTRTFT 1994 and ProCoS 1994. LNCS, vol. 863. Springer, Heidelberg (1995)
[24] Parrow, J.: Fairness properties in process algebra with applications in communication protocol verification. PhD thesis, Uppsala University (1985)
[25] Davis, J., Schneider, S.: An Introduction to Timed CSP. PRG- 75, PRG Programming Research Group, Oxford (1989)
[26] Abadi, M., Lamport, L.: An old-fashioned recipe for real time. ACM Transactions on Programming Languages and Systems 16(5), 1543–1571 (1994)
[27] Gössler, G., Sangiovanni-Vincentelli, A.L.: Compositional Modeling in Metropolis. In: Sangiovanni-Vincentelli, A.L., Sifakis, J. (eds.) EMSOFT 2002. LNCS, vol. 2491, pp. 93–107. Springer, Heidelberg (2002)
[28] Ptolemy II Website (2008), http://ptolemy.eecs.berkeley.edu
[29] Zhou, Y., Lee, E.A.: Causality interfaces for actor networks. Transactions on Embedded Computing Systems 7(3) (2008)

2 Model-Based Integration

Holger Giese[1], Stefan Neumann[1], Oliver Niggemann[2], and Bernhard Schätz[3]

[1] Hasso Plattner Institute at the University of Potsdam, Germany
{holger.giese,stefan.neumann}@hpi.uni-potsdam.de
[2] Fraunhofer IOSB - Competence Center Industrial Automation, Lemgo, Germany
oliver.niggemann@iitb.fraunhofer.de
[3] fortiss GmbH, München, Germany
schaetz@fortiss.org

Abstract. The integration of different development activities and artifacts into a single coherent system is a major challenge for the development of complex embedded real-time systems. For complex software the functional integration alone is a major undertaking, in the case of embedded real-time systems we in addition have to cope with all the affected system characteristics such as real-time behavior, resource consumption, and behavior in the case of failures.

In this chapter we will discuss the state-of-the-art of model-based integration. Therefore, we will clarify the terminology concerning integration, provide a classification of the integration challenges for complex embedded real-time systems, and outline the fundamental techniques employed to cope with the integration challenges. This framework is then used to explain the current standard practice concerning integration of hardware and software for functional development as well as function integration. Furthermore, a number of advanced proposal how to address some of the remaining integration challenges such as AUTOSAR and MECHATRONIC UML using model-based concepts are presented using the framework.

2.1 Introduction

One of the major challenges for the development of complex embedded real-time systems is the integration of different development artifacts into a single coherent system. The integration problems we have to face are exacerbated even further as today's advanced embedded real-time systems tend to contain more functionality than in former times, are often expected to exhibit adaptive behavior and take advantage of wireless or local networking, or even have to be classified as system of systems rather than systems alone.[1]

For complex software systems integrating the functional aspects alone is often already a major challenge. However, in the case of embedded real-time systems we usually cannot restrict our attention to an abstract view on the software only. In addition we have to cope with real-time behavior, resource consumption, and behavior in the case of failures. All these system characteristics have

---

[1] See [1] for a discussion of the resulting integration efforts.

to be covered to ensure a proper integration. Therefore, the integration of complex embedded real-time systems has not only to cover the integration of highly complex functionality, but also to take care of additional relevant system characteristics such as hard real-time constraints, a proper use of resources, and dependable operation with respect to severe reliability, availability and safety requirements. Usually, it is not sufficient to only consider the software. Effects of the hardware and the run-time environment as well as low-level design and implementation decisions may be relevant.

When developing complex systems besides the technical aspects also process issues as well as organization aspects become relevant (cf. [2]). However, we will further restrict our discussion to the technical integration of development artifacts looking in particular at model-based development and ignore process and organizational issues of integration. In [3] several approaches are characterized which support the model-based development of embedded control systems and cope with several integration problems. Selected domain-specific modeling languages (like AADL [4]) and tool sets (like Fujaba [5]) are characterized concerning the integration problems which are solved by the particular approach. In [6] several effects are discussed which result from the partitioning of a system in subsystems or features, the separate development and the later combination and integration. In this work different types of integration properties are discussed, as well as what kinds of effects exist concerning these properties and how to cope with them. In [6] the focus is set to shared resources, communication features and interacting control concerning the partitioning and later composition of the overall system. This chapter in contrast focuses on the underlying problem and integration concepts and provides a characterization of existing integration problems and fundamental integration techniques. It presents their role by discussing the standard approach to hardware integration and function integration as well as several advanced model-based integration approaches.

In essence, the integration problem we have to face is related to the fundamental fact that we as humans somehow have to divide the problem into less complex elements that can be handled independently or with limited dependencies. While from a system engineering perspective [2] the technical integration problem can be restricted to the problem of integrating complete components, for software and embedded real-time systems the integration can also happen at the level of abstract software components or even abstract software or hardware models. Therefore, integration is usually related to some form of combining artifacts, which is often proceeded by some form of separation that divides the development task into the later combined artifacts.

This need for integration also at the level of software and hardware components rather than complete systems has several reasons. First the reuse of already developed artifacts either as developed or including some adaption in order to reduce costs is one major driver. In addition, the complexity of today's embedded real-time systems often makes it necessary to separate the development of subsystems within a single organization or across a network of suppliers, sub-suppliers and the manufacturer (in the case of the automotive domain called

OEM). While the following discussed aspects and characteristics concerning the integration are fundamental for several types of embedded systems the discussed examples focus on the automotive domain. However, the assumptions and conclusions made are also valid for the other domains. An example for the latter case is the automotive domain, where the traditional model of division of labor is that the OEM integrates complete subsystems including software and hardware developed by their suppliers. However, today this traditional model of division of labor where only complete subsystems combining software and hardware are integrated is no longer valid and therefore also integration scenarios where different software has to be integrated on the same hardware are relevant.

In this chapter we will present a framework to discuss how model-based integration can in particular facilitate the outlined problem of the technical integration of development artifacts. Therefore, we will first clarify the terminology concerning integration, provide a classification of the integration challenges for complex embedded real-time systems, and review the main techniques employed to cope with the integration challenge in Section 2.2. Then, this framework will be employed in Section 2.3 to explain the current standard practice concerning integration of hardware and software for functional development as well as function integration. A number of advanced proposals for how to address some of the remaining integration challenges such as AUTOSAR and MECHATRONIC UML using model-based concepts are discussed also by means of the framework in Section 2.4. In this section we also review other approaches and discuss which integration problems of our classification are covered and which fundamental techniques are employed. Finally, we discuss which challenges have been addressed and which open problems remain and provide our final conclusions and outlook on expected further research in Section 2.5 and close the chapter with some final remarks.

2.2 Integration

No uniform definition for the term integration can be found in the literature (cf. [2]). Therefore, we will at first outline the terminology employed in this chapter, a classification for the integration challenges and the fundamental techniques employed to support integration. These elements provide a conceptual framework that is then used throughout the rest of the chapter to explain how different approaches address integration.

2.2.1 Terminology

The counterpart of the integration is the division or decomposition of a system into subsystems. At the more abstract level discussed here it relates to the division of development artifacts and not necessarily complete subsystems. Note that the division can happen explicitly when a development artifact is used to plan and document the decomposition or implicitly when either ad hoc or traditionally certain parts of a solution are developed separately. It is also worth mentioning that any such decomposition usually goes hand in hand with a breakdown and refinement of requirements.

The division used to derive development artifacts of reduced complexity consequently leads to the need to integrate the results of the separate tasks to derive the originally intended complete development artifact at the required level of detail. The integration itself therefore consists of the activities required to achieve a proper composition when combining development artifacts.[2]

Therefore, consequently, integration activities can be found during the decomposition of a system into subsystems as well as during its composition from the subsystems, and when related subsystems are developed in parallel. This can include preventive activities such as the definition of abstract interfaces or analytic activities such as integration testing which check system requirements that could not be addressed at the subsystem level. In addition, all activities to solve problems during the composition are included; this includes problems encountered during the system division or during the parallel development.

Integration essentially happens when in a development task multiple development artifacts serve as input and the specific combination of versions and variants of these development artifacts has not been considered beforehand. Many of these development artifacts are in fact models that represent some abstract view on specific aspects of the envisioned or already existing version and/or variant of a system or subsystems. Therefore, also in the case of classical integration usually models are already of paramount importance. If these models are furthermore not only paper-and-pencil models, but are employed to derive other development artifacts or facilitate specific integration steps, we consider this to be one form of model-based integration.

Given a development task and a fixed level of detail, the decomposition step results in a number of components with reduced dependencies. At a coarse grain level the resulting structure is referred to as architecture which decomposes the problem into components connected via links. The overall architectures capture the different components and their structure in the form of links between them.

For our discussion it is useful to further distinguish two cases of architectural decomposition: In the first case of a hierarchical architecture the components are modules at the same conceptual level (e.g., an architecture with separate components for motor and steering control) while in a layered architecture the decomposition provides layers where one layer is operating on top of the underlying one (e.g., application, operating system, hardware). When developing using such layers sometimes the models abstract from the layer beneath by means of modeling concepts and thus no explicit initial decomposition is visible.

Taking the many different facets of composition into account, we can define *integration* as the development activities which are employed to ensure a proper composition when multiple conceptually development aspects are combined which usually results in new or changed development artifacts.

---

[2] We do not consider the problem here that the different artifacts have been defined using different models of communication, computation, or causality as discussed in Chapter 1.

The composition of different software modules into a larger module is an example of combined conceptual development aspects being used as concrete development artifacts. Here decomposition as well as composition is done explicitly.

On the other hand, the combination of a software module with its underlying hardware relies on an implicit decomposition from the underlying layer. Afterwards, aspects are composed implicitly when enriching the model to include more information of that layer. This enrichment happens usually in several steps where additional conceptual development aspects such as limited precision, execution times or memory consumption are integrated with development artifacts that beforehand were not included in these aspects (by initially abstracting from them).

2.2.2 Classification of Integration Problems

Several problems are encountered when integrating different development artifacts. However, most of the relevant kind of conflicts can be covered by the classification of integration problems depicted in Figure 2.1.

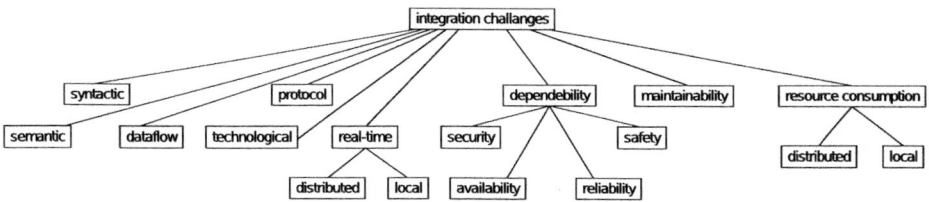

Fig. 2.1. Classification of Integration Problems

The outlined classification includes technological aspects refering to the ability to integrate at all on the basis of some technology and the syntactical aspect covering that the parts have to be integrated concerning the exchanged or shared data as well as the offered or required operations. Furthermore, we have the full semantic, which ensures that the encoding of data and side effects of the offered operations are compatible, the protocol, which addresses issues such as non-uniform service availability the synchronization and control flow between different parts, and dataflow, which covers the specific needs for composing dataflow computations as required for most control algorithms, are included.

While technological and syntactical integration are often addressed upfront during decomposition (e.g., AUTOSAR in Section 2.4.1), more demanding integration problems such as semantic, protocol or dataflow are today often only handled when it comes to composition.

The dependability resp. quality-of-service aspects include that the composition has to fulfill given reliability goals, availability requirements, exclude unacceptable safety or security problems, and still ensures maintainability. The real-time behavior may include the local behavior but also distributed real-time compatibility. For the resource consumption, either node limitations such as CPU time, network limitations such as bandwidth or even system limitations such as the overall power consumption may be relevant.

Embedded systems differ from standard software systems in particular when it comes to integration problems related to real-time or resource requirements. As real-time requirements, severe resource constraints, and more demanding dependability requirements have to be fulfilled by an integrated system with often rather limited hardware, a large fraction of the development efforts have to be spent on addressing them explicitly. In standard software development, developers use resources rather excessively and remain at a higher level of abstraction, and thus avoiding having to optimize their solution for a specific hardware. In the embedded world in contrast this luxury is not possible and a sufficient behavior has to be achieved with rather constraint resources and thus all kinds of hardware-related constraints such as execution times and resource consumption have to be addressed explicitly.

2.2.3 Fundamental Integration Techniques

To exemplify the rather general considerations presented so far, we will now look into a number of fundamental techniques for the proper integration. As the steps undertaken to divide the labor and the effort for the final integration of the independently achieved results can only be understood in combination, we will introduce the basic principles for both aspects at once and then discuss their interplay.

Explicit Horizontal Decomposition & Composition

The first fundamental concept operates at a constant level of abstraction and looks into an explicit decomposition into subsystems at the same level (horizontal decomposition). In order to achieve a suitable decoupling of the separately considered parts some form of separation of concerns [7] as well as information hiding [8] are usually employed, too. For example, architectural aspects (concerns) covered by modules or components which provide interfaces. Using an interface allows to decompose the architecture and hide implementation details (information hiding) at the same horizontal abstraction level.[3]

This form of horizontal decomposition of the system usually permits the subsystems to be developed in parallel and work on disjoint sets of development artifacts. In addition to the description of the system, this decomposition also happens for the requirements which are broken down from the system into its subsystems.

The explicit composition brings together subsystems which have been developed in parallel. In the ideal case all relevant system or subsystem characteristics are captured during the decomposition and are guaranteed when doing the composition. However, often this is not the case. For example, when using separation

---

[3] Some techniques (e.g., information hiding) also support vertical abstraction like described later.

of concerns several aspects are often not covered during decomposition but become relevant when doing the composition (potentially in a later development stage) or when the composition not only exhibits the characteristics of its components but also characteristics which are determined by the composition (sometimes call emergent) itself. It is particularly relevant for the integration that all system requirements that have not been broken down into subsystems requirements are checked for the composition result. This includes that characteristics such as deadlocks which can often not be predicted when doing the decomposition have to be addressed when doing the composition. Therefore, depending on the question of which characteristics are compositional or not resp. which requirements have been broken down to local properties of the subsystems more or fewer characteristics of the composition have to be checked at composition time to ensure a proper integration.

The standard case for composition is that the individual constituent parts are simply combined by some generic form of composition (e.g., scheduling in the case of processes on an operating system). More advanced cases employ declarative constraints contained in the specification of the components to ensure that the composition behaves properly (e.g., scheduling with guaranteed deadline in case of processes on a real-time operating system).

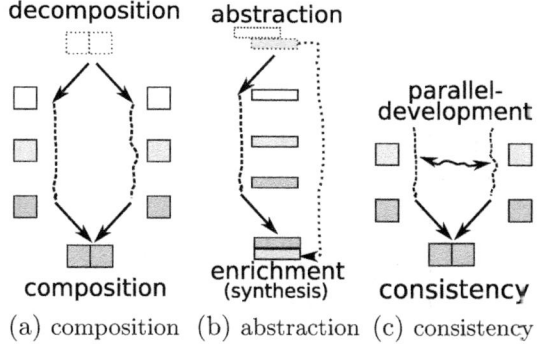

Fig. 2.2. Fundamental techniques employed to approach integration

The resulting interplay of decomposition and composition is depicted in Figure 2.2 (a). At a rather high level of abstract the system is decomposed into two or more subsystems that are developed in parallel. These subsystems, which are then further elaborated in parallel, are composed later on according to the decomposition done upfront.

It has to be noted that in contrast to more restricted interpretations of the term integration (cf. [9]) we do not limit integration to the case where checks at composition time are employed, but also include the case where upfront activities such as defining interfaces ensure a proper composition.

Module interfaces are the classical example for the decomposition / composition. If done properly as in many programming languages, the definition of the

Table 2.1. Coverage of integration aspects by modules with syntactical interfaces

Integration problems	Technique	Explanation
syntactical	D/(S)	Is checked already during decomposition guarantee proper integration during composition. Synthesize syntactical correct implementation of interfaces.
technological	S	try to solve problem using synthesis, e.g., generating compatible C-Code for the implementation of interfaces.
semantically	C	Is checked only late during composition.

(legend: D = during decomposition – C = during composition – S = by synthesis).

syntactical interfaces at the time of the decomposition guarantees that during the later composition no syntactical integration problem can result. In Table 2.1 this result is depicted using a classification that focuses on the question when the integration problem becomes visible. The case *during decomposition* (D) implies that the integration is guaranteed upfront, while the *during composition* case (C) results in a risk that an integration problem is detected rather late. The case *by synthesis* (S) refers to automated techniques that can generate a solution that solves the integration problem in principle. However, usually no guarantee can be given that the synthesis will thus be able to find a solution as there might be not resolvable conflicts.

Vertical Abstraction & Enrichment

Another fundamental concept to separate details during the development of a system is vertical abstraction and enrichment. Compared to the previously described horizontal decomposition and composition, where the abstraction level is the same when doing the composition or decomposition, vertical abstraction and enrichment change the abstraction level itself.

In the abstraction step we omit details of the envisioned system and focus on the characteristics that are relevant for the current development step. The abstraction can be seen as a form of implicit separation by omitting the details for a certain time. These omitted details are then later added to the development artifacts when enriching them.

The abstraction concept can be employed to ease development when there is only a unidirectional dependency between the upfront-addressed details and the omitted ones. Often architectural layers are employed to realize the abstraction independent of the concrete omitted details (application, operating system, hardware). The fact that the lower architectural layers do not depend on the higher layers in combination with standardized interface for the lower-level layers allows that the higher layers could be developed more or less independent, and omitted details are either filled in by the lower levels or can be considered later on.

The design characteristics that are affected by the combination are then usually considered later. E.g., first the end-to-end timing constraints are considered and the specific timing resulting from the integration with the operating system and the hardware layer is considered later. As the real-time scheduling can only be considered in combination with the concrete real-time operating system, the analysis of the real-time characteristics are only addressed when both are integrated. When abstraction is used in this manner, dependencies between the separated layers often seriously constrain the decoupling of the development activities for the subsystems. The development artifacts of the different activities depend on each other such that the ordering of the development activities reflects the usage of development artifacts of other activities. E.g., details are stepwise added to the development artifacts during function development (see Section 2.3.1) where in each step the effects of other layers (runtime system, hardware, ...) is added.

Vertical enrichment is the counterpart of abstraction where new characteristics are added to a development artifact. We have to further distinguish two fundamentally different forms of how enrichment can occur. Either the detail adds new characteristics to the development artifact. In this case no constraint between the abstract development artifact and the added detail exists (e.g., a logical untimed model is enriched by timing information). On the other hand the development artifact may already contain some abstract information about some system characteristic which constraints the added details (e.g., an idealized model equal to some set of differential equations has to be developed into a model equal to a set of difference equation scheme). Often the development starts with such an abstract artifact and abstraction is used rather than explicitly applied. In this case you either want to have refinement when the more detailed structure and behavior is included in the more abstract one or approximation such that the more detailed structure and behavior is similar to the idealized abstract one.

In the case of refinement the abstraction already includes the refined behavior as one possibility and thus checking crucial properties for the abstraction can guarantee these properties also for the refinement. In contrast, for an approximation it holds that the abstraction is an idealization and that the behavior which could be observed for the enrichment should be somehow similar. Therefor, here the opposite observation holds that only in the case a required property does not hold for the approximation it can also not be expected to hold for any enrichment (even though this is not necessarily always the case). Thus, refinement can be used to guarantee the absence of failures upfront, while approximation can be employed to detect possible integration failures upfront. A fully inconsistent enrichment which is neither a refinement nor an approximation would mean to redo all the construction work contained in the abstract development artifact and thus is usually not intended. However, it must be mentioned that different characteristics of a development artifact may be enriched differently and thus some may be refined while others are approximated.

In both cases the enrichment is some form of implicit (vertical) composition as a new or yet only insufficient covered development aspect is now considered. The initially not considered development aspect is then brought back into the picture and thus the related information about the related layer beneath is implicitly composed with the model that beforehand used abstraction to omit that information. An example for such a case is the usage of abstraction layers, e.g., hiding communication details or hardware properties like in case of the different layers of the AUTOSAR architecture (see Section 2.4.1).

Like in the case of (horizontal) composition again synthesis can be used to automatically apply enrichment. Depending on the applied form of enrichment respectively the previously applied abstraction, synthesis does not guarantee in any case that all desired properties are fulfilled (e.g., no schedule for a set of tasks can be synthesized). In many cases enrichment is automatically applied using synthesis, e.g., in the case of automatic C-code-generation supported for embedded systems.

As depicted in Figure 2.2 (b), the initial abstraction allows to omit a development aspect and later consider it when enriching the model in that respect.

Table 2.2. Coverage of integration aspects by the different approaches

Integration problems	Technique	Explanation
syntactical	A	Abstraction guarantees that the composition is syntactically correct.
technological	A/E(S)	Abstraction and enrichment (potentially by synthesis) provides some technological compatibility.
real-time local	E	The initial abstraction does not provide any guarantees.

(legend: A = during abstraction – E = during enrichment – S = by synthesis).

The explicit consideration of real-time constraints for a software function in a subsequent development step is an example for an abstraction and enrichment step. Upfront, the developer abstracts from the timing issues and instead focuses on the functional aspect of the solution. Then, in a later step the derived functional solution is enriched with timing information in the form of deadlines etc. As outlined in Table 2.2 the initial abstraction step does not provide any guarantee for the later enrichment and thus the integration problem has to be addressed late when the enrichment happens.

Often horizontal decomposition & composition, where parts of the system are decomposed at a specific abstraction level and vertical abstraction & enrichment, where the level of abstraction changes, are used in combination. An example for such a situation is when different parts or subsystems are developed by different stakeholders and one has only an abstract view on a subsystem provided by a supplier while other parts are available on a more detailed level.

Consistency & Synchronization

A third fundamental concept to handle integration issues is to not only decompose the problem initially and resolve integration problems later, but somehow reflect the dependencies between the different artifacts throughout the parallel development.

The first approach is to check the consistency of the models and resolve the issue immediately or at least in the near future rather than waiting for the time of integration. This case of consistency refers to horizontal consistency [10] and takes care that no conflicts arise when the models developed in parallel are later integrated.

Another option which provides a higher degree of automation is model synchronization [11] where the equivalent parts of two models are automatically kept consistent. Like for consistency, the relevant case here is only horizontal synchronization of parallel-developed models (analogous to horizontal model transformations [12]).

The main benefit of both approaches is that in contrast to the two former ones the independently developed model can more freely evolve without resulting in harm later on. In the case of decomposition and composition in contrast somehow the basis for the separation is fixed after doing the upfront decomposition. Also in the case of abstraction and enrichment the separation is somehow a-priori fixed when doing the abstraction and it thus too only provides limited degrees of freedom when enriching the model later on. However, this higher degree of flexibility can only be preserved as long as the consistency rep. synchronization is keeping track of the dependencies to prevent integration problems later on.

On the other hand, unless fully automated as in the case of model synchronization there is the permanent need to resolve inconsistencies during the parallel development and therefore both parallel development activities might be slowed down considerably. Therefore, the sketched benefit comes with the drawback that no "fully" parallel and independent development is really possible.

If during the parallel development the consistency is checked as depicted in Figure 2.2 (c), integration problems during the later composition can be prevented. Co-simulation of different models, which are developed independently or for analysis purposes (like in the case of plant-models), is one example where different development activities are checked for their consistency (see Table 2.3). It is important to note that this also enables consistent changes of the interface between the initially separated subsystems. Without consistency checks interface changes would endanger the proper composition later on.

Table 2.3. Coverage of integration aspects by the different approaches

Integration problems	Technique	Explanation
semantic	P	The co-simulation helps to find inconsistent behaving development artifacts.

(legend: P = during parallel development).

Combinations in Practice

It is important to note that in practice instead of these pure cases of parallel and sequential processing you will encounter partially ordered activities that are coupled by the production and use of different versions of development artifacts depending on the employed decompositions and abstractions.

A frequently employed approach which combines horizontal decomposition with vertical abstraction is interfaces. When planning the decomposition, e.g., when horizontally decomposing a system into subsystems, interfaces are used to capture at a more abstract level the dependencies between the components. Additionally such interfaces can be also defined between layers at different level of abstraction. If the dependencies are properly designed in the interfaces this prevents that related integration problems will be encountered during composition. However, interfaces usually only cover a very restricted subset of the component characteristics and they only prevent integration problems for that restricted set of characteristics. Examples where different sets of characteristics are covered by interfaces presented later in the chapter in the case of AUTOSAR and MECHATRONIC UML.

Using an abstract model of the environment or the other subsystems is another technique used in engineering which combines decomposition and abstraction. This is particularly useful when a simple interface will not capture all required properties of the environment properly. Please note that such an environment model together with the subsystem model can be checked during the parallel development against requirements of the system which could not have been broken down into subsystem requirements due to their non-compositional nature (like the reactive interplay between the plant and implemented control functionality). In the case of control engineering so-called plant models are employed to capture that part of the environment which is relevant. Simulation runs check that the given control requirements are met. The function development described in the next section is a typical example where environment models play a prominent role.

Another approach to derive a valid composition at a more detailed level is a dedicated manual or automated synthesis step. The synthesis step generates a solution which fulfills the constraints (e.g., fixed schedule for a dedicated hardware and software stack) configuring an underlying layer or determining an additional glue component. To fulfill several resource constraints the synthesis can also target to minimize the resulting resource consumption (e.g., synthesizing a minimal runtime kernel which only includes the necessary modules/functionality). Alternatively, an online solution is often employed (e.g., real-time scheduling in the case of processes on a real-time operating system). In this case usually an additional check at integration time is required that evaluates whether the constraints can be met (e.g., schedulability checks). Examples for synthesis approaches for the local real-time integration problem are MECHATRONIC UML (see Section 2.4.2) and TDL (see Chapter 5).

Depending on the specific domain and the severity of the encountered conflicts quite different means for the resolution of the integration problems are applicable. A technique can only be employed when the resulting solution adheres to the specific constraints of the domain (e.g., in a domain with high cost pressure such as the automotive domain using more powerful hardware and additional abstraction layers are often not affordable). Also the development efforts as well as the scalability of composition techniques are important factors that have to be considered. E.g., in the automotive domain the high cost pressure does not allow the intensive use of more powerful hardware and additional abstraction layers. However, in domains where safety issues prevail, like in the case of avionics systems, more advanced concepts such as IMA exist allowing the modular verfication of decomposed system parts [13]. Also the development efforts as well as the scalability of composition techniques are important factors that have to be considered.

2.3 State-of-the-Art Approach

The construction of current complex embedded systems – as found, e.g., in the avionics or automotive domain – is characterized by two sources of complexity: First, these systems are composed of *interacting distributed components*. Second, these components are developed *in parallel by different suppliers*. Thus, integration of these components becomes a core issue during development, especially if individual components of different suppliers are combined by the equipment manufacturer on a single electronic control unit.

To overcome the *problem of late integration*, often a model-based integration approach is chosen, allowing the development process to be modularized. Here, the techniques of *decomposition* and *enrichment* as described in Section 2.2.3 are used to obtain two *orthogonal dimension* of development: By using *functional decomposition*, braking the system down into separate functions or components, these functions can be constructed, validated, and verified independently. By using *incremental enrichment*, going from functional via logical to technical models, these components can then be safely integrated on a common platform.

To effectively support such a development process, two prerequisites are necessary: On the one hand, the approach must support the *description of the system at different levels of abstraction*, to support a stepwise enrichment of the models of an individual function. This ensures that the more detailed model respects the limitations of the more abstract model. On the other hand, the approach most support the *combination of the models of all functions* at each level of abstraction, to enable a safe integration of the overall system. This ensures that the combined functionality implements the intended overall behavior.

To support the different levels of abstraction, increasingly model-based approaches are used, especially for control functionality. Typically, here the functional, logical, and technical level are realized by function-oriented models (e.g., MATLAB/Simulink or ASCET-MD), by software-oriented models (e.g., TargetLink or ASCET-SD), and prototyping or pre-production platforms.

Currently, parallel engineering is achieved by decomposing the system into several components at the functional level. Thereafter, the different functionalities are developed separately. Composition is achieved mainly at the platform level by defining the realization of the joint interfaces (e.g., the exchanged bus messages).

Obviously, due to increasing dependencies between formerly independent functionalities such an approach requires to support a combination of the top-level functions. Furthermore, since obviously the late integration can cause inconsistencies (e.g., when fixing different discretizations of joint interface signals during the construction of the software model), current approaches specifically provide support for a safe integration at earlier levels (e.g., Intecrio or SystemDesk). Additionally, platforms (e.g., IMA[4], AUTOSAR[5]) and a corresponding development environment eliminate the need for the manual integration at the platform

---

[4] Integrated Modular Avionics.
[5] Automotive Open System Architecture.

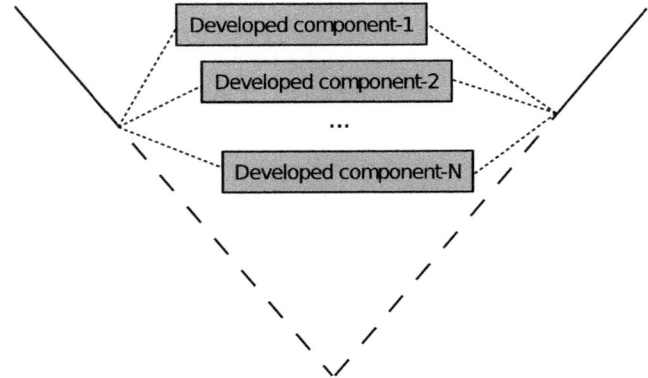

Fig. 2.3. Parallel development within the V-model (according to [14])

level, lifting the integration to the software level. Besides the functional aspects, such an approach requires to include system wide properties like real-time requirements.

As mentioned above and illustrated, e.g., in [14], the standard engineering process combines parallel and incremental development, allowing to develop components or functions in parallel while especially taking into account platform and other restrictions (e.g., real-time and resource restrictions) in a stepwise manner. The parallel development of multiple components is illustrated in Figure 2.3, showing that parallel development is achieved by forking the development for each component or function in the design phase, joining these components in the integration phase. Each component or function, as described in the multiple V-model shown in Figure 2.4, is iteratively developed with increasing level of detail via a simulation, prototyping, and pre-production stage.

As indicated in Figure 2.5, the development of functions at different levels of abstraction affects not only the design of the functions via the different models used at these levels; it furthermore affects the different validation and verification phases in the multiple V-model, e.g., concerning the used models of the environment or test cases. However, it also requires providing an integration of the

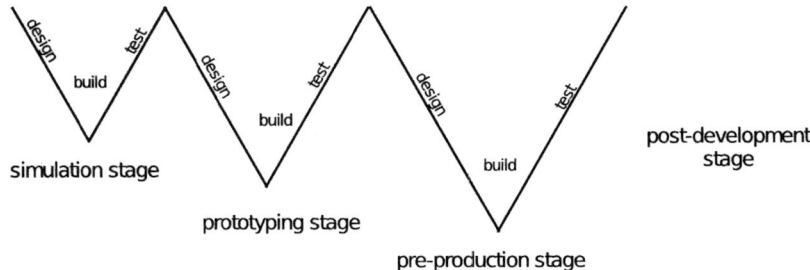

Fig. 2.4. Stages of the multiple V-model (according to [14])

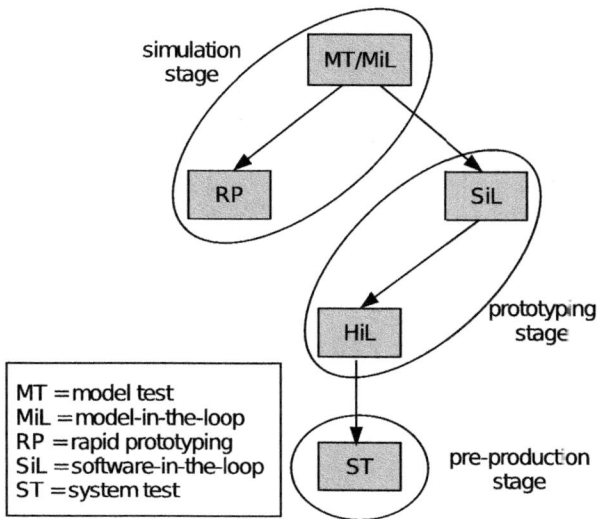

Fig. 2.5. Test and simulation activities within the different stages (according to [14])

functions at these different levels to support the early validation or verification of the system under development.

In the following, in Subsection 2.3.1 the development of an individual function at different levels of abstraction is described in more detail, while Subsection 2.3.2 illustrates the possibilities of integrating several functions at these different levels of abstraction.

2.3.1 Function Development

During function development a complex functionality in the form of a control algorithm or reactive behavior controlling a physical process is developed. This functionality is incrementally developed by adding constraints imposed by the plant (environment) as well as the platform in a stepwise fashion. This stepwise integration is accompanied by validation and verification techniques at the different stages. Figure 2.5 shows an overview of these different levels of abstractions as well as the corresponding techniques.

The simulation and prototyping stages use abstracting assumptions to eliminate details both from the environment as well as from the platform. Examples for such assumptions are unlimited HW resources, e.g., eliminating the need to consider execution times or memory consumption, or simplified plant models, e.g., eliminating the need to deal with failures of sensors and actors.

Simulation Stage
At this stage, purely functional models build the basis for development. Commonly, data flow models in form of block diagrams (e.g., ASCET, Simulink) or control flow models in the form of state diagrams (e.g., Statecharts) are used.

Functionality is developed independent from platform and its interfaces to the environment (e.g., A/D and D/A converter). Therefore, these models ignore properties like WCET limitations, characteristics of the HW (e.g., register size), or memory consumption. Typically, these models use values and signals of types that differ from these used within the real system (e.g., floating-point data types instead of fix-point, abstract messages instead of CAN messages). Due to the focus on the logical execution order and the data flow, for verification and validation often simulation of the models is used, using either no plant or a plant model as environment. The goal of this stage is a first proof of concept and the verification and validation of the overall design and control law. Using a plant model during the simulation stage in parallel supports the validation of the developed functionality concerning semantically as well as dataflow integration problems like described in Section 2.2.2.

For validation and verification, at the simulation stage model tests (MT), model-in-the-loop tests (MiL) and rapid control prototyping (RCP) are used.

For MT, one-way simulation of the models – also called one-shot simulation – is used. Here, all input and output values are generated and analyzed for a single execution of the system, abolishing the need for a dynamic interaction with the environment. For MiL, the model of the function is simulated back-to-back with virtual models of the dynamics of the environment in the form of a plant model. The plant model is an abstracted and simplified representation of the real environment, allowing a validation of the functionality. Due to the abstraction of the plant model, issues like the calibration of the functionality generally cannot be treated at his level. For a more accurate validation of the models, RCP can be used together with the models of the simulation stage. To that end, typically high-performance RCP-HW is used, allowing platform restrictions to be ignored (e.g., computations times, floating point vs. fix point). By replacing the plant model by the actual plant (or a close equivalent), RCP allows the functionality to be validated against the real plant including the real-time behavior of the plant, access special-purpose HW, and use actual actuators and sensors.

Prototyping Stage

At this stage, many aspects like real-time properties and resource restrictions removed by the platform abstraction in the simulation stage are taken into account. The focus of this stage is the implementation of the designed functionality and its validation and verification. Depending on the level of abstraction, only software aspects (e.g., modularization, used data types) or also hardware aspects (e.g., computation times and storage restrictions) can be addressed.

On the software level, software models are executed on a host computer. Unlike to MT and MiL, the software models additionally consider implementation aspects like discretization of the functionality in the value and the time dimension, e.g., by using fix-point arithmetics and task schedules. In practices, software models are often generated from the functional models via parameterized autocoders (e.g., ASCET-SD, TargetLink, or Embedded Coder) and in such a way that at the technological level the integration is supported via synthesis. An essential part of the prototyping stage consists in the verification that the models from this stage are an enrichment of those of the simulation stage and intended semantics are fulfilled, as mentioned in Section 2.2.3. Here, enrichments dealing with refinement or approximation are often semi-automatically provided by a target code generator during discretization. By using a specific host platform, some aspects of the final platform are still ignored. Typically, resource restrictions are not examined or only in a simplistic fashion. Using such host platforms, verification and validation via MiL can be covered by corresponding techniques via software in the loop (SiL) test.

To again include more platform restrictions, software models to be executed on the target processor can also be used in this stage. Such an integration of software and target hardware can be done on different levels. The target processor can be used to run a processor in the loop (PiL) simulation, providing extended debugging or calibration functionalities. Therefore, in PiL simulation often the used platform is different to the final one, and often Evaluation Boards are used providing additional interfaces for debugging and calibrating. Alternatively, the real hardware (the ECU) can be used for HiL simulation, allowing reliable results to be obtained, e.g., also w.r.t. execution times, which may be affected by the additional debugging and calibration functionalities. In both cases, the environment is simulated via the use of a simplified and abstract model of the real plant. However, in the PiL approach often simplified models are sufficient, while HiL approaches in general use fine-grained models often requiring the use of real-time systems for execution. Furthermore, within the HiL simulation additional system parts can be included, e.g., legacy ECUs or even mechanical parts (e.g., the throttle of an otherwise simulated engine). By using more detailed execution platforms (e.g., in the case of PiL simulation or in the case the real ECU is used) additional properties like execution times and the resource consumption can be evaluated at least for the local case by using the concept of enrichment as introduced in Section 2.2.3.

Pre-Production Stage
Within the pre-production test the system is tested against external influences of the environment. This includes the effects on the system due to environmental conditions (like temperature, shock or vibration). The system under test is built of prototyping HW fulfilling the required specifications for the end product. The goal of the tests is to identify and fix problems and to measure the robustness of the system as early as possible.

2.3.2 Function Integration

The function integration happens at the latest when the functions are integrated during the pre-production stage. However, in contrast to such a big bang integration usually whole functional groups are integrated beforehand using replacements for the missing rest of the system in order to ease the integration testing.

The current practice for integrating multiple software modules on one node is characterized by the following stepwise partially manual process: (1) *Specification:*

The interface and decomposition of the software into modules are specified on a high abstraction level while mainly functional properties are targeted (if at all), then (2) *Partitioning:* The software is partitioned into concurrent modules resp. logical threads with appropriate periods to make it run on a real-time operating system or kernel (usually without adequate analysis), (3) *Implementation:* The software is implemented (often manually), (4) *Integration:* The threads are combined using either static schedules or concurrent threads in a RTOS or kernel and it is verified that the software fulfills all real-time constraints in its given environment. In the case the implemented software is combined with an RTOS, at least at the functional level the technological aspect concerning the integration is supported by composing the software with a standardized execution platform (e.g., an OSEK RTOS). Additionally scheduling analysis of the used RTOS tasks can be applied to analyze and validate the real-time behavior, at least for the local case. If the real-time constraints do not hold, partitioning, implementation and integration have to be repeated. Repeating this cycle a number of times is usually very costly but often unavoidable.

Prototyping Stage

The outlined early integration of functional groups might be addressed already at the prototyping stage when integrating the control algorithm with more appropriate substitutes for the final hardware. This could happen using the prototypes of other functions as well as their final version depending on the availability.

Pre-Production Stage

In the last stage of the multiple v-model (pre-production stage) the real system is build including the real plant.

The system is tested within the real-life environment to ensure that all requirements are met, including conformance to relevant standards like industrial or governmental ones. The build system is close to the later product and some calibration and configuration can be done. Tests concerning functional and nonfunctional properties are possible but fixing problems concerning properties of one of the earlier stages could not or only with extensive effort be done. E.g., to change the system architecture or the design of the control functions is rarely possible at this stage.

Encountered Problems

When the system is initially decomposed several properties (e.g., real-time properties or needed resources) are not considered. When composing the developed parts these properties can lead to crucial problems, potentially leading to extensive changes related to the earlier development stages. Furthermore, the composition of the developed parts can result in characteristics which are caused only by the composition and not by the characteristics of the components itself. For example, if several independent developed components have to use the same communication channel (e.g., a shared bus) problems can occur which could be hardly detected when components are tested and simulated individually.

Decomposition at the system level is almost done at the same architectural level (while additionally layered architectures play an important role, e.g., in case of the integration of operating system properties like scheduling).

2.3.3 Discussion

The outlined process of decomposition, functional development and system integration represents a nearly optimal solution for systems with more or less independent functions and functions which require only a usual control law. Table 2.4 and 2.5 summarize the integration aspects concerning function development and function integration, which are somehow supported by the described standard approach. Unfortunately today's embedded real-time systems often include much more sophisticated designs where an overwhelming number of functions exist that have to interact in complex ways to achieve the envisioned overall functionality. Therefore, this style of development and integration often result in severe problems perhaps being detected rather later during system integration. These can be true for

- interface compatibility problems,
- protocol compatibility problems,
- dependability issues,
- real-time behavior and
- resource consumption.

In all these cases, the rework required to fix such problems if encountered during system integration can be quite costly.

Table 2.4. State-of-the-Art: coverage of integration aspects during FD

Integration problems	FD	Explanation
technological	A/E(S)	via code generation, standardized tools (e.g., MATLAB)
syntactical	D/C	define interface to sensors and actuators of the plant
semantic	A/E(S)/P	enrichment at each stage using (initial abstract) refined SW (potentially synthesized) and HW models; simulation of the SW models in combination with the model of the plant
dataflow	A/E(S)/P	enrichment at each stage using (initial abstract) refined SW (potentially synthesized) and HW models; simulation of the SW models in combination with the model of the plant
real-time compatibility		
local	E/P	enrichment at each stage using refined SW and HW models; consistency with HW checked during parallel development by means of simulation techniques, e.g., PiL simulation
distributed		
resource consumption		
local	E/P	enrichment at each stage using refined SW and HW models; consistency with HW checked during parallel development by means of simulation techniques, e.g., prototyping stage
distributed		

(legend: FD = function development – A = during abstraction – C = during composition – D = during decomposition – E = during enrichment – P = during parallel development – S = by synthesis).

Table 2.5. State-of-the-Art: coverage of integration aspects during FI

Integration problems	FI	Explanation
technological	A/C	later integration on standard platforms such as OSEK allow upfront abstraction
syntactical	A/(D)/C	upfront definition of components and their interfaces (not standard)
semantic	C/(P)	checked during integration testing – potentially using functional groups in parallel with replacement for the rest of the system
protocol	C	checked during integration testing
dataflow	C/(P)	checked when programming and compiling the integrated code, checked during integration testing – potentially using functional groups in parallel with replacement for the rest of the system
real-time compatibility		
local	C	scheduling analysis of the integrated tasks
distributed	C/(P)	using simulation techniques (later in the development lifecycle) – potentially using functional groups in parallel with replacement for the rest of the system
resource consumption		
local	C	resource analysis of the integrated tasks and their code
distributed	C	analysis of the bus allocation for the integration of multiple nodes

(legend: FI = function integration – A = during abstraction – C = during composition – D = during decomposition – P = during parallel development –).

2.4 Advanced Model-Based Solutions

A number of model-based approaches for the development of complex embedded real-time systems have been proposed to avoid some of the problems encountered for the standard approach for system development as outlined in the preceding section. We will first review an industrial approach developed in the automotive domain. The approach deals with some integration problems present in complex and highly heterogeneous systems. In addition, we will discuss a particular academic proposal and a number of related approaches that try to address other challenging integration problems.

2.4.1 AUTOSAR

AUTOSAR has been founded 2003 as an industrial standardization body for the automotive industry and is now supported by all major car manufacturers, their suppliers, semiconductor producers, and tool suppliers (see also [15, 16]). AUTOSAR aims at the improvement of the software development and integration for Electronic Control Units (ECUs) by providing standards for software architectures and software modeling techniques.

AUTOSAR responds to an increasing number of problems with the development of ECUs, especially software development and software integration:

Traditionally manufacturers bought dedicated ECUs for dedicated functionalities from ECU suppliers, e.g., an engine ECU or a left-door ECU. The increase of functionalities has led therefore to an increase of the number of ECUs – reaching numbers of up to 80 ECUs in some vehicles. A further increase of the number of ECUs is hardly feasible: Function integration in such settings means integrating ECUs via communication busses; a feat which becomes more and more difficult when the number of functionalities increases and functionalities need to communicate intensively with each other.

This dilemma is solved by breaking with the tradition of equating functionalities and hardware modules (i.e. ECUs). Instead software – not hardware – becomes the means for implementing functionalities. This decouples the number of ECUs from the increasing number of functionalities. But on the other hand this entails changes in the development process: Manufacturers will not continue to buy production-ready ECUs from suppliers. Instead they will buy software modules, i.e. functionalities, and (generic) hardware platforms separately. So function integration then means software integration.

In the automotive industry, software has so far not been a product; just finding a pricing schema for software will therefore be a challenge in itself. Furthermore, in the future several software functionalities must be integrated on one hardware platform consisting of one or more interconnected ECUs. This will either be done by traditional suppliers or by manufacturers which try to extend their fields of competence – and try to become more independent from suppliers.

Another problem addressed by AUTOSAR is the testing and quality assurance of the developed ECUs. Would it be just for the increasing number of ECUs, the problem could be solved by also increasing the number of ECU tests. But nowadays most functionalities communicate with other functionalities (which may be on other ECUs) or high-level functionalities are even implemented by combining already existing functionalities—being an example for the emerging behavior described in Section 2.2.3.

Testing such distributed functionalities is difficult since all possible communication combinations between the building functionalities must be covered – causing in the worst-case an exponential increase of necessary test cases. Traditional engineering fields such as electrical engineering have faced this problem for decades and have mainly come up with three main solutions: (i) Systems are decomposed into separate components, in doing so dependencies inbetween are minimized – corresponding to the ideas from Section 2.2.3. (ii) Components are tested separately. Tested and used components are then reused in later projects – thus avoiding unnecessary and error-prone new implementations. (iii) Errors are taken into consideration, e.g., by means of safety margins and diagnosis functions.

AUTOSAR is trying to apply these ideas – to some extent – to automotive software architectures. Generally speaking AUTOSAR comprises three main activities:

(1) Introducing Software Component Models
(2) Standardizing Basic Software Modules
(3) Standardized Application Interfaces

Compared to the integration problems discussed in Section 2.2.2 in the current release version 3.1 of AUTOSAR primarily syntactical aspects are supported in the form of standardized interfaces for components and modules. Other key issues like semantics, protocol integration or real-time properties are not covered.

How other aspects can be included into the AUTOSAR standard is discussed for the case of real-time properties in [17].

(1) Software Component Models

AUTOSAR introduces a software component model for automotive application software. By this, AUTOSAR wants to facilitate (i) software reuse and (ii) software exchange between different parties. Software components can either be atomic (i.e. components implemented as C Code) or compositions which are formed by interconnected software components. In AUTOSAR all applications of a vehicle are part of one overall software composition – i.e. independent of their distribution on several ECUs. This top-level composition forms the vehicle's software architecture. From a software developer's point of view, AUTOSAR software components mainly introduce standardized C-APIs to communicate with other software components, either on the same ECU or via the communication bus. No direct calls to C code contained in other software components or to C code of basic software modules such as drivers are allowed anymore. This makes the software independent of other components, hardware and basic software modules – and hence reusable. By using APIs in form of C-APIs not only syntactical aspects but also technological ones are treated.

These standardized C-APIs are based on a classical port concept. I.e. a software component A will not send a signal directly to software component B anymore; e.g., it will not directly call a method $B\_receiveSignal()$. Instead A sends this signal to one of its own ports; ports being just proxies for other software components. A software architect later on connects (normally in a modeling tool) A's port to the corresponding receiving port of B. The same concept is used to connect software components to basic software modules such as I/O drivers, the COM stack, or error management modules.

The usage of standardized C-APIs to communicate with a component's environment has several advantages: Applications become reusable, hardware and basic software dependencies are eliminated, communications become clearly visible. But on the other hand, standardized C-APIs also mean that (i) existing C code must be wrapped or modified and (ii) existing tool chains such as code generators must be adapted. These changes to the development process may well delay the introduction of new AUTOSAR concepts.

This leaves one key question unanswered. How are A and B connected on the C code level? Who generates the gluing code to implement the signal transfer from A's sending port to B's receiving port? In AUTOSAR this is done by a middleware layer, the so-called Run-Time Environment (RTE). The RTE comprises the C code for the definitions of the C-API commands used within software components. Since resource consumption and process usage are important for ECU software development, this RTE code is generated for each ECU individually.

Here two key concepts of software integration can be seen implemented: (i) Software components and compositions are an example for composition and decomposition (see Section 2.2.3). (ii) The standardized C-API generated by the RTE abstracts the underlying hardware platform and is therefore an example of abstraction (see Section 2.2.3). Thus, the AUTOSAR framework supports the composition and decomposition at the same conceptual level using software components and compositions, while the RTE is an example where abstraction is used in case of a layered architecture to allow the decomposition of the system at different conceptual levels like discussed in Section 2.2.1.

From a process point of view, application software components and compositions are modeled first and then mapped onto ECUs – ECUs and their basic software can also be described by AUTOSAR. The developer then connects the application software components to basic software modules such as I/O drivers or the operating system. These basic software modules also have to be configured appropriately. E.g., tasks have to be defined for the operating system. Then the RTE can be generated which connects software components to other software components and to basic software modules.

(2) Standardized Basic Software
AUTOSAR also standardizes the C-API and the configuration files for ECU basic software modules such as I/O drivers, COM stack, operating system, error manager, mode management, and network management. One goal was to make basic software modules interchangable, i.e. a COM stack from provider 1 should be used with an operating system from provider 2. Furthermore configuration settings should become more reusable, i.e. the configuration files from an older project (where provider 1 was used), should also be usable for a newer project (where provider 2 is used). The introduction of standardized basic software modules also eases the implementation of the RTE because the to-be abstracted basic software becomes more uniform.

The basic software modules are organized into several layers, making them an example for the vertical abstraction described in Section 2.2.3 supporting syntactical as well as technological aspects.

(3) Standardized Application Interfaces
Another activity from AUTOSAR is the standardization of the APIs for application software components. E.g. the interior light control software used by different manufacturers should have the same API. By this, manufacturers and suppliers hope for fewer redundant implementations of the same functionality by different software providers and for easier integration processes. Of course, only commodity modules are standardized—no standardization is planned in competitive areas.

As model-based integration is the key issue here, it is worth reviewing the AUTOSAR approach with regard to improvements of the ECU integration process. First of all, several integration steps exist: basic software integration, ECU integration, and ECU system integration. In the following, AUTOSAR's contribution to these integration steps are assessed using the categories from Figure 2.1.

Basic Software Integration: Basic software is composed of different software modules: operating system, drivers, services, and communication stack – most of these modules are further decomposed into different submodules. AUTOSAR makes the integration of these modules into one basic software layer easier by means of syntactical, technological and data flow agreements: *(i)* The decomposition is standardized including the C-APIs between modules and *(ii)* configurations are expressed using a standardized set of parameters – being an example of the "Decomposition & Composition" principle of Section 2.2.3.

AUTOSAR does not address the key issue of semantic and protocol integration: In the standard, the precise behavior of the modules is not defined in a formal way, leading to integration problems such as incorrect emerging behavior – especially since the basic software is implemented by different software suppliers. This problem is worsened by the large number of interacting and behavior-influencing parameters. Furthermore real-time issues are not modeled in a satisfying manner, leading to problems with the temporal features of the integrated software system.

All these drawbacks lead to situations where, e.g., one basic software layer from software supplier A behaves differently to an equally configured basic software layer of supplier B – a situation rendering the reuse of configurations and the exchange of software modules almost useless. A solution could be a precise, executable model of the basic software behavior including the effects of parameter settings.

ECU Integration: In this step, the application software components on one ECU are integrated with each other and with the basic software layer. Unlike with the basic software, the decomposition and the C-API cannot be standardized in most cases – except for AUTOSAR's limited " Standardized Application Interfaces" activity explained above. This make the application of the "Decomposition & Composition" principle harder, in fact the interface (and port) principle from Section 2.2.3 must be used: modules (i.e. components) do not refer directly to each other but refer indirectly to each other via ports and interfaces. The data flow between modules is modeled by means of connections between ports; at run-time these connections are implemented by the RTE middleware (see above). So again, AUTOSAR solves to some extent the syntactic, technological and data flow integration problem (see Figure 2.1).

The introduction of the component/interface software engineering pattern causes the need for explicit software architectures; which in turn causes the need for software architects, for a separate software design step in the development process and for appropriate tools. This significant change to the development process is one of the challenges when AUTOSAR is introduced: Software architecture models must be synchronized with existing models (e.g., behavior models), new tools must be tested, and new development teams must be established.

Just like with the basic software integration, semantic, protocol, real-time, and resource consumption integration problems are not addressed sufficiently. I.e. predictions about the functional and real-time behavior of the integrated ECU cannot be made. Since AUTOSAR does not cover algorithmic models, a solution to the semantic and protocol problem cannot be expected.

Predicting the precise resource consumption of the ECU (i.e. processor and memory usage) of the integrated software is another unsolved but highly relevant issue. Estimation techniques and simulation approaches might help in the future to ease this problem.

ECU System Integration: In this final step, the ECUs are integrated into the overall ECU network. The main agreement or contract between the ECUs is the communication configuration, i.e. the messages and signals used to transport information on the communication network. In traditional development processes, this configuration is defined first – and it is defined manually. In AUTOSAR, this configuration is derived automatically from the mapping of the software architecture on the hardware topology. This eases the integration since software architecture and network configuration are therefore synchronized automatically – an example of the synchronization principle of Section 2.2.3.

Real-time problems such as too high message delays on the network are not addressed. Neither are resource consumption problems such as too-high network loads. Again, due the lack of behavior models in AUTOSAR, problems concerning the dynamic interaction between ECUs (semantic and protocol problems) cannot be expected to be solved by AUTOSAR. Dependability issues such as redundancy are also relevant but are currently not solved satisfyingly by AUTOSAR.

AUTOSAR is continuing to extend the standard (see, e.g., [18]). Currently AUTOSAR works on topics such as variant management, MultiCore support, functional safety, and the modeling of timing information such as end-to-end timing on the application level – this may ease the real-time integration problems.

This short overview of AUTOSAR's role in the automotive software integration process shows that AUTOSAR helps mainly with statical, functional integration problems such as syntactic, data flow, or technology issues. Dynamical problems such as semantic and protocol issues are not solved, neither are non-functional issues such as the estimation of resource consumptions. So AUTOSAR is not the "golden bullet" for integration but only a first step towards a software-aware development process in the automotive industry.

Several studies have shown that AUTOSAR requires significant changes of the development processes and of current business models: Software becomes a product, software models must be created, new roles – e.g., a software architect – must be established, manufacturers try to become software integrators, and new tools must be introduced.

This leads to a problem that goes beyond simple missing features of AUTOSAR such as insufficient support for dynamic, real-time or non-functional integration aspects like in the case of needed resources: AUTOSAR's approach to software engineering has been, from the very beginning on, based on the component-oriented software engineering paradigm – mainly influenced by the EAST project (see [19]). This paradigm requires an explicit software architecture defined as inter-communicating software components. And it requires therefore an explicit software architect, an explicit tool chain for software architectures, explicit verification and testing strategies for software architectures, and especially an explicit software architecture design step in the development process.

Table 2.6. Coverage of integration aspects using AUTOSAR

Integration problems	AUTOSAR	Explanation
technological	D/E/S	C-APIs; provides standardized platform and supports code generation; code generation for the implementation of the RTE provided by tools
syntactical	D/A/E	AUTOSAR standardized APIs and means to define components and ports; virtual function bus provides realization; provides layered architecture with interfaces between
semantic	C	checked during integration testing
protocol	C	checked during integration testing
dataflow	C	checked when programming and compiling the integrated code, checked during integration testing
maintainability	D/A	decomposition the software architecture; abstraction via standardized interfaces between different layer
real-time compatibility		
local	C	scheduling analysis of the integrated AUTOSAR/OSEK tasks
distributed	C	using simulation techniques later in the development lifecycle
resource consumption		
local	C	resource analysis of the integrated tasks and their code
distributed	C	analysis of the bus allocation for the integration of multiple nodes

(legend: A = during abstraction – C = during composition – D = during decomposition – E = during enrichment –

S = by synthesis)

While these requirements can be met in classical computer science domains such as business software or telecommunication, this must not be true for the automotive software development. This domain possesses an established development process based on ideas from control theory and signal processing – and it possesses an adequate established tool chain and adequately trained developers. So one might ask whether AUTOSAR should have chosen a software architecture paradigm leveraging established procedures. And one might ask whether, instead of choosing a software architecture approach from a technical (computer science) point of view, AUTOSAR should have chosen an approach which would minimize changes to existing development processes and which would exploit strengths of automotive's long-term and successful software development history.

To give an example: Data-centric software engineering approaches (see [20, 21] for details) couple software components via a signal repository. Components may either write or read signals in the repository. Unlike with component-oriented approaches, no explicit software architectures are required – and therefore no separate tool chains and fewer changes to development processes are needed. And such an approach also resembles the existing automotive development process where ECUs communicate via communication buses, i.e. via a common pool of bus signals. Of course, this does not mean that a data-centric approach would solve all problems. But it may serve as an example that fundamentally different alternatives would have existed and might have demanded fewer changes to the established development process.

2.4.2 MECHATRONIC UML

As outlined in Section 2.3.2, the current practice for model-based development of software components with hard real-time constraints – whether AUTOSAR is employed or not – is characterized by the following step-wise partially manual process (1) *Specification*, (2) *Partitioning*, (3) *Implementation*, and (4) *Integration* which has to be repeated when the integration is not able to fulfill the required real-time constraints.

Consequently, it would be attractive to extend the idea of model-driven architecture (MDA) [22, 23] to design software for embedded hard real-time systems. When using MDA for such systems, the developer would have to specify the so-called *Platform-Independent Model (PIM)* which describes the system behavior including the real-time constraints which must be met. Ideally, a tool would then automatically partition the specification and map it to the *Platform-Specific Model (PSM)*, based on a *Platform Model (PM)* that provides details about the target platform. The PSM describes the active objects and their scheduling parameters which are required to implement the system behavior, specified by the PIM. In the next step, the PSM would be compiled automatically into the platform-specific implementation which guarantees a correct implementation of the PIM's semantics. The implementation would guarantee the real-time constraints by construction and thus, no verification of the real-time constraints is required. This would make the above mentioned manual steps *(3) Implementation* and *(4) Integration* unnecessary. However, the UML standard as well as proposed extensions for embedded real-time systems [24, 25, 26, 27, 28, 29, 30, 31] fail to provide a proper basis for this as the suggested models are not sufficient to talk about platform-independent real-time behavior.

The MECHATRONIC UML approach (MUML) [32] in contrast provides the missing platform-independent real-time models and also supports MDA for embedded real-time systems [33]. Therefore, by applying MUML the sketched iterative manual process often followed today in practice can be avoided by using the automatic mapping of a PIM to a PSM that is appropriate for real-time systems. In addition to (1) MDA for embedded real-time systems, MUML provides support for two particular problematic cases for integration embedded real-time systems: (2) the real-time coordination of embedded real-time systems and (3) their safety analysis. Tool support for (MUML) is provided in the form of the Fujaba real-time tool suite, which offers a wide range of UML based diagrams, the appropriate extension for the specification of real-time properties as well as modelchecking and consistency analysis support [34].

(1) MDA for Embedded Real-Time Systems

The structure of embedded real-time systems consist of a complex architecture of components. UML [35] despite it shortcomings can be considered as the standard to model complex software systems even in the real-time domain [28, 29, 30, 31]. MUML therefore supports to specify the architecture and complex real-time communication between the components by UML component diagrams and patterns respectively [36].

The semantics of the UML State Machines assumes the transitions to be fired within zero-time cannot be realized in practice and the pragmatic interpretation that zero-time means *fast enough* is only helpful in simple systems where a single periodic deadline can characterize for the whole state machine and it states what *fast enough* means. Therefore, in MUML *Real-Time Statecharts* (RTSC) [32, 37] extend UML State Machines to allow the explicit specification of the really required timing. Transitions are not assumed to fire *infinitely fast*, which is unrealistic on real physical devices (especially when considering the execution of the actions attached to the transitions), but it is possible to specify deadlines for each transition which in turn determine what *fast enough* really is. Similar to the notion in timed automata [38, 39] clocks and clock invariants are employed to describe when transitions are enabled and what the minimum time and the maximum time (d_0, t_{ans}) for finishing the execution of a transition has to be (more details see [33]).

Generating a PSM, consisting of active objects and deadlines, that guarantee the real-time constraints as specified in the model is of course only possible, when the model does not contain any conflicts between the declarative elements such as time guards and time invariants. A possible conflict is, for example, when multiple real-time constraints are contradicting and thus no behavior exists which fulfills them (time-stopping deadlock). To exclude such conflicts, the full state space of a Real-Time Statechart model has to be checked in the general case. As outlined in [33], model checking with UPPAAL and static analysis techniques can be employed to exclude such conflicts.

In order to generate the PSM, WCETs are required for all actions (side effects, *entry()*, *exit()*, and *do()*- operations) and for the elementary instructions that build the code fragments realizing the Real-Time Statechart behavior (e.g. checking guards, raising events, etc.).

As the WCETs are platform-dependent, we first deploy our components (whose behavior is each specified by a Real-Time Statechart) by a UML deployment diagram. In such a deployment diagram, we assign the component instances of our systems to dedicated nodes and the cross node links to available network connections in form of busses or direct communication links. Given such an assignment, we can further look into the specific characteristics of the different nodes as described in the platform model.

To analyze the resulting model with platform-specific annotations, we extend our timed automata model for model checking as well as our static analysis technique such that it also reflects the WCET behavior of the side effects of the transitions (cf. [33]).

After modeling and analyzing the PIM with components and Real-Time Statecharts and specifying the platform-specific WCET information in the PM and the deployment, we have to map the components and links to active objects and to network and communication links to come up with the final platform-specific model. In our case the PSM can be described by the UML Profile for Schedulability, Performance, and Time [29], as it allows the specification of priorities, periods, and deadlines for active objects. We use it as a platform-*specific* model, as these values, which we derive automatically from the platform-*independent* model, are different for different platforms. For such a PSM we can derive code that guarantees the in the PIM and PSM specified timing constraints for Real-Time Java and C++.

While MUML has been developed in the context of a research project different case studies have been realized like described in [40] using an evaluation platform equipped with a 40 Mhz Power PC processor. For the derivation of WCETs the tool Bound-T[6] has been employed within the evaluation example described in [41].

Table 2.7. Coverage of integration aspects using MUML

Integration problems	MUML	Explanation
technological	D*	platform-independent model and code generation for map those the a specific platform (*but only realized for one)
syntactical	D/C	models capture components and ports; mapping to code provides realization
semantic	(D)	model checking of the models prevent some semantic integration problems
protocol	D	model checking of the models exclude protocol-related integration problems
dataflow	D	dataflow part of the interface and modular syntax checks guarantee proper dataflow specification
dependability		
safety	D	compositional hazard analysis of the models enable up-front guarantees; requires HW reliability data
real-time compatibility		
local	S	generated task periods and scheduling analysis guarantee correct timing
distributed	D	generated local tasks plus model checking guarantee correct distributed timing
resource consumption		
local		
distributed		

(legend: A = during abstraction − C = during composition − D = during decomposition − S = by synthesis).

(2) Correct Real-Time Coordination

As MUML further provides a compositional verification approach for the real-time coordination of systems of systems with reconfiguration [36, 42, 43, 44]. It further allow the model-based analysis of interoperability problems for the functional and real-time behavior. By extending UML components the syntactical compatibility (data, operations, ...) and semantic compatibility (data, operations, ...) is guaranteed while a run-time environment guarantees technological compatibility. In [45] is outlined how MUML interfaces also take care of the execution order of dataflow computations employed for evaluating control algorithms. The extended port specifications by means of RTSC together with the mentioned verification further ensure protocol compatibility (non uniform service availability, synchronization) and real-time compatibility.

(3) Safe Real-Time Systems

In addition, an approach for a compositional safety analysis [46, 47] permits to do a model-based upfront analysis of the resulting system safety when decomposing

---

[6] http://www.tidorum.fi/bound-t/

the systems into components in the form of an architecture. Therefore, safety issues have not to be addressed later when integrating the components into the overall system.

To sum-up, Table 2.7 provides an overview of the integration problems more or less covered by the MUML approach. Like mentioned before MUML has been developed in the context of a research project. While several studies have shown the applicability of the approach a coherent professional tool chain currently does not exist.

2.4.3 Other Approaches

A number of other approaches using models that also address several of the integration problems outlined in Figure 2.1. We will provide only a sketch of their benefits in the following text and refer to the referenced literature resp. chapters for more information.

Infrastructure Abstraction

Several approaches address like MUML the problem that real-time issues can in the traditional approach only be addressed rather late and that the resolution of related integration problems can become quite costly. The time-triggered approach [48] addresses this problem at the hardware and network level and provides a platform where the different real-time communication issues can be clearly separated with respect to time and dependability. An approach which similar to MUML address the timing problem at a single node is Giotto [49] as well as its successor TDL (see Chapter 5). Here a virtual machine guarantees that time constraints specified in the specification are guaranteed by the execution environment also allowing the higher level abstractions to be analyzed to detect protocol compatibility problems.

Other approaches try to synthesize a proper task allocation from a given software model [50] in order to meet the timing requirements. In addition, besides avoiding the integration problem a model-based analysis of the composition of distributed real-time embedded system may also be simply beneficial by enabling an earlier analysis [51].

An approach targeting to an earlier analysis concerning the later used HW infrastructure, including multiple nodes and the communication path between them is described in [52]. Virtual execution platforms, which represent the later used HW infrastructure, are used to provide an execution environment for simulation purpose, taking characteristics like the execution time into account.

An approach somehow in the middle is platform-based design [53] where no full abstraction is provided but instead the stepwise realization of higher-level abstractions by means of underlying platform components are the main design step. This often allows to reduce the integration problems as designs are derived by proper combinations of components with some degree of built-in compatibility as they together represent a platform and not a ragtag group of components. In some cases even correctness-by-construction can be achieved [54] by means of a platform-based approach.

The problem is also related to problem of heterogeneity (technology as well as semantics) which is in particular problematic when the artifacts to be integrated have not been decomposed upfront with the same model of computation (see Chapter 1).

Interfaces and Component Models for Integration
Like advocated in AUTOSAR and the MUML approach, interfaces and component models are a suitable concept to address integration issues upfront when decomposing a system. Related approaches propose extended interface for component-based design [55, 56] also covering stateless and stateful protocol behavior as well as real-time behavior. In [57], like in [45] for MUML, interfaces that take care of the execution order of dataflow as employed for computing control algorithms are presented.

One more component-oriented approach is the rich component model [58] that has been proposed mainly targeting reuse but also support early checking to avoid integration problems. Another is the Behavior-Interaction-Priority (BIP) component framework [59, 60] that can ensure the proper deadlock-free composition using a much simpler check of the resulting dependency graph rather than the complete component synchronization and thus permits to do it upfront when decomposing a system.

In [3] several approaches are discussed that provide specific DSLs for component models (EAST-ADL and AADL): For example, AADL [4] is a DSL for the development of embedded real-time systems which supports the description and analysis of the system architecture addressing the integration of SW and HW parts which can be developed by different stakeholders. EAST-ADL is a DSL and architecture description language which is based on UML and SysML. One key aspect of EAST-ADL is the usage of abstraction and an according system model is structured with several abstraction layers. The EAST-ADL language can be used within a tool, like it is done in the form of Papyrus for EAST-ADL.

Integrated Model-Based Development
Another thread of work focuses on a proper representation of all required system characteristics by means of models and their consistent further elaboration. In [3] several approaches are discussed that provide tool support (Fujaba [5], GeneralStore [61], ToolNet [62] and IDM [63]) for the integrated model-based development of embedded systems.

For keeping the different models, potentially used in different tools consistent, model-transformation and model-synchronization techniques can be used. In [64] the authors describe how modelsynchronization is used to keep AUTOSAR and SysML models consistent.

In addition in [65] model-integrated computing (MIC) as a paradigm to address the integration problems for embedded real-time systems. In [66] it is advocated that the MIC approach employing a number of domain-specific languages (DSL), supporting the proper consistency of the different model, doing frequent model analysis by mapping these models to available analysis tools, and generating refined models as well as code (synthesis) can help to substantially reduce the later experienced integration problems. In [67] an implementation of model-based integration for the development of avionics systems is evaluated. Within this evaluation the benefits of the Model-Based Integration of Embedded Software (MoBIES) development process[7] and a special DSL (ESML) for the development of avionic systems are evaluated.

2.5 Summary

If we review the presented results, we can conclude that a number of promising approaches for different problems exist, while no solution for the overall problem seems available. This impression is also confirmed by the coverage of the integration problems summarized in Table 2.8.

Table 2.8. Coverage of integration aspects by the different approaches

Integration problems	FD	FI	AU-TOSAR	mUML	Other approaches
technological	A/E(S)	A/C	D/E/S	(D)	
syntactical	D/C	A/(D)/C	D/A/E	D(UML)	D [55, 56, 59, 60]
semantic	D/E(S)/P	C/P	C	D(UML)	D [55, 56, 59, 60]
protocol		C	C	D	D [55, 56, 59, 60]
dataflow	A/E(S)/P	C/(P)	C	D	D [57]
dependability/ quality of service					
reliability		(C)			
availability		(C)			
safety		(C)		D	D [58]
security		(C)			
maintainability			D		
real-time compatibility					
local	E/P	C	C	S	S [49], see TDL in Chapter 6
distributed		C/(P)	C	D	D [56, 58], S [48, 50]
resource consumption					
local	E/P	C	C		
distributed		C	C		D [48]

(legend: FD = function development – FI = function integration A = during abstraction – C = during composition – D = during decomposition – E = during enrichment – P = during parallel development – S = by synthesis).

If we review the summary of the findings depicted in Table 2.8, we can make the following specific observations:

- It seems that the need for support of the functional integration problem at the technological and syntactical level has been identified also in industry and AUTOSAR or related approaches start to address them in a standardized manner.

---

[7] A project funded by the Defense Advanced Research Projects Agency (DARPA).

- In contrast semantic, protocol or dataflow issues are at first addressed by academic research projects but in practice they are addressed rather late if at all (compare Section 2.3). Here it seems beneficial if the existing research results could be transferred into industrial strength solutions to minimize the integration costs by addressing these issues earlier in the development life cycle. However, as these issues related to some extent to formal modeling, it is not clear whether such a transfer is really possible taking the existing workforce and their educational background into account.
- The integration of non-functional dependability resp. quality-of-service aspects is besides safety not very well covered either by industrial nor research approaches. This is probably due to the fact that these are often system properties which could not be easily established using in a compositional manner and indicates that these topics require much more attention from the research community.
- Concerning real-time compatibility we can observe that several well-suited research results have been achieved and some of them are in a transition phase to industrial praxis (e.g., see TDL in Chapter 5). These solution promises to ease the integration efforts considerably as they permit to exclude that the integration problems are detected rather late resulting in enormous costs due to the required rework of the integrated solutions.
- Finally, the resource consumption is a currently rather superficially covered aspect. However, the importance of hardware costs in fields like the automotive domain as well as the increasing importance of energy efficient and resource-aware solutions will make this aspect another highly relevant research topic. Here also the problem seems to be that resource consumption is a system property that is not easily addressed in a compositional manner.

If we take a look at the overall picture, we can see that handling an integration problem at composition time (C) as advocated in the traditional functional development and functional integration is in principle always possible. However, there is a clear trend that model-based integration results in a front loading where instead of costly efforts to handle integration problems after the fact these problems are upfront addressed by decomposition/composition (D), abstraction/enrichment (A) or parallel development & consistency (P). The more mature approaches are those ones where instead of checking integration problems late when combining the different system constituents, the seperation in the form of decomposition or abstraction provides already the basis to exclude or limit most of these problems upfront (see also [9] for a related observation based on several industrial studies).

However, it also became apparent that besides MUML and rich components [58] most approaches provide a rather isolated solution to one integration problem. Therefore, the main challenge for integration seems to be establishing a comprehensive solution that covers not only the rather simple problems such as syntax and technology compatibility but also most of the challenging aspects such as protocol compatibility, dataflow compatibility and real-time behavior. As most proposals for these advanced integration concepts have not yet been employed thoroughly in industrial practice and may have contradicting constraints, it is not clear whether such an "integration of advanced integration concepts" is really feasible.

Therefore, the current challenge is not only to develop better solutions of the outlined separate integration problems (cf. Figure 2.1) but also to combine the existing solutions into overall integration approaches that provide a coherent solution that covers all required integration problems. It can be expected that suitable overall integration approaches have to be tailored for the specific domain of embedded real-time systems such as AUTOSAR while its ingredients will often be applicable in several domains.

Acknowledgements

We thank Ingolf Krüger and Florence Maraninchi for their feedback on earlier versions of the paper.

References

[1] Lane, J.A., Boehm, B.: System of systems lead system integrators: Where do they spend their time and what makes them more or less efficient? Systems Engineering 11(1), 81–91 (2008)
[2] Sage, A.P., Lynch, C.L.: Systems integration and architecting: An overview of principles, practices, and perspectives. Systems Engineering 1(3), 176–227 (1998)
[3] Chen, D., Torngren, M., Shi, J., Gerard, S., Lonn, H., Servat, D., Stromberg, M., Arzen, K.E.: Model integration in the development of embedded control systems - a characterization of current research efforts. In: 2006 IEEE International Symposium on Computer-Aided Control Systems Design, October 4-6, pp. 1187–1193 (2006)
[4] Feiler, P., Gluch, D., Hudak, J.: The architecture analysis & design language (aadl): An introduction. Technical Report CMU/SEI-2006-TN-011, Software Engineering Institute, Carnegie Mellon University, Pittsburgh, PA, USA (2006)
[5] Burmester, S., Giese, H., Niere, J., Tichy, M., Wadsack, J.P., Wagner, R., Wendehals, L., Zündorf, A.: Tool Integration at the Meta-Model Level within the FUJABA Tool Suite. International Journal on Software Tools for Technology Transfer (STTT) 6(3), 203–218 (2004)
[6] Mosterman, P.J., Ghidella, J., Friedman, J.: Model-based design for system integration. In: Second CDEN International Conference on Design Education, Innovation, and Practice, Kananaskis, Alberta, Canada, July 18-20 (2005)
[7] Dijkstra, E.W.: On the role of scientific thought, pp. 60–66. Springer, New York (1982)
[8] Parnas, D.L.: On the Criteria To Be Used in Decomposing Systems Into Modules. Communications of the ACM 15(12), 1053–1058 (1972)
[9] Bosch, J., Bosch-Sijtsema, P.: From integration to composition: On the impact of software product lines, global development and ecosystems. Journal of Systems and Software 83(1), 67–76 (2010) (SI: Top Scholars)
[10] Küster, J.M., Engels, G.: Consistency management within model-based object-oriented development of components. In: de Boer, F.S., Bonsangue, M.M., Graf, S., de Roever, W.-P. (eds.) FMCO 2003. LNCS, vol. 3188, pp. 157–176. Springer, Heidelberg (2004)

[11] Giese, H., Wagner, R.: From model transformation to incremental bidirectional model synchronization. Software and Systems Modeling 8(1) (March 2009) (online first: 3/2008)
[12] Mellor, S., Scott, K., Uhl, A., Weise, D.: MDA Distilled: Principles of Model-Driven Architecture. Addison-Wesley, Reading (2004)
[13] Watkins, C.B.: Modular Verification: Testing a Subset of Integrated Modular Avionics in Isolation. In: 25th Digital Avionics Systems Conference, 2006 IEEE/AIAA, Portland, OR, IEEE Xplore (2006)
[14] Broekman, B., Notenboom, E.: Testing Embedded Software. Addison-Wesley, Reading (2003)
[15] AUTOSAR: Web page, http://www.autosar.org/
[16] Fennel, H., Bunzel, S., et al.: H.H.: Achievements and Exploitation of the AUTOSAR Development Partnership. In: Convergence, Detroit, USA (2006) (SAE 2006-21-0019)
[17] Richter, K.: On the Complexity of Adding Real-Time Properties to the AUTOSAR Software Component Model. In: Proc. of the 4th Workshop on Object-oriented Modeling of Embedded Real-Time Systems (OMER 4), Paderborn, Germany (October 2007)
[18] Fürst, S.: AUTOSAR - A World Wide Standard is on the Road. In: 14th International VDI Congress Electronic Systems for Motor Vehicles, Baden-Baden, Germany (October 2009)
[19] Lönn, H.: Far east: Modeling an automotive software architecture using the east adl. In: ICSE 2004 workshop on Software Engineering for Automotive Systems, SEAS (2004)
[20] Oki, B., Pfluegl, M., Siegel, A., Skeen, D.: The information bus: an architecture for extensible distributed systems. In: SOSP 1993: Proceedings of the fourteenth ACM symposium on Operating systems principles, pp. 58–68. ACM, New York (1993)
[21] Pardo-Castellote, G.: OMG Data-Distribution Service: Architectural Overview. In: International Conference on Distributed Computing Systems Workshops, p. 200 (2003)
[22] Allen, P. (ed.): The OMG's Model Driven Architecture. Component Development Strategies, The Monthly Newsletter from the Cutter Information Corp. on Managing and Developing Component-Based Systems, vol. XII (January 2002)
[23] Object Management Group: MDA Guide Version 1.0, Document omg/2003-05-01 (May 2003)
[24] Selic, B., Gullekson, G., Ward, P.: Real-Time Object-Oriented Modeling. John Wiley & Sons, Inc., Chichester (1994)
[25] Awad, M., Kuusela, J., Ziegler, J.: Object-Oriented Technology for Real-Time Systems: A Practical Approach Using OMT and Fusion. Prentice Hall, Englewood Cliffs (1996)
[26] Douglass, B.P.: Real-Time UML: Developing Efficient Objects for Embedded Systems, 2nd edn. The Addison-Wesley Object Technology Series. Addison-Wesley, Reading (October 1999)
[27] Gomaa, H.: Designing Concurrent, Distributed, and Real-Time Applications with UML. Addison-Wesley, Reading (January 2000)
[28] Bichler, L., Radermacher, A., Schürr, A.: Evaluation uml extensions for modeling realtime systems. In: Proc. on the 2002 IEEE Workshop on Object-oriented Realtime-dependable Systems, WORDS 2002, San Diego, USA, pp. 271–278. IEEE Computer Society Press, Los Alamitos (2002)

[29] Object Management Group: UML Profile for Schedulability, Performance, and Time Specification. OMG Document ptc/02-03-02 (September 2002)
[30] Gu, Z., Kodase, S., Wang, S., Shin, K.G.: A Model-Based Approach to System-Level Dependency and Real-Time Analysis of Embedded Software. In: The 9th IEEE Real-Time and Embedded Technology and Applications Symposium, Toronto, Canada (2003)
[31] Masse, J., Kim, S., Hong, S.: Tool Set Implementation for Scenario-based Multithreading of UML-RT Models and Experimental Validation. In: The 9th IEEE Real-Time and Embedded Technology and Applications Symposium, Toronto, Canada (May 2003)
[32] Burmester, S., Giese, H., Tichy, M.: Model-Driven Development of Reconfigurable Mechatronic Systems with Mechatronic UML. In: Aßmann, U., Aksit, M., Rensink, A. (eds.) MDAFA 2003. LNCS, vol. 3599, pp. 47–61. Springer, Heidelberg (2005)
[33] Burmester, S., Giese, H., Schäfer, W.: Model-driven architecture for hard real-time systems: From platform independent models to code. In: Hartman, A., Kreische, D. (eds.) ECMDA-FA 2005. LNCS, vol. 3748, pp. 25–40. Springer, Heidelberg (2005)
[34] Burmester, S., Giese, H., Hirsch, M., Schilling, D., Tichy, M.: The Fujaba Real-Time Tool Suite: Model-Driven Development of Safety-Critical, Real-Time Systems. In: ICSE 2005: Proceedings of the 27th International Conference on Software Engineering, pp. 670–671. ACM Press, New York (2005)
[35] Object Management Group: UML 2.0 Superstructure Specification. Document: ptc/04-10-02 (convenience document) (October 2004)
[36] Giese, H., Tichy, M., Burmester, S., Schäfer, W., Flake, S.: Towards the Compositional Verification of Real-Time UML Designs. In: Proc. of the European Software Engineering Conference (ESEC), Helsinki, Finland. ACM Press, New York (2003)
[37] Giese, H., Burmester, S.: Real-Time Statechart Semantics. TechReport tr-ri-03-239, University of Paderborn (2003)
[38] Larsen, K., Pettersson, P., Yi, W.: UPPAAL in a Nutshell. Springer International Journal of Software Tools for Technology 1(1) (1997)
[39] Henzinger, T.A., Nicollin, X., Sifakis, J., Yovine, S.: Symbolic Model Checking for Real-Time Systems. In: Proc. of IEEE Symposium on Logic in Computer Science (1992)
[40] Tichy, M., Giese, H., Seibel, A.: Story Diagrams in Real-Time Software. In: Giese, H., Westfechtel, B. (eds.) Proc. of the 4th International Fujaba Days, Bayreuth, Germany. Volume tr-ri-06-275 of Technical Report. University of Paderborn, pp. 15–22 (September 2006)
[41] Henkler, S., Oberthur, S., Giese, H., Seibel, A.: Model-Driven Runtime Resource Predictions for Advanced Mechatronic Systems with Dynamic Data Structures. In: Proc. of 13th International Symposium on Object/component/service-oriented Real-time distributed Computing (ISORC), May 5-6. IEEE Computer Society Press, Los Alamitos (accepted 2010)
[42] Burmester, S., Giese, H., Hirsch, M., Schilling, D.: Incremental Design and Formal Verification with UML/RT in the FUJABA Real-Time Tool Suite. In: Proceedings of the International Workshop on Specification and vaildation of UML models for Real Time and embedded Systems, SVERTS 2004, Satellite Event of the 7th International Conference on the Unified Modeling Language, UML 2004 (October 2004)

[43] Becker, B., Beyer, D., Giese, H., Klein, F., Schilling, D.: Symbolic Invariant Verification for Systems with Dynamic Structural Adaptation. In: Proc. of the 28th International Conference on Software Engineering (ICSE), Shanghai, China. (2006)
[44] Becker, B., Giese, H.: On Safe Service-Oriented Real-Time Coordination for Autonomous Vehicles. In: Proc. of 11th International Symposium on Object/component/service-oriented Real-time distributed Computing (ISORC), May 5-7, pp. 203–210. IEEE Computer Society Press, Los Alamitos (2008)
[45] Burmester, S., Giese, H., Gambuzza, A., Oberschelp, O.: Partitioning and Modular Code Synthesis for Reconfigurable Mechatronic Software Components. In: Bobeanu, C. (ed.) Proc. of European Simulation and Modelling Conference (ESMc 2004), Paris, France, pp. 66–73. EOROSIS Publications (2004)
[46] Giese, H., Tichy, M., Schilling, D.: Compositional Hazard Analysis of UML Components and Deployment Models. In: Heisel, M., Liggesmeyer, P., Wittmann, S. (eds.) SAFECOMP 2004. LNCS, vol. 3219, pp. 166–179. Springer, Heidelberg (2004)
[47] Giese, H., Tichy, M.: Component-Based Hazard Analysis: Optimal Designs, Product Lines, and Online-Reconfiguration. In: Górski, J. (ed.) SAFECOMP 2006. LNCS, vol. 4166, pp. 156–169. Springer, Heidelberg (2006)
[48] Kopetz, H., Bauer, G.: The time-triggered architecture. Proceedings of the IEEE 91(1), 112–126 (2003)
[49] Henzinger, T., Horowitz, B., Kirsch, C.: Giotto: a time-triggered language for embedded programming. Proceedings of the IEEE 91(1) (January 2003)
[50] Wang, S., Shin, K.G.: Task construction for model-based design of embedded control software. IEEE Trans. Software Eng. 32(4), 254–264 (2006)
[51] Madl, G., Abdelwahed, S.: Model-based analysis of distributed real-time embedded system composition. In: EMSOFT 2005: Proceedings of the 5th ACM international conference on Embedded software, pp. 371–374. ACM, New York (2005)
[52] Krause, M., Bringmann, O., Hergenhan, A., Tabanoglu, G., Rosentiel, W.: Timing simulation of interconnected AUTOSAR software-components. In: DATE 2007: Proceedings of the conference on Design, automation and test in Europe, San Jose, CA, USA, EDA Consortium, pp. 474–479 (2007)
[53] Sangiovanni-Vincentelli, A.: Defining platform-based design. EEDesign of EE-Times (February 2002)
[54] Horowitz, B., Liebman, J., Ma, C., Koo, T., Sangiovanni-Vincentelli, A., Sastry, S.: Platform-based embedded software design and system integration for autonomous vehicles. Proceedings of the IEEE 91(1), 198–211 (2003)
[55] de Alfaro, L., Henzinger, T.A.: Interface theories for component-based design. In: Henzinger, T.A., Kirsch, C.M. (eds.) EMSOFT 2001. LNCS, vol. 2211, pp. 148–165. Springer, Heidelberg (2001)
[56] Thiele, L., Wandeler, E., Stoimenov, N.: Real-time interfaces for composing real-time systems. In: EMSOFT 2006: Proceedings of the 6th ACM & IEEE International conference on Embedded software, pp. 34–43. ACM, New York (2006)
[57] Zhou, Y., Lee, E.A.: A causality interface for deadlock analysis in dataflow. In: EMSOFT 2006: Proceedings of the 6th ACM & IEEE International conference on Embedded software, pp. 44–52. ACM, New York (2006)
[58] Damm, W., Votintseva, A., Metzner, A., Josko, B., Peikenkamp, T., Böde, E.: Boosting re-use of embedded automotive applications through rich components. In: Proc. of Foundations of Interface Technologies 2005, FIT 2005 (2005)
[59] Gössler, G., Sifakis, J.: Composition for component-based modeling. Sci. Comput. Program. 55(1-3), 161–183 (2005)

[60] Bliudze, S., Sifakis, J.: The algebra of connectors: structuring interaction in bip. In: EMSOFT 2007: Proceedings of the 7th ACM & IEEE international conference on Embedded software, pp. 11–20. ACM, New York (2007)

[61] Reichmann, C., Markus, K., Graf, P., Müller-Glaser, K.D.: Generalstore - a casetool integration platform enabling model level coupling of heterogeneous designs for embedded electronic systems. In: ECBS 2004: Proceedings of the 11th IEEE International Conference and Workshop on Engineering of Computer-Based Systems, Washington, DC, USA, p. 225. IEEE Computer Society, Los Alamitos (2004)

[62] Altheide, F., Dörr, H., Schürr, A.: Requirements to a Framework for sustainable Integration of System Development Tools. In: Stoewer, H., Garnier, L. (eds.) Proc. of the 3rd European Systems Engineering Conference (EuSEC 2002), Toulouse, AFIS PC Chairs, pp. 53–57 (2002)

[63] Karsai, G., Lang, A., Neema, S.: Design patterns for open tool integration. Software and System Modeling 4(2), 157–170 (2005)

[64] Giese, H., Hildebrandt, S., Neumann, S.: Towards Integrating SysML and AUTOSAR Modeling via Bidirectional Model Synchronization. In: 5th Workshop on Model-Based Development of Embedded Systems, MBEES (2009)

[65] Sztipanovits, J., Karsai, G.: Model-Integrated Computing. Computer 30(4), 110–111 (1997)

[66] Karsai, G., Sztipanovits, J., Ledeczi, A., Bapty: Model-integrated development of embedded software. Proceedings of the IEEE 91, 145–164 (2003)

[67] Schulte, M.: Model-based integration of reusable component-based avionics systems - a case study. In: ISORC 2005: Proceedings of the Eighth IEEE International Symposium on Object-Oriented Real-Time Distributed Computing, Washington, DC, USA, pp. 62–71. IEEE Computer Society, Los Alamitos (2005)

Part II

Language Engineering

3 Metamodelling
State of the Art and Research Challenges

Jonathan Sprinkle[1], Bernhard Rumpe[2], Hans Vangheluwe[3], and Gabor Karsai[4]

[1] University of Arizona, Tucson, AZ, USA
sprinkle@ECE.Arizona.Edu
[2] RWTH Aachen University, Germany
http://www.se-rwth.de
[3] McGill University, Montreal, Canada
hv@cs.mcgill.ca
[4] Vanderbilt University, Nashville, TN, USA
gabor.karsai@vanderbilt.edu

Abstract. This chapter discusses the current state of the art, and emerging research challenges, for metamodelling. In the state-of-the-art review on metamodelling, we review approaches, abstractions, and tools for metamodelling, evaluate them with respect to their expressivity, investigate what role(s) metamodels may play at run-time and how semantics can be assigned to metamodels and the domain-specific modeling languages they could define. In the emerging challenges section on metamodelling we highlight research issues regarding the management of complexity, consistency, and evolution of metamodels, and how the semantics of metamodels impacts each of these.

3.1 Metamodelling: State of the Art

Models are powerful tools to express the structure, behavior, and other properties in mathematics, each of the hard sciences and in all areas of engineering. While models are very common, an explicit definition of a modelling language and an explicit manipulation of its models is tightly connected to computer based tools. Additional power can be gained by explicit definition and computer based manipulation of models e.g. in CAD, control engineering, algebraic mathematics and of course computer science. To be able to manipulate models, their language needs to be specified as model of these models—metamodels. In this section, we describe the state of the art for metamodelling, including the metamodelling of data structures, as well as the metamodelling of languages systems where appropriate.

3.1.1 Concepts in Metamodelling

Metamodelling (literally, "beyond Modelling") is the Modelling of models. In their most common use, metamodels describe the permitted structure to which models must adhere [1]; although out of the scope of this chapter, meta-meta-models formally describe metamodels, as they define the core abstractions permitted in metamodelling. In fact, some metamodelling languages are self-descriptive [2, 3]. A metamodel therefore describes the syntax of the models [4]. Through various extension mechanisms and additional rules with this representation of the syntax of models metamodels can also help to define the semantics of models, as we discuss later. The layered approach to modelling (through metamodelling) is depicted in Figure 3.1.

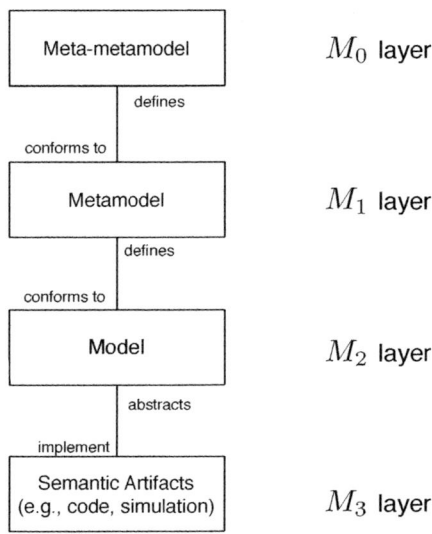

Fig. 3.1. The Four-layer metamodelling Architecture [5, 2, 6]. Some publications reverse the numbering of these layers, but we use this numbering, as platform transformations may create layers of arbitrary depth (e.g., M_3 may in fact be models defined in another layer M_2).

In the four-layer approach (generalized to n-layers in the MDA, discussed next) artifacts in each layer conform to, and/or are abstracted by the more abstract layer adjacent (in this case, the layer with a lower subscript). Thus, semantic artifacts in the M_3 layer are abstracted in models from the M_2 layer, which in turn conform to the metamodels from the M_1 layer. As these layers of abstraction are traversed, the role of each abstraction layer changes.

For example, a model is an encoding of some application or design in a different abstraction. Metamodels constrain the structure (and perhaps behavior) of models, but metamodels are relevant to all designs, not just a specific design. A widely-known example of this is that an XML document conforms to some type definition (either a DTD or XSD schema), but given only a schema, it is not possible to recover a particular XML file. When modelling languages, a metamodel basically needs to be considered a model of the abstract syntax of a language.

In Figure 3.1 we mention semantic artifacts. These are data, running programs, files, etc. that have some *meaning* in another context, e.g. by the user. They are artifacts in that they are produced through the design process. The running programm is regarded as a semantic artifact by a number of approaches, if it is generated from high-level models, such as a state chart or dataflow model, while others regard code mainly as another (and final) syntactic representation of the system to be developed. For a metamodelling approach to have significant impact, some of those artifacts must eventually be produced in the design process; else, the modelling process is best classified as sophisticated documentation.

Metamodelling: A Design Process
Generally, a design process that utilizes metamodelling first involves abstraction of the concepts of some domain or application into the appropriate meta-types (these are defined by the meta-metamodel M_0), using the metamodelling tools working on the M_1 layer. Metamodels can apply various archetypal concepts to constrain how models models are built, as shown in Table 3.1. The informed reader will see a dramatic similarity of these concepts with class modeling for software design. In fact, the visual representation chosen for many object-oriented modelling environments, and of metamodelling languages, is most commonly that of UML class diagrams.

Table 3.1. Archetypal abstractions used in metamodelling (adapted from [2, 4])

Archetypal Concept	Description
Class	Specific classes of entities that exist in a given system or domain. Domain models are entities themselves and may contain other entities. Entities are instances of classes. Classes (thus entities) may have attributes.
Association	Binary and n-ary associations among classes (and entities).
Specialization	Binary association among classes denoting an IS-A relation.
Hierarchy	Binary association among classes denoting "aggregation through containment". Performs encapsulation and information hiding.
Constraint	An expression that defines the (statically computable) correctness of part of the model: only if all these constraints evaluate to true, the model is called "well formed".

A visual depiction of the metamodelling design process can be seen in Figure 3.2. In this figure, the metamodelling Interface corresponds to tools on the M_1 layer, and the Modeling Environment corresponds to tools on the M_2 layer, while the Application Domain corresponds to tools in the M_3 layer. As the application evolves, changes are made not to the M_3 layer, but to the more abstract layer (M_2). Similarly, as the Modeling Environment requires new types, they are modified in the model of the modeling environment (the Metamodel Specifications). As we discuss in Section 3.1.5, metamodels that specify languages can also denote the concrete syntax for language elements, and constraints for the language, in the metamodel.

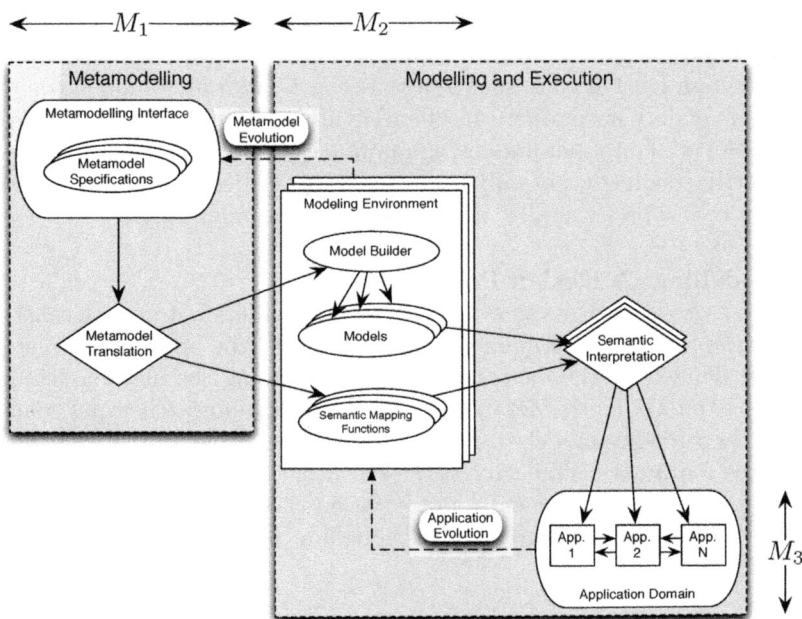

Fig. 3.2. The metamodelling design process

A concrete example may help to better understand the metamodeling design process. Let us consider the application domain (M_3) of electronic control units (ECUs) for automotive applications. The tool (M_2) should permit components to be connected to one another, and for components to be defined in terms of mathematical operations on their multiple input/output connections. There should also be constraints that prevent the outputs of two components to be connected to one another, for example. The metamodelling interface (M_1) can be used to create the tool, M_2, through the specification of the abstract syntax that permits these kinds of applications to be modeled. Metamodel translation synthesizes the tool, M_2, and semantic interpretation of models built using M_2 generates, for example, the embedded code for each component, a schedule for execution of components on a real-time operating system, logging functions for debugging purposes, and other necessary features in the application domain of ECUs.

In the event that a particular design (M_3) should be changed, the models created in M_2 should be modified, and then the semantic interpretation should be performed again. This is called *application evolution*. If the domain somehow changes, perhaps through additional constraints, new types that become available, or changes in design philosophy, then *metamodel evolution* must be performed, in order to change the design environment M_2. As shown in Figure 3.2, M_2 should be evolved by changing the metamodel specifications (M_1) and performing metamodel translation again.

Archetypal Metamodelling Abstractions

All major metamodelling approaches permit some significant subset of the basic abstractions shown in Table 3.1 [3, 7], though various tools may use a slightly different nomenclature [8]. As an example, the fundamental abstraction of "information hiding" is usually implemented using containment (as in hierarchical states).

Once the metamodels are defined in the design process, some transformation process generates the semantic artifacts necessary to continue in the design. This may be the generation of software skeletons that implement a class diagram, or the synthesis of configuration files that permit the use of a generic modeling tool. Textual modelers may use the metamodels to generate configurations for parsers and lexers to operate on text files that conform to the defined metamodels.

This synthesis process (metamodel translation, in Figure 3.2) maps types defined in the metamodel to concrete abstractions that an end-user will utilize to abstract their model-based design. Once the design of the application or system design is encoded in terms of meta-types (using tools from the M_1 layer), some transformation from the instances of these meta-types into the semantic domain is performed. In the remainder of this section, we will discuss some significant state-of-the-art approaches to metamodelling.

3.1.2 Meta Object Facility (MOF)

Central to the design and implementation of the UML 2.0 infrastructure and superstructure is the concept of model transformations between varying layers of abstraction. In order to permit these transformations (model to model, or as instances, object to object), some additional specification must be used to describe the structure of these objects—and this is termed the *Meta-Object Facility* (MOF).

The purpose of metamodelling through MOF is to describe the models in these various layers using common Modelling abstractions. This permits homogeneous access to models at all layers using reflection, standardizes access across tools through a common API, and permits serialization of models through the XMI standard. The specification of the MOF standard itself is well-described in the OMG document governing MOF (see especially [9]), and we do not attempt to fully describe those formalisms and terms here. However, we will describe the modelling concepts used by MOF to perform metamodelling, and we will do this from the perspective of the *Essential* MOF (EMOF), as described next.

3.1.3 Essential MOF (EMOF)

One major benefit of Modelling languages is to include concrete formalisms that makes modelling of particular concepts easy. For metamodelling, however, a significant amount of freedom in specification can lead to complexity when various models need to be updated. Thus, the use of an essential subset of a metamodelling language can insulate created models from changes to the metamodel.

This is the concept behind the Essential MOF (EMOF), which is a subset of the Complete MOF (CMOF) [9]. We present here the key concepts to EMOF, so that they can be compared to those of other metamodelling languages. In essence, what we present here is the metamodel of EMOF.

Reflection, Identifiers and Extension

Basic assumptions for using EMOF (and CMOF) include the ability to utilize reflection, extension, and identifiers. *Reflection* is the ability of an object to determine its type (class, or "metaobject" in the EMOF vocabulary, and its associated metadata). *Extension* is the ability of an object to be dynamically annotated with name-value pairs. This permits some amount of runtime Modelling of particular objects, enabling an object to create new data fields which it could later use, without creating a new type. Note that when using extension, only that object receives these new name-value pairs, they should not propagate to all objects of that type. Finally, *identifiers* are a way for objects to maintain uniqueness regardless of any values with which it is instantiated or any extensions with which it is annotated.

In essence, two of these three concepts correspond to key attributes of Object-Oriented languages: an object is unique, an object knows its type. The other (extension) is a novel introduction to models when compared to textual languages, as in order to extend a class in a textual language requires creation of a new type.

EMOF Classes

The fundamental metamodelling archetypes are easily visible in Figure 3.3 (compare to Table 3.1). However, as this is the MOF metamodel, such concepts are rewritten slightly. MOF permits Class objects (which inherit from *Type*). These Class objects will be able to contain Property and Operation features. In turn, the Property and Operation features that belong to a class are further made up of Parameter objects, or associated with Property objects.

It should be clear to the reader from examining the kinds of attributes in this metamodel that one major goal of MOF is Modelling *software models*. That is, the Class object has a specific data member isAbstract, which is an attribute commonly associated with software architecture. Not all metamodelling techniques are used simply as abstractions for software models, as we discuss in Section 3.1.5.

Given the wide acceptance of EMOF as a metamodelling framework, there are some key features and benefits to EMOF. It is possible to serialize EMOF models using the accompanying XMI standard (which provides mapping rules from EMOF to XML). There are also mappings from EMOF to Java, so as to generate software architectures and APIs from the models. Using these transforms, it is also possible to generate reflective operations in software, to permit manipulation of metamodel elements.

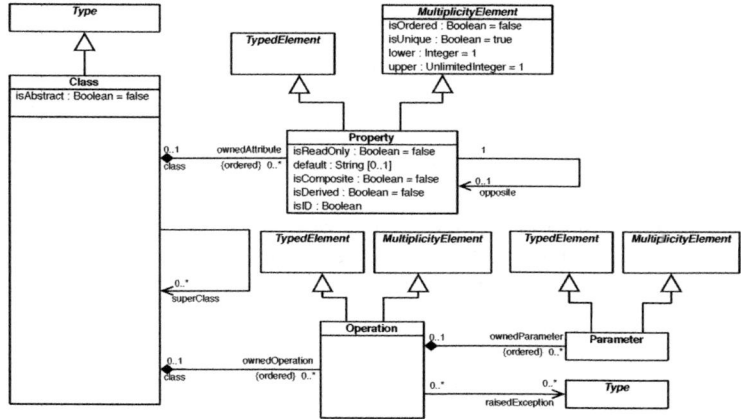

Fig. 3.3. The EMOF metamodel [10]

3.1.4 Eclipse Modelling Framework (EMF)

Similarly to EMOF, the Eclipse Modelling Framework (EMF) is a facility for building models of data structures. EMF, as it is tied to a particular tool (Eclipse), presents some additional benefits in that it can generate refined tools and applications that are tailored for Eclipse. At root, it is still quite closely tied to creating software models. EMF is a restricted subset of UML class diagram concepts, namely the definitions of *classes*, *attributes* belonging to those classes, and *relations* between classes.

Accompanying the EMF toolsuite is a set of plugins that permit reuse of EMF models. Among the most significant are tools that permit editing EMF models (and customizing EMF editors), and synthesizing software from EMF models. These are discussed further in metamodelling-languages surveys, such as [11], as well as the EMF documentation.

The popularity of the Eclipse toolsuite brings with it a plethora of Eclipse, and EMF, plugins and tools that use the serialization that comes with the EMF use of XMI standards, and the implicit tool interchange that is possible through popular acceptance of the Eclipse platform.

There is a companion modeling framework for the visualization of EMF models, called the Graphical Modeling Framework (GMF), which is part of the Eclipse toolsuite. GMF uses the Graphical Editing Framework (GEF) in order to interface with domain models graphically, and can leverage existing EMF metamodels to bootstrap the visual language definition in GMF. Some features of domain-specific modeling, such as constraint specification and multi-aspect visualization, are not yet part of the GMF toolsuite, but it is nonetheless a strong tool for modeling of Java-based applications.

3.1.5 Metamodelling of Languages

The domain-specific modeling approach using metamodelling treats metamodels as language specifications, not software structure specifications. This difference from the common uses of EMOF and EMF distinguishes MetaGME [12], MetaEdit+ [13], and AToM³[14, 15] tools. A common use of language generators is to synthesize domain-specific languages [16], and a rich legacy of application in this area can be found in the proceedings of [17, 18, 19, 20], and also in [21].

Language metamodelling defers representation and usage to a language or model editor. Meta-configurable editors such as GME, AToM³, and MetaEdit+, are capable of reusing the same editor framework for many different languages. Eclipse's GEF also permits single editor multi-configuration reuse. However, in language use (alternatively, modeling environment use), the issues of concrete syntax and visualization must be addressed. We discuss this in Section 3.1.7.

An additional property found in the metamodelling of languages is the ability to specify selective visualization (also known as *aspects* or *viewpoints*). These properties permit filtering of the visualization space for an intuitive subset of the design, as partitioned at design time. An example of these properties is seen in Figure 3.4, where subsets of each object are visible, depending on the aspect in use. In this particular example from the signal processing domain, certain computational blocks may share parameters, but these blocks are not functionally connected. For the purposes of design it can be convenient to see what computational blocks share the same parameters, but if this information were shown in the same screen as the functional connections, it would be difficult to understand the diagram.

The final property we discuss with respect to language modelling not regularly found in data Modelling is that of constraint specification within the metamodel. Constraints may exist for certain data Modelling applications, but at the language level, such constraints can *prevent* or *restrict* the ability of a modeler to create certain constructs that are known *a priori* to have no well-defined semantic interpretation (or perhaps a disallowed, but known, semantic interpretation). These constraints may be specified in terms of the OCL (Object Constraint Language) [22].

A common use of constraints is to permit simple metamodel specifications, with small exclusions from their use. For example, a metamodel may define connections between ports of container objects (such as that shown in Figure 3.6a). However, a constraint can prevent the connection of two output ports to one another *unless* those output ports are at different levels of hierarchy (i.e., passing a value on to a parent's output port). Such a constraint can be written in OCL as:

```
OutPort.attachingConnections( BufferedConnection )->forAll( c |
    c.connectionPoints( "src" )->theOnly( ).target( ).parent( ).parent( ) =
    c.connectionPoints( "dst" )->theOnly( ).target( ).parent( ) )
```

This concisely states that if an `OutPort` object participates in an association of kind `BufferedConnection`, that the grandparent of the `src` must be the parent of the `dst`. This prevents two `OutPort` objects of the same `Component`

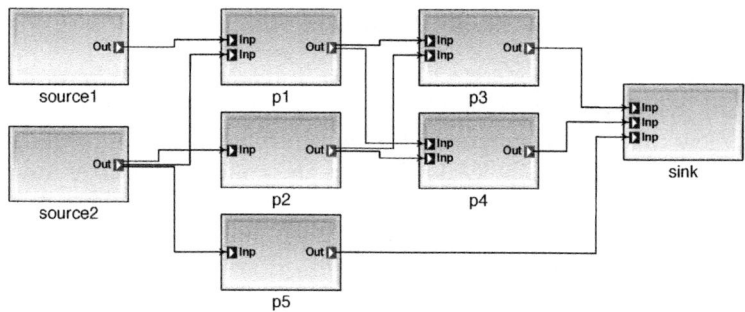

(a) Signal flow structure.

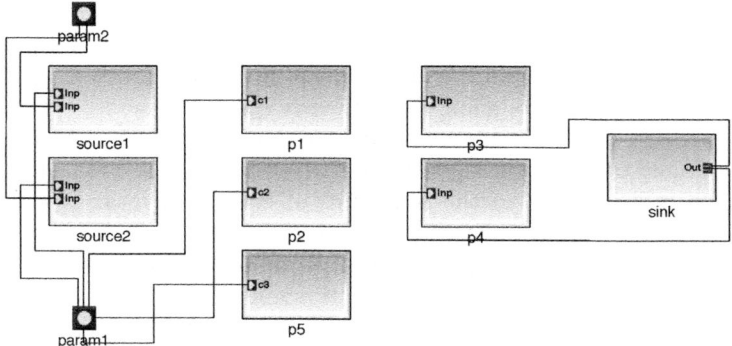

(b) Shared parameters of the system.

Fig. 3.4. These two figures show how the same structural elements can be shown in different aspects to visualize elements of the system effectively. In (a), the structure as related to the signal flow is given. In (b) the parameters shared between components are easily seen—and easily changed.

from connecting through this kind of association. Constraints provide a powerful means to restrict the modeler from creating ill-formed models, while maintaining a language that is easy to compile.

3.1.6 Textual Metamodelling

Metamodelling is useful in textual, as well as graphical/visual, languages. In fact, there are certainly cases where a textual language is preferred [23]. When considering the traditional methods of specifying grammars (e.g., Backus Naur Form [24], and Extended BNF), it is apparent that such specifications do define the abstract syntax of a language. Tools that generate parsers and lexers for such grammars (such as antlr, bison, etc.) are the textual analogs to the abstract syntax tree generators found in modeling environments [25]. The application of programming language types to their semantics is well-studied [26], and rigorous treatment of their specification can permit subtle understandings.

3.1.7 Concrete and Abstract Syntax

The differences between concrete and abstract syntax are well known, and well-studied [26, 27]. However, their application to new modelling languages brings into question how to specify concrete syntax best during the metamodelling phase of language design, as well as how the abstract and concrete syntaxes are used.

For textual modelling languages, some concrete syntax is required in order to streamline model construction. Although visual language developers have been among the most vocal proponents of modelling, there is significant research in textual domain-specific languages, because of their better efficiency, both in language definition and use [28].

For graphical modelling languages, a (default) concrete syntax can be synthesized directly from the metamodel, as long as a default concrete syntax is provided for each archetypal type. The GME tool e.g. has generic types as defined by the meta-metamodel, which provide a default concrete syntax if not overridden at the metamodelling level. Overriding the concrete syntax is fairly straightforward (many tools such as Simulink and LabVIEW permit this as well), it can be done at the M_2 or M_1 level. There are important questions that must be resolved with regards to the semantics of concrete syntax changes at any level, especially for tools that are domain-specific in nature, and depend on an intuitive understanding of visual models; we discuss these issues below. Other tools such as DiaGen [29] utilize concepts similar to that of GME to attach visualization attributes to the nodes and edges graph that encodes the model.

Metamodelling Level Concrete Syntax Specification.
Defining a specific concrete syntax is possible for graphical languages by specifying a glyph or glyph-generator that will provide a visualization (perhaps context-specific) for a particular type in the language. Then, for every instance of this type that is visualized, this image (or the imaged produced by the glyph-generator) replaces the default value. This is very useful for simple domain-specific visual languages, where concrete domain items can be composed easily with other domain items.

Modelling Level Concrete Syntax Specification.
Redefining the appearance of a model, namely the concrete syntax, at the modelling level is also possible, though not as widespread as overriding at the metamodelling level. Whereas redefinition at the metamodelling level operate for each created instance, redefinition at the modelling level overrides just for one particular instance thus allowing allows individual shapes for each model element. Such overrides are to some extent also questionable due to the fact that (for some reason) the metamodel designer chose a different concrete syntax. Why is this concrete syntax being overridden? Will this confuse other modelers using this model? Any confusion in these areas will reduce the positive impact seen in the utilization of modelling languages to specify a design, as new users will be unable to distinguish between semantics of the language, and visual preferences of another modeler.

Concrete Syntax

Concrete syntax is carefully chosen to represent domain concepts (for domain-specific languages), as "syntactic sugar" (for DSLs as well as general-purpose languages), or to otherwise make programming or Modelling easier. For the semantic interpretation of a language, however, there is the possibility that the concrete syntax *could* be used in semantics definitions, or that different variants of concrete syntax make the visual representation of a model ambiguous to developers. The inability to distinguish ambiguous representation in a screenshot of the model is shown in Figure 3.5.

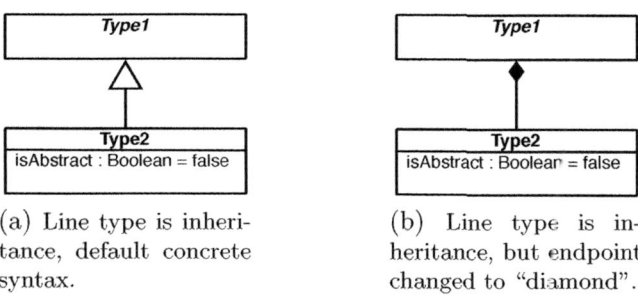

(a) Line type is inheritance, default concrete syntax.

(b) Line type is inheritance, but endpoint changed to "diamond".

Fig. 3.5. Changing the concrete syntax at modelling time can lead to confusion. In (a) the default concrete syntax for inheritance is used. In (b) a modeler has changed the appearance of the line, but the semantic interpretation will still be inheritance.

It is a good practice, generally, to *not* use the concrete syntax details in the mapping of semantics, but to depend entirely on the abstract syntax tree. An interesting research challenge would be general-purpose tools that could identify issues such as these in completed models (or perhaps in their semantic mapping) as potential design flaws in the model, language, or language compiler.

3.1.8 Type System

Type systems in traditional programming languages are established at language-design time. In (typed) programming languages as well as in math, a type is basically a description for a set of values together with a set of operations to manipulate these values. In the metamodelling setting this approach needs to be adapted to meta-type structures. Different to the programming language approach, metamodelling approaches tend to merge typing and the meta-level. The challenge of defining a type within the model while applying it at the same time can then be met through the use prototypes and clones as discussed below.

For modelling language types the most common mode of type definition is through specification in a metamodel (as described in Section 3.1.1). Using this mode of definition, the traditional object-oriented abstractions of type definitions can be leveraged into the modelling language. These include notions of inheritance, containment, and association. Propagation of model features (i.e., subclasses have all features of parent classes) through generalization/ specialization relationships in the metamodel provide a means to effect polymorphic behaviors at model execution or interpretation time. Please note that it is rather convenient to lift the type infrastructure of the object-oriented realization of the metamodel into the defined modelling language. However, one could build an entirely different type system, and when the oo style of typing doesn't fit we are even forced to do so.

There are the following phases of type system use and specification:

- Meta-metamodelling time: specification of the fundamental meta-types, which define how types permit containment, association, attribute values, etc.
- metamodelling time: specification of the metamodel, using meta-metamodel types, in order to define model types (e.g., domain-specific concepts), and the abstract syntax of the language for language-generating metamodels.
- Modelling time: specification of certain clones and clone structures as templates for further instantiation.

New patterns and structures not conceived at language-design time, however, may emerge *after* the metamodels have been designed. Many modelling environments permit the Modelling-time specification of new type systems, which permits a modeler to develop a new "type" out of composition and association of instances of domain types. In this case, the new "type" may be reused, reinstantiated multiple times, and may propagate changes made to the type to any instances of that type.

In order to distinguish easily between types (defined at metamodelling time), and Modelling-time types, we use the following nomenclature (from [30]):

- **prototypes**: modelling-time types; and
- **clones**: instantiations of prototypes.

We explore this modelling-time types behavior through an example.

Types and Clones at Modelling Time

Consider the language defined by the metamodel shown in Figure 3.6a[1], and models built using this language shown in Figure 3.6b. Now, let us consider that model C2, contained in Component1 is a *clone* of the *prototype* Component2. This would mean that for each object contained in Component2, there would be a corresponding object of the same type (and participating in corresponding *internal* associations) in C2. By *internal* associations, we mean to say that the association is contained by that model (and is not an association that resides outside the type).

Regarding the attribute values of these models, whenever a clone of a prototype is created, it receives the attribute values of the prototype. From this time on, there are several semantic issues which must be addressed by the modelling environment.

---

[1] This metamodel is reused in Chapter 9 in order to discuss the evolution of models.

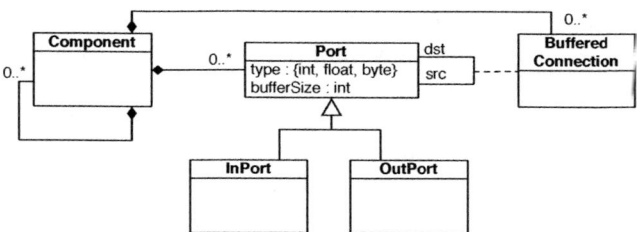

(a) The metamodel allows objects of kind Port, which is specialized as InPort and OutPort.

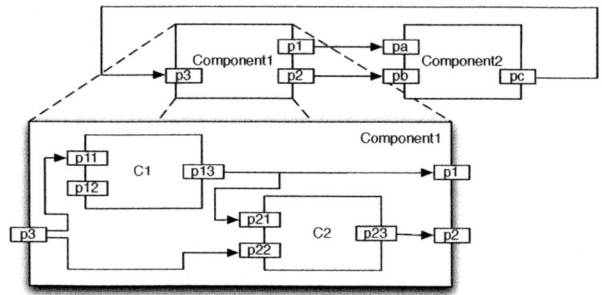

(b) A model built using the metamodel in (a). The contents of Component1 are shown to display the additional associations in which its Port objects play a role.

Fig. 3.6. (a) A metamodel allowing port interconnection between components. (b) A model built using the metamodel in (a). The "arrow" end of the connections represent the dst role.

(1) Are attribute values of the clones permitted to be modified?
(2) If an attribute value of a clone is modified, and the prototype attribute value is modified, what then should be done for the clone model's attribute values?
(3) If the attribute values of the prototype change, should unmodified attributes of the clones be updated?

Tools and environments that permit prototypes and clones adopt a fairly consistent view of these questions. Both GME [12] and Ptolemy II [31] permit attribute value modifications of clones. In the event of changes to the original prototype, an attribute-specific copy-on-write behavior is utilized, where unmodified attribute values reflect the prototype values, rather than maintaining the values at instantiation-time. We discuss in the next section how selective permission to contain new objects in clones and prototypes can create some confusion.

It is generally up to the tool developer to determine how to visually depict prototypes and clones. If a separate browser that permits searching for or displaying only prototype and clone hierarchies is given, it is not necessary to even have a visual cue that a particular object is a clone.

A final restriction on clone models is that they cannot contain objects that are not contained in the prototype object (i.e., the correspondence function is bijective). Similar to attribute propagation, then, new models created inside an prototype propagate to *all* clones of that prototype. If an object contained within an prototype is deleted, all clones remove their corresponding object (and any associations to that model, as appropriate).

Prototypes and Subprototypes at Modelling Time
Subprototypes have a subset of the restrictions and constraints of clones. As a class diagram permits subclasses to specialize the structure of their superclass, a subprototype can add to the features of an prototype at modelling time. Thus, the restriction that there does not exist any object in the instance that does not correspond to a type-contained object is not necessary (i.e., the correspondence is injective from prototype to subprototype).

3.1.9 Merging of Metamodels

Given the ability of models to apply hierarchy and refinement as abstractions, and the fact that metamodels are models themselves, the ability to merge metamodels (others say "compose metamodels" structurally) is somewhat trivial. The semantics of this merge, however, deserves some discussion.

Consider two metamodels, $\mathcal{M}_1, \mathcal{M}_2$, such that $\mathcal{M}_1 \cap \mathcal{M}_2 = \emptyset$ (i.e. they do not share any meta-class). Now, consider that some elements from each of these metamodels can be related in a new, merged, metamodel, $\mathcal{M}_3 = \langle \mathcal{M}_1 \cup \mathcal{M}_2, f \rangle$. When merging the metamodels, some elements from each of the two metamodels must somehow be associated with one another. We can use this function f to define appropriate relations between metamodel elements.

These relations can be considered as mappings for identity, or new properties. As discussed in [30, 32], the identity equivalence maps two types (one from each metamodel) as identical, and thus permits the associations and attribute values of those types to be a union of the definition in the two metamodels. More subtle is the desire to transfer only some of the associations and attribute values of a certain type. These are created as new metamodel types (found only in $\mathcal{M}_3$) which can inherit either the interface of the existing types, or the implementation of the existing types (meaning that containment and other relations are, or are not, transferred). For these subtleties, we refer the reader to [30, 32] for a full explanation with examples.

A short example of merging metamodels is given in Figure 3.7. In Figure 3.7a we see a simple modeling language for discrete systems. In this language, the behavior of the system is obtained by firing the Behavior model(s) in the current state. A simple modeling language for continuous time systems is shown in Figure 3.7b, with the capability to assign values to Variable objects through algebraic and differential (Flow) equations. In order to create a new language,

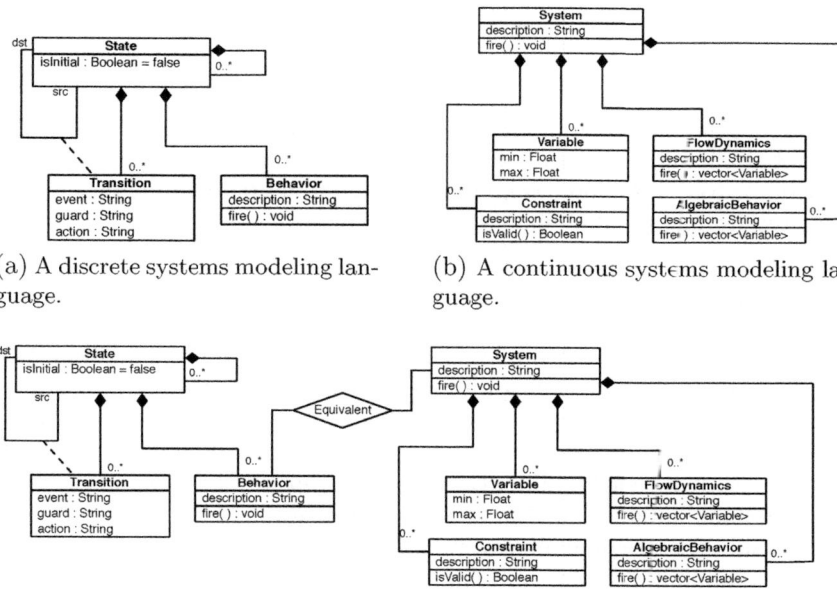

Fig. 3.7. Elements from the discrete, and continuous, domains are merged/composed in a new domain, and an equivalence relationship is used to indicate identity of one element in each metamodel

capable of modeling hybrid systems (those systems where each discrete state has a continuous dynamics), we can merge the two metamodels, and indicate an equivalence relationship between `Behavior` in the discrete systems language, and `System` in the continuous systems language. In this new modeling language, it is then possible to create new objects of kind `System` inside of a `State`, even though this is not explicitly shown through containment relations between `System` and `State`. It is possible to further assign relations between objects, using containment, association, or other relations.

An additional, metamodelling, concern is the propagation of constraints when metamodels are merged. These, and especially the issues of semantics, are research issues, and discussed in the following Section 3.2.

3.2 Metamodelling: Research Challenges

Metamodelling as a technology provides significant power to designers and users, and it has been thoroughly explored in terms of modeling data, software, and languages. Although many of the properties, semantics, and uses of metamodelling are now "solved" problems, there are significant research challenges still outstanding, regarding usability, evolution, intuitive representation, etc. We discuss these research challenges (in brief) in this section.

A Unifying Issue: Semantics

A unifying characteristic of nearly all outstanding research issues we discuss with respect to metamodelling is the issue of semantics. The meaning of composed, cloned, evolved, etc., models and metamodels may be unclear, depending upon the circumstances under which these operations are performed.

3.2.1 Semantic Attachment

We recall that metamodels are basically pure syntactic representations of the models they describe [4]. Significant strides have been made in attaching additional information to these metamodels, which in many approaches is called "semantic attachments". The significant issue in semantic attachment is not "can it be done?" but rather "what methods are appropriately efficient and intuitive?" A traditional compiler that traverses an abstract syntax tree to produce artifacts in the semantic domain can be readily produced (and tools to significantly automate the parsing and traversal have been developed [33, 28]).

Methods to ground semantics between metamodels to a common semantic domain show promise [34, 35] or through explicit definition of the semantics domain [36, 37, 38]. Utilizing those techniques, lossless, bijective, semantically-correct mappings between a metamodel and a semantic anchor such as the abstract state machine language (ASML) could foster semantically-correct interchange between tools, or from one semantic domain to another. One issue that deserves further research is the specifications of these mappings for complex metamodels, and their intuitive representation.

The expected use cases for attaching semantics to metamodels include:

- for the purpose of documentation/precise definition;
- facilitation of automatic verification of some property; and
- automated translation between tools.

3.2.2 Inference between Metamodels

As described in Section 3.1.9 it is possible to merge metamodels into a new metamodel, and (by design) mark certain metamodel elements as equivalent. The automation of this identification between two related (but separately specified) metamodels is an interesting research challenge. Issues are are present include:

- Semantic equivalence of inferred equivalent types in each metamodel;
- Visualization issues; and
- Propagation of constraints.

Among these, the semantic equivalence may require user interaction to determine. The propagation of constraints, however, presents a few interesting issues.

It may, for example, be possible to evaluate all models to determine whether constraints are violated *prior* to performing type inference. Perhaps, then, type inference is predicated on constraint satisfaction. On the other hand, issues such as selective propagation of containment (or containee) relationships may enable constraint satisfaction, so an intelligent approach to utilizing implementation and interface inheritance may permit some equivalence inference, while not violating any of the (union) of constraints.

3.2.3 Evolution of Models Driven by Metamodel Evolution

This issue presents tremendous challenge in the preservation of structure, constraints and semantics. As the metamodel evolves, e.g. because the tools are updated to a new version, it may be that models built using the metamodel will no longer conform to the evolved metamodel. In this case, evolution of the models (to conform to the new metamodel) may be required. This issue has been studied for visual languages [39], but further research is necessary to determine the best way to intuitively (and accurately) portray such evolutionary transformations. An interesting extension is the automation of such transformations *based on* changes to the metamodel in its evolution. More discussion is devoted to this complex topic in the Chapter of Model Management.

3.3 Conclusions

Despite the many tools available for metamodelling, the underlying mechanism of object-orientation has fostered that most of the metamodelling tools use a common set of abstractions with only slight variations Using these abstractions, it is possible to raise the specification of a language and its tooling far above the implementation layer. Additional capabilities increase the power of metamodelling by permitting the synthesis of languages as well as automated or semi-automated analysis and synthesis techniques. Run-time modelling tools permit users to define their own prototypes, and leverage new patterns not anticipated at metamodel-design time, and the visualization of models can be carefully specified to ensure that information is appropriate presented to modelers. Multiple metamodels can be merged to specify new languages that appropriately integrate the concepts into one big metamodel .

Model-based engineering and in particular modelling of embedded systems benefits heavily from metamodelling due to the structure that metamodelling gives to models, and the semantics that can be attached to metamodels. Given this structure, the specification of semantics is easier, correspondences between metamodels can be denoted, parsers/lexers can be synthesized, and constraints can be evaluated. These capabilities are the foundations for raising the level of specification of systems to models, rather than low-level implementation.

References

[1] Vangheluwe, H., de Lara, J.: Xml-based modeling and simulation: meta-models are models too. In: WSC 2002: Proceedings of the 34th conference on Winter simulation, Winter Simulation Conference, pp. 597–605 (2002)

[2] Karsai, G., Nordstrom, G., Ledeczi, A., Sztipanovits, J.: Specifying graphical modeling systems using constraint-based meta models. In: IEEE International Symposium on Computer-Aided Control System Design, CACSD 2000, pp. 89–94 (2000)

[3] Sprinkle, J., Karsai, G., Lédeczi, A., Nordstrom, G.: The new metamodeling generation. In: Eighth Annual IEEE International Conference and Workshop on the Engineering of Computer Based Systems, April 2001, pp. 275–279 (2001)

[4] Harel, D., Rumpe, B.: Meaningful modeling: What's the semantics of "semantics"? Computer 37(10), 64–72 (2004)

[5] Peltier, M., Bézivin, J., Ziserman, F.: On levels of model transformation. In: XML Europe 2000, pp. 1–17 (2000)

[6] Sprinkle, J.: Model-integrated computing. IEEE Potentials 23(1), 28–30 (2004)

[7] Vangheluwe, H., de Lara, J.: Foundations of multi-paradigm modeling and simulation: computer automated multi-paradigm modelling: meta-modelling and graph transformation. In: WSC 2003: Proceedings of the 35th Conference on Winter Simulation, Winter Simulation Conference, pp. 595–603 (2003)

[8] Weisemöller, I., Schürr, A.: A comparison of standard compliant ways to define domain specific languages. In: ATEM 2007: 4th International Workshop on (Software) Language Engineering, in conjuction with MoDELS (2007)

[9] Object Management Group: Meta Object Facility 2.0 (January 2006)

[10] Object Management Group: Unified Modeling Language 2.1.2: Superstructure and Infrastructure (November 2007)

[11] Emerson, M., Neema, S., Sztipanovits, J.: 33. In: Metamodeling Languages and Metaprogrammable Tools. CRC Press, Boca Raton (2008) ISBN: 9781584886785

[12] Ledeczi, A., Bakay, A., Maroti, M., Volgyesi, P., Nordstrom, G., Sprinkle, J., Karsai, G.: Composing domain-specific design environments. Computer 34(11), 44–51 (2001)

[13] Tolvanen, J.P., Rossi, M.: Metaedit+: defining and using domain-specific modeling languages and code generators. In: OOPSLA 2003: Companion of the 18th Annual ACM SIGPLAN Conference on Object-Oriented Programming, Systems, Languages, and Applications, pp. 92–93. ACM, New York (2003)

[14] Mosterman, P.J., Vangheluwe, H.: Computer automated multi-paradigm modeling: An introduction. Simulation: Transactions of the Society for Modeling and Simulation International 80(9), 433–450 (2004); Special Issue: Grand Challenges for Modeling and Simulation.

[15] de Lara, J., Vangheluwe, H., Alfonseca, M.: Meta-modelling and graph grammars for multi-paradigm modelling in AToM^3. Software and Systems Modeling 3(3), 194–209 (2004)

[16] Kurtev, I., Bézivin, J., Jouault, F., Valduriez, P.: Model-based dsl frameworks. In: OOPSLA Companion, pp. 602–616 (2006)

[17] Gray, J., Sprinkle, J., Rossi, M., Tolvanen, J.P. (eds.): 8th OOPSLA Workshop on Domain-Specific Modeling (DSM 2008), University of Alabama at Birmingham, OOPSLA (October 2008), ISBN: 978-0-61523-024-5

[18] Sprinkle, J., Gray, J., Rossi, M., Tolvanen, J.P. (eds.): 7th OOPSLA Workshop on Domain-Specific Modeling (DSM 2007), University of Jyväskylä, Jyväskylä, Finland, OOPSLA (October 2007), ISBN: 978-951-39-2915-2
[19] Tolvanen, J.P., Gray, J., Sprinkle, J. (eds.): 6th OOPSLA Workshop on Domain-Specific Modeling (DSM 2006), University of Jyväskylä, Jyväskylä, Finland, OOPSLA (October 2006), ISBN: 951-39-2631-1
[20] Tolvanen, J.P., Sprinkle, J., Rossi, M. (eds.): 5th OOPSLA Workshop on Domain-Specific Modeling (DSM 2005), University of Jyväskylä, Jyväskylä, Finland, OOPSLA (October 2005), ISBN 951-39-2202-2
[21] Gray, J., Tolvanen, J.P., Kelly, S., Gokhale, A., Neema, S., Sprinkle, J.: Domain-specific modeling. In: Fishwick, P.A. (ed.) Handbook of Dynamic System Modeling. Chapman & Hall/CRC, Boca Raton (2007), ISBN: 1584885653
[22] Warmer, J., Kleppe, A.: The Object Constraint Language: Precise Modeling With UML. Addison-Wesley, Reading (1999)
[23] Whitley, K.: Visual programming languages and the empirical evidence for and against. Journal of Visual Languages and Computing 8(1), 109–142 (1997)
[24] Knuth, D.E.: backus normal form vs. backus naur form. Commun. ACM 7(12), 735–736 (1964)
[25] Rekers, J., Schürr, A.: Defining and Parsing Visual Languages with Layered Graph Grammars. Journal of Visual Languages and Computing 8(1), 27–55 (1997)
[26] Pierce, B.C.: Types and Programming Languages. The MIT Press, Cambridge (2002)
[27] Winskel, G.: The Formal Semantics of Programming Languages. Foundations of Computing Series. The MIT Press, Cambridge (1993)
[28] Krahn, H., Rumpe, B., Völkel, S.: MontiCore: Modular development of textual domain specific languages. In: Paige, R.F., Meyer, B. (eds) Proceedings of the 46th International Conference Objects, Models, Components, Patterns (TOOLS-Europe), pp. 297–315. Springer, Heidelberg (2008)
[29] Minas, M.: Visual Specification of Visual Editors with VisualDiaGen. In: Pfaltz, J.L., Nagl, M., Böhlen, B. (eds.) AGTIVE 2003. LNCS, vol. 3062, pp. 473–478. Springer, Heidelberg (2004)
[30] Karsai, G., Maroti, M., Ledeczi, A., Gray, J., Sztipanovits, J.: Composition and cloning in modeling and meta-modeling. IEEE Transactions on Control Systems Technology 12(2), 263–278 (2004)
[31] Eker, J., Janneck, J., Lee, E., Liu, J., Liu, X., Ludvig, J., Neuendorffer, S., Sachs, S., Xiong, Y.: Taming heterogeneity–the Ptolemy approach. Proceedings of the IEEE 91(1), 127–144 (2003)
[32] Ledeczi, A., Nordstrom, G., Karsai, G., Volgyesi, P., Maroti, M.: On metamodel composition. In: Proceedings of the 2001 IEEE International Conference on Control Applications (CCA 2001), pp. 756–760 (2001)
[33] Nordstrom, S., Shetty, S., Chhokra, K.G., Sprinkle, J., Eames, B., Lédeczi, Á.: Anemic: Automatic interface enabler for model integrated computing. In: Pfenning, F., Smaragdakis, Y. (eds.) GPCE 2003. LNCS, vol. 2830, pp. 138–150. Springer, Heidelberg (2003)
[34] Jackson, E., Sztipanovits, J.: Formalizing the structural semantics of domain-specific modeling languages. Software and Systems Modeling 8(4), 451–478 (2009)
[35] Chen, K., Sztipanovits, J., Abdelwahed, S., Jackson, E.: Semantic anchoring with model transformations. In: Hartman, A., Kreische, D. (eds.) ECMDA-FA 2005. LNCS, vol. 3748, pp. 115–129. Springer, Heidelberg (2005)

[36] Broy, M., Cengarle, M.V., Rumpe, B.: Semantics of UML – Towards a System Model for UML: The Control Model. Technical Report TUM-I0710, Institut für Informatik, Technische Universität München (February 2007)
[37] Broy, M., Cengarle, M.V., Rumpe, B.: Semantics of UML – Towards a System Model for UML: The State Machine Model. Technical Report TUM-I0711, Institut für Informatik, Technische Universität München (February 2007)
[38] Broy, M., Cengarle, M.V., Rumpe, B.: Semantics of UML – Towards a System Model for UML: The Structural Data Model. Technical Report TUM-I0612, Institut für Informatik, Technische Universität München (June 2006)
[39] Sprinkle, J., Karsai, G.: A domain-specific visual language for domain model evolution. Journal of Visual Languages and Computing 15(3-4), 291–307 (2004); Special Issue: Domain-Specific Modeling with Visual Languages

4 Semantics of UML Models for Dynamic Behavior
A Survey of Different Approaches

Mass Soldal Lund[1], Atle Refsdal[1], and Ketil Stølen[1,2]

[1] SINTEF ICT, Norway
{Mass.S.Lund,Atle.Refsdal,Ketil.Stolen}@sintef.no
[2] Department of Informatics, University of Oslo, Norway

Abstract. Models are used for a number of different purposes, from the requirements capture and design of a new system, to the testing of an existing system. Many different modeling languages are available, and the semantics given for the languages vary from informal natural language descriptions to various kinds of mathematical or logical definitions. When choosing a modeling language and accompanying semantics, a number of things need to be taken into consideration, such as who are the users of the models, what is the purpose of the models, what kind of application is being modeled, and what are the essential features that must be captured.

When modeling embedded systems, an essential aspect is the interaction between hardware and software. Hence, we need to capture the behavior of the hardware and software components. For capturing the dynamic behavior of components, modeling languages like UML sequence diagrams, state machines and similar notations are often used. This paper surveys different approaches to formally capturing the semantics of models expressed using languages of this kind.

4.1 Introduction

In the context of development of embedded systems, a model is a description of a computer system, possibly including its human users, controlled process or environment, in some modeling language. Modeling plays an increasingly important role and is used for a number of purposes throughout the lifetime of a system, from initial requirements capture and design to testing and maintenance of the running system. Some models are intended to be processed automatically, for example by code generators or model checkers, while other models are used as an aid in communication between for example system developers and client representatives.

A large number of languages for modeling computer systems are available. The semantics given for the languages range from natural language explanations of modeling language constructs and examples to highly formal mathematical or logical definitions. When choosing a language and accompanying semantics, a number of issues need to be taken into consideration. One question is: Who will use the models, and what level of training do they have? Clearly, the language need to have a notation that is understandable by the users of the models, at least at an intuitive level. As an example, mathematical and logical formulas may be well understood by computer scientists and some developers, but will be incomprehensible for most client representatives.

Another question is: What is the purpose of the models? If the models are to be used for giving formal proofs of system properties, then the language must be supported by a formal semantics defined in clear mathematical or logical terms. If the models will be used for code generation or automatic model checking then we need to ensure that the semantics can also be processed by a computer. On the other hand, if the models are intended for communicating with client representatives then a natural language explanation of the language features may be appropriate.

A third question is: What kind of system is being modeled, and what are the essential features or properties that need to be captured by the models? For example, capturing real-time requirements may be essential when modeling an emergency communication network, but of little importance when designing a chocolate automaton.

For embedded systems, the interaction between software and hardware components is an essential feature. This means that we need to model the behavior of hardware as well as software components, and in particular their mutual interaction. For capturing dynamic component behavior, modeling languages like UML [1] sequence diagrams and state machines are currently the most highly profiled.

This paper surveys approaches to giving formal semantics to models expressed in UML sequence diagrams, state machines or similar notations, such as MSC [2], LSC [3], the Statecharts language [4], SDL [5], etc. An overview is given of different types of semantics and their strong and weak points. The survey is not exhaustive, but covers the most common variants. The survey does not address semantics for hybrid models which is a field in its own [6, 7].

The rest of this paper is organized as follows: Sect. 4.2 characterizes the scope of the survey and defines more carefully notions like "model", "semantics", and "embedded system". Furthermore, the semantic challenges related to embedded systems are discussed and summarized as a set of criteria against which the different approaches should be evaluated. In Sect. 4.3, the different types of semantics and their strong and weak points are discussed. Sect. 4.4 presents a survey of semantics approaches for UML sequence diagrams and similar notations. Section 4.5 is similar to Sect. 4.4, except that we now consider semantic approaches for UML state machines and similar notations. In Sect. 4.6 we evaluate the semantic approaches surveyed in the previous two sections with respect to the evaluation criteria formulated in Sect. 4.2. Summary and conclusions are given in Sect. 4.7.

4.2 Characterization of Scope, Main Notions, and Criteria for Evaluation

Embedded systems can be defined as "combinations of computer hardware and software, and perhaps mechanical or other parts, designed to perform dedicated functions"[1] or "programmable, electronic (often in combination with mechanical) systems that control and determine the functioning of devices (machines, appliances, instruments, constructions)."[2] MP3 players, routers, sensors, copying machines, and cars are examples of embedded systems. The fact that embedded systems, unlike general purpose computers, are dedicated to specific tasks, means that they can be optimized with respect to for example performance or reliability.[3]

A model is a description of a system in some modeling language, such as the UML. The semantics of a model explains what the model means. More exactly, the semantics of a model is a function mapping the syntactically well-formed models of the modeling language into syntactically well-formed expressions in a language that is well understood. What is a well-understood language depends on the intended users of the semantics. It often makes sense to define several equivalent semantics for the same modeling language; for example, an axiomatic semantics for logical deduction, a denotational semantics for mathematical reasoning, an operational semantics for building tools, and a natural language semantics to explain the language to its end-users. If the expressions of the modeling language is mapped into a mathematical or logical domain so that the semantic representation can be manipulated and analyzed using well-established mathematical and logical techniques, we say that the language has a formal semantics.

When modeling and developing embedded systems, several considerations need to be taken into account. One is that the close interplay between dedicated hardware and software components means that there is less room for corrections and refactoring during the development process than for conventional computer systems. A formal approach to model analysis and incremental development is therefore highly desirable when developing embedded systems. Hence, modeling languages should be supported by formal semantics, as well as definitions of refinement characterizing what it means for a more concrete or detailed model to "implement" or fulfill the requirements of a more abstract model. This reduces ambiguity and facilitates rigorous, and possibly automated, mathematical or logical proof of system properties.

An essential requisite for an incremental development process is the ability to leave some decisions open for later development steps. Consequently, we need

---

[1] From Netrino embedded systems glossary:
http://www.netrino.com/Embedded-Systems/Glossary
[2] From Embedded Systems Institute:
http://www.esi.nl/frames.html?/institute/research.html
[3] Other examples are characteristics such as size and power usage, but this is outside the scope of this paper.

a modeling language that has the ability to express underspecification or implementation freedom. By this we mean that a model may explicitly provide alternative ways of fulfilling a task, so that the choice is left open to those responsible for implementing or further refining the specification. Moreover, in an incremental development process one cannot describe all the relevant system behavior in a single step. Thus we want to be able to produce models that are incomplete in the sense that not all system behavior has been considered and categorized as either positive (acceptable, desirable) or negative.

Finally, there is the issue of the kinds of features or properties that can be captured by the modeling language. Properties can be categorized according to the basis on which they are falsified: Properties that can be falsified on the basis of a single trace are called *trace properties*, while properties that are falsified on the basis of a set of traces are called *trace-set properties* [8].[4] Examples of trace properties are safety and liveness [9, 10], while permissions often used in relation to policies and many information flow properties are examples of trace set properties. Most modeling languages are well-suited to capture trace properties, but only some allow us to specify trace-set properties as something distinguishable from underspecification. Distinguishing trace-set properties from underspecification is necessary since trace-set properties should be preserved under refinement while this is not the case for underspecification.

Performance and reliability requirements are usually of high importance for embedded systems. This is for example the case for routers and sensors. Indeed, in many cases the motivation for building a dedicated embedded system is to achieve high performance and reliability. Performance and reliability requirements are typically expressed in terms of time and/or probability. Therefore, modeling languages for embedded systems should ideally have the ability to capture real-time requirements (a special kind of trace properties) and probabilistic requirements (a special kind of trace-set properties). These requirements should be fully integrated in the semantics of the models in order to ensure that they are taken into account when analyzing the models.

Based on the above considerations, we have identified the following questions that we will use to evaluate the surveyed semantic approaches:

- What kind of semantics is given?
- Can underspecification be represented?
- Can trace-set properties be represented?
- Can incomplete models be represented?
- Is the approach supported by definitions of refinement?
- Can real-time requirements be captured by the semantics?
- Can probabilistic requirements be captured by the semantics?

In the following we survey and evaluate a number of semantic approaches with respect to these questions. But first we give an overview of main categories of semantics of relevance.

---

[4] In [8], the term *possibilistic properties* is used instead of trace set properties.

4.3 Main Categories of Semantics

At an overall level, semantics of modeling languages can be categorized based on whether they are formal or not, i.e. whether the expressions of the modeling language are mapped into a mathematical or logical domain, or explained in natural language. An advantage with natural language explanations is that they can be understood by anyone, without requiring specialized training. However, natural language explanations tend to be ambiguous and often contain inconsistencies. For example, this is the case with the UML semantics provided by the Object Management Group (OMG) [11, 12, 13]. Formalizing the semantics of a language will help uncover ambiguities and inconsistencies. Moreover, formal semantics allows models to be analyzed with mathematical and logical tools and techniques, thus allowing system properties to be explored in a rigorous manner before the implemented system even exists. Being able to perform this kind of analysis as early as possible is particularly important when developing embedded systems, as the cost of redesigning dedicated components at a late stage typically will be high. Hence, a formal semantics is needed for the development process. There are, however, different styles of formalizing semantics, each with their strong and weak points. For the graphical modeling languages we are concerned with in this paper, denotational and operational semantics are the most relevant styles. We now look at these two styles of semantics and their strong and weak points.

David A. Schmidt [14] provides the following explanation for a denotational semantics:

> The *denotational semantics* method maps a program directly to its meaning, called its *denotation*. The denotation is usually a mathematical value, such as a number or function. No interpreters are used; a *valuation function* maps a program directly to its meaning.

This corresponds well with the explanation given by Andreas Prinz [12, p. 149]

> The basic idea is to give a denotation to every element of the language. This means to map the syntactical expressions of the language to a well-known domain.

Denotational semantics typically allows a fairly abstract system description. As they also build on known domains, they are well suited for mathematical reasoning and formal proof of properties. On the negative side, a denotational semantics provides little guidance for tool developers and will typically be too complex for users. Expressing states and operations is usually difficult with a denotational semantics.

For operational semantics, [14] suggests the following definition:

> The *operational semantics* method uses an interpreter to define a language. The meaning of a program in the language is the evaluation history that the interpreter produces when it interprets the program. The evaluation history is a sequence of internal configurations [...]

As a methodology for language development he suggests that "a denotational semantics is defined to give the meaning of the language" and that "the denotational definition is implemented using an operational definition" [14, p. 4].

Table 4.1. Different styles of semantics

Type of semantics	Advantages	Disadvantages
Informal	– Easy to communicate – Does not require specialized training	– Tends to be ambiguous – Often contains inconsistencies – Cannot be formally analyzed
Denotational	– Allows a fairly abstract system description – Builds on known domains – Well suited for mathematical reasoning and formal analysis of properties	– Provides little guidance for tool developers – Too complex for users – Expressing states and operations is usually difficult
Operational	– Provides good formalization of implementation – Well suited for building tools – Expressing states and operations is usually easy	– Tends to be very detailed – It is often difficult to derive formal proofs – Relies on the underlying semantics of the abstract computer

Hoare and He [15, p. 258] describe more explicitly the notion of an operational semantics:

> An *operational* semantics of a programming language is one that defines not the observable overall effect of a program but rather suggests a complete set of possible individual steps which may be taken in its execution. The observable effect can then be obtained by embedding the steps into an iterative loop [...]

Taken together, these two descriptions suggest that formalizing an operational semantics of a language is to define an interpreter for the language. The formal definition of the interpreter describes every step that can be made in the execution of the language in such a way that the executions are in conformance with the meaning of the language as defined by a denotational semantics.

Major advantages of operational semantics is that such semantics provides good formalization of implementation and is well suited for building tools. It is also typically well suited for state-based languages. On the other hand, operational semantics tends to be very detailed, and it is often difficult to derive formal proof from operational semantics. Besides, an operational semantics relies on the underlying semantics of the abstract computer on which the interpreter is assumed to run [12]. Table 4.1 summarizes the strong and weak points of the different styles of semantics discussed above.

In the next three section we present and discuss a number of approaches to giving semantics to models. We concentrate on two categories of models, models expressed in a sequence diagram style and models expressed in a state machine style, and two main categories of semantics, denotational and operational. Making a complete and exhaustive presentation of every existing approach is an impossible task, and it has therefore been necessary to make a selection. With respect to sequence diagrams, we focus on UML sequence diagrams and Message Sequence Charts (MSC), and with respect to state machines, we focus on statecharts, UML state machines and SDL. Furthermore, we have aimed at making a representative selection of the approaches that exist.

In Sect. 4.4 we present and discuss approaches to giving semantics to sequence diagrams and similar notations, and in Sect. 4.5 we do the same with respect to state machines and similar notations. In Sect. 4.6 we evaluate and compare the approaches using the evaluation criteria identified in Sect. 4.2.

4.4 Sequence Diagrams and Similar Notations

In this section we present different approaches to defining formal semantics to models expressed in UML sequence diagrams and similar notations. This presentation cannot, however, be seen independently of the history of sequence diagrams. The various approaches of defining semantics have emerged at different points in this history, and are clearly influenced by the state of the language(s) at the time of their emergence.

Sequence diagrams is a graphical specification language defined in the Unified Modeling Language (UML) $2.x^5$ standard [1]. Sequence diagrams as defined in the UML 2.x standard are the last of a sequence of languages that have evolved over the last 15 to 20 years. Both UML sequence diagrams and their predecessor Message Sequence Charts (MSC) [2] are specification languages that have proved themselves to be of great practical value in system development.

An early version called Time Sequence Diagrams was standardized in the 1980s (see [17, 18]). Better known are MSCs that were first standardized by ITU in 1993 (see e.g. [19]). This standard is usually referred to as MSC-92, and describes what is now called basic MSCs. This means that MSC-92 did not have high-level constructs such as choice, but merely consisted of lifelines

---

[5] The UML standard exists in versions 1.3, 1.4, 1.4.2, 1.5, 2.0 and 2.1.1. The for us relevant changes occurred in the transition from version 1.5 to version 2.0. Hence, in this paper we will operate with UML 1.x and UML 2.x with versions 1.4 [16] and 2.1.1 [1] as representatives.

and messages. MSC-92 had a lifeline-centric textual syntax[6], and was given a semantics formalized in process algebra.

In 1996, a new MSC standard was defined, called MSC-96 [20]. In this standard, high-level constructs and high-level MSCs were introduced, a kind of diagrams that show how control flows between basic MSCs. Further an event-centric textual syntax[7] and a new semantics were defined [21]. This semantics is also a kind of process algebra, but holds substantial differences from the MSC-92 semantics. Finally, the MSC-96 standard was revised in 1999 and became MSC-2000 [2], but kept the MSC-96 semantics. A further discussion on the MSC semantics is found below.

The first versions of the Unified Modeling Language (UML 1.x) [16] included a version of sequence diagrams similar to MSC-92, i.e., consisting of lifelines and messages but no high-level constructs. An important difference, however, was that the sequence diagrams of UML 1.x did not have the frame around the diagram, which in MSC-92 allowed messages to and from the environment of the specified system.

Sequence diagrams in UML 2.x may be seen as a successor of MSC-2000, since many of the MSC language constructs have been incorporated in the UML 2.x variant of sequence diagrams. UML 2.x sequence diagrams are, however, neither a subset nor a superset of MSC-2000; there are both similarities and differences between the languages [22]. Most notably MSCs do not have any notion of negative behavior.

The UML standard defines the semantics of sequence diagrams informally. Most notably, this is a trace-based semantics:

> **Basic trace model:** The semantics of an Interaction[8] is given by a pair $[P, I]$ where P is the set of valid traces and I is the set of invalid traces. $P \cup I$ need not be the whole universe of traces.
> A trace is a sequence of event occurrences denoted $\langle e1, e2, ..., en \rangle$. [1, pp. 479–480]

The UML standard [1] defines four timing concepts: Duration observation, duration constraint, time observation and time constraint. The timing concepts of the UML Testing Profile [23] are a combination of the timing concepts from the UML standard and the timers from MSC. In the UML Profile for Schedulability, Performance, and Time [24] timing is specified by timestamps on events. This UML

---

[6] Lifelines represent the time-lines of communicating parts or components in a sequence diagram. In "MSC-terminology", lifelines are called *instances* or *instance lines*. A lifeline-centric syntax means that each lifeline is characterized by itself and a diagram as a collection of lifelines.

[7] In an event-centric syntax events, as opposed to lifelines, are the basic building blocks of a diagram. The event-centric syntax of MSCs is more general than the lifeline centric-syntax in that all diagrams expressed in the lifeline-centric syntax can be expressed in the event-centric syntax, but not the other way around.

[8] In the UML standard, *Interaction* is used as the common name for diagrams specifying interaction by sending and receiving of messages. Sequence diagrams are then one kind of Interaction [our note].

profile also has the notion of a timer, and a notion of a system clock that can produce interrupt events. The MSC standard defines three timing concepts: Timer, relative time constraints or relative time delays, and absolute measure or timing.

In the following we present briefly different denotational and operational semantics of sequence diagrams and similar notations. We start by presenting denotational semantics, then denotational semantics with time, and then denotational semantics with probabilities. Then we present operational semantics following the same structure.

4.4.1 Denotational Semantics

In [25] Katoen and Lambert define a denotational semantics for MSCs over sets of partially ordered multisets. They define two translations, one for basic MSCs and one for high-level MSCs. The former is defined over the instance oriented textual syntax of MSCs and therefore have one rule for strict sequencing of events on a single lifeline and one rule for co-region and parallel composition of lifelines. These two levels seem to be unnecessary. If the rules for lifeline composition is combined with the rules for sequential and parallel composition, the semantics can be defined directly over the HMSC syntax and we then get a more general approach. A similar denotational semantics for both basic MSCs and high-level MSCs is given in [26].

In [27], Krüger defines a variant of Message Sequence Charts that is supported by formal definitions of the semantics, as well as refinement relations. The semantics is defined in terms of streams, which consist of a sequence of system channel valuations and a sequence of state valuations. A system is represented semantically by a set of streams, and the existence of more than one stream indicates nondeterminism. The MSC variant proposed in [27] has some features that go beyond standard MSC. For example, a trigger composition operator allows us to specify that that the occurrence of an interaction sequence always causes the occurrence of another, thus providing a way of specifying liveness properties. In addition, [27] defines four different interpretations of MSCs: an existential interpretation, an universal interpretation, an exact interpretation and a negative interpretation. Four different refinement relations are defined: binding of references, which allows references to empty MSCs, property refinement, which reduces the set of possible behaviors of the system, message refinement which allows a single message to be replaced by a whole interaction sequence, and structural refinement, which allows a single lifeline to be replaced by a set of lifelines thus allowing decomposition.

The STAIRS semantics [28, 29] is a trace based formalization of sequence diagrams based on an extension of the semantic model of the UML standard, and hence distinguishes between positive, negative and inconclusive traces. But instead of a single pair (p, n) of positive and negative traces the semantic model of STAIRS is a set of pairs $\{(p_1, n_1), \ldots, (p_m, n_m)\}$. Such a pair of sets of traces (p_i, n_i) is referred to as an *interaction obligation*. The word "obligation" is used in order to emphasize that an implementation of a specification is required to fulfill every pair captured by the specification. This semantic model makes it possible to define trace-set properties. Refinement is defined as refinement of each interaction obligation, and refinement of interaction obligations is defined as reducing the set of positive traces by making them negative and reducing the set of inconclusive traces by making them positive or negative.

Störrle [30, 31, 32] defines a denotational trace based semantics for UML 2.x sequence diagrams that is quite similar to the STAIRS semantics. Among the notable differences are that Störrle does not treat choices as underspecification. Further, Sörrle gives a different treatment of negative behavior where sequence diagrams are not allowed to be inconsistent and the negative operator can indirectly specify positive traces. Refinement is defined, but are more restricted as there is no treatment of underspecification in the semantics.

Cengarle and Knapp [33] defines denotational semantics for UML 2.x sequence diagrams. Their denotational semantics is trace based and similar to STAIRS and the semantics of Störrle with respect to the positive parts of sequence diagrams. In difference from STAIRS and Störrle, they make a prefix closure of negative traces, but does not allow inconsistent sequence diagrams. Their refinement relation differs from STAIRS in that the set of inconclusive traces may be increased, something which is a problem with respect to the monotonicity of the composition operators.

In [34], Küster-Filipe gives an LSC inspired denotational semantics of UML 2.x sequence diagrams based on partially ordered sets. The partially ordered sets of sequence diagrams is used to build event structures, and modal logic constraints over these event structures are used to express negative behavior, as well as must and may behavior.

4.4.2 Denotational Semantics with Time

In [35, 36], the semantics of basic MSCs given in [26] is presented in a timed version. A timing function assigns time stamps to the events of a MSC, and the MSC can be annotated with timing constraints in the form of minimum and maximum time intervals between events. In addition, algorithms are given for checking the realizability of MSCs and whether or not there exists a timing function that is consistent with the timing constraints of an MSC.

In [37], Zheng et al. give a semantics with time for MSC-2000. The semantics is based on labeled partially ordered sets and defines semantics for both basic MSCs, high-level operators of MCSs and high-level MSCs. Time is represented by a function mapping each event in a diagram to a set of time values, giving the absolute time interval in which the event should occur. Relative timing constraints are expressed by a function mapping pairs of events to intervals of time values. In [38], horizontal and vertical refinement of their timed MSCs are defined.

Timed STAIRS is an extension to STAIRS defined in [39, 40]. In timed STAIRS there is a distinction between syntactic and semantic events: in the syntax an event is a triple of a kind (transmit, receive, or consumption), message and time-stamp tag, while in the semantic events the time-stamp tags are mapped to timestamps represented by real numbers. A requirement is placed on traces to ensure that time increases monotonically in every trace. Time constraints are defined as Boolean expressions over the time-stamp tags of the events of a diagram. If the mapping of timestamps to time-stamp tags in a trace satisfies the constraint, the trace is interpreted as positive, otherwise it is interpreted as negative.

4.4.3 Denotational Semantics with Probabilities

Performance Message Sequence Chart (PMSC) [41, 42] extends MSC with syntactic constructs for expressing performance requirements. The aim is to integrate performance characteristics, such as response time and throughput, in functional specifications. Of particular interest is the new operator altprob for probabilistic choice that is introduced in [42]. This operator allows exact probabilities to be assigned to the alternatives represented by its operands. This means that underspecification with respect to probability cannot be captured by this operator. Apart from mentioning instance decomposition, refinement is not discussed, and no definition is given of what it means for a system to comply with a PMSC specification. The semantics of PMSC is explained at a purely intuitive level.

Probabilistic STAIRS (pSTAIRS) [43, 44, 45] generalizes timed STAIRS in order to allow probabilistic requirements, including soft real-time requirements, to be captured. Sets of acceptable probabilities, rather than a single probability, can be assigned to alternatives. Hence, it is possible to express requirements such as "the probability of receiving a reply within 5 seconds after sending a request should be at least 0.9" or, for a machine simulating a coin toss, "the probability of getting a heads outcome should be between 0.4 and 0.6". Semantically, probabilistic STAIRS extends the semantic model of timed STAIRS by assigning probability sets to each interaction obligation, thus yielding so-called *p-obligations*. Refinement is defined in a similar way as for timed STAIRS, with the additional constraint that the probability set of the refined p-obligation must be a subset of the original p-obligation, thus narrowing the range of acceptable probabilities.

4.4.4 Operational Semantics

In 1995 a formal algebraic semantics for MSC-92 was standardized by ITU [46, 47]. MSC-92 has a lifeline-centric syntax and its semantics is based on characterizing each lifeline as a sequence (total order) of events. These sequences are composed in parallel and a set of algebraic rules transforms the parallel composition into a structure of (strict) sequential composition and choice. The causality of messages is obtained by a special function that removes from the structure all paths that violate the invariant. In a way this semantics is not a proper operational semantics since a diagram first has to be transformed into the event structure before runs can be obtained. This transformation replaces parallel composition with choice and hence creates an explosion in the size of the representation of the diagram. In addition, the lifeline-centric syntax is not suitable for defining nested high-level constructs. In [48], similar semantics for UML 1.x sequence diagrams is given.

MSC-96 got a standardized process algebra semantics in 1998 [21, 49, 50]. This semantics is event-centric and has semantic operators for all the syntactic operators in MSC-96. Further, these operators are "generalized" to preserve the causality of messages by coding information about messages into the operators in the translation from syntactical diagrams to semantic expressions. Runs are characterized by inference rules over the semantic operators. Compared to UML semantics, the most notable thing about this semantics is that it has no notion of negative behavior, and therefore also makes no distinction between negative behavior and inconclusive behavior (behavior that is neither positive nor negative). This is no surprise since MSC does not have the negative operator of UML 2.x. The only available meta-level is a flat transition graph, and this does not give sufficient strength to extend the semantics with negative behavior. Nor is it possible to define trace-set properties over this transition graph. The semantics has no explicit communication medium; the communication model is "hard-coded" in the semantics by the "generalized operators" and does not allow for variation. Even though MSC has timing concepts, these are not given proper treatment in the semantics.

Another process algebra semantics for MSC is presented in [51]. This semantics may in some respects be seen as more general than both the MSC-92 and the MSC-96 semantics. A simple "core semantics" for MSCs is defined and this semantics is then inserted into an environment definition. Varying the definition of the environment allows for semantic variability and extendibility, e.g., with respect to the communication model. However, the semantics is heavily based on synchronization of lifelines on the entry of referenced diagrams and combined fragments and diverges in this respect from the intended semantics of MSCs and UML sequence diagrams. Further, the same strategy as for the MSC-92 semantics is applied; interleaving is defined by means of choice, and the message invariants obtained by removing deadlocks. This results in an unnecessary amount of computation, especially in the cases where we do not want to produce all traces but rather a selection of the traces that a diagram defines.

Realizability of MSCs is the focus of both [36, 52] and [53]. They define synthesis of MSC to concurrent automata and parallel composition of labeled transition systems (LTS), respectively. (Each lifeline is represented as an automaton or LTS; the lifelines are then composed in parallel.) Further they define high-level MSCs as graphs where the nodes are basic MSCs. In addition, [53] defines both syntax and semantics for negative behavior. In both approaches the translation of high-level MSCs to concurrent automata/LTSs removes the semi-global nature of choices in a specification, and the high-level MSC graphs are non-hierarchical, disallowing nesting of high-level operators. In [53] communication is synchronous.

Various attempts at defining Petri-net semantics for MSCs have been made [54, 55, 56, 57]. In [54, 56] only basic MSCs are considered. In [57], high-level MSCs are defined as graphs where each node is a basic MSC. As with the above mentioned semantics, it is then possible to express choices and loops, but the approach does not allow for nesting of high-level operators. In [55], a Petri-net translation of the choice operator is sketched, but no loop defined. In [58] a Petri-net semantics for UML 1.x sequence diagrams is presented, but as with the Petri-net semantics of basic MSCs it has major limitations.

Jonsson and Padilla [59] present a semantics for MSC which is based on syntactic expansion and projection of diagram fragments during execution. Each lifeline is represented by a thread of labels where the labels refer to events or diagram fragments. The threads are executed in parallel and when a label referring to a fragment is reached the fragment is projected and expanded into the threads. Expansions may happen at arbitrary points since there are no rules in the semantics itself for when to expand. This creates a need for execution strategies, and the approach may be seen as having an informal meta-level where ad hoc strategies are described. However, if completeness is to be ensured, or if the semantics is to be extended with negative behavior or trace-set properties, this meta-level must be formalized. The semantics requires explicit naming of all diagram fragments and this yields an unnecessary complicated syntax. It does not have an explicit communication medium; the communication model is "hard-coded" into the semantics and does not allow for variation.

In [60, 61] an operational semantics for UML 2.x sequence diagram is given. The semantics is defined as the combination of two transition systems, which are referred to as an *execution system* and a *projection system*. The projection system is used for finding enabled events at each stage of the execution and is defined recursively. These two systems work together in such a way that for each step in the execution, the execution system updates the projection system by passing on the current state of the communication medium, and the projection system updates the execution system by selecting the event to execute and returning the state of the diagram after the execution of the event. The execution system can be configured with different communication models, and the semantics also provides a formal meta-level for specifying execution strategies and for handling of negative behavior and trace-set properties. The semantics is proved to be sound and complete with respect to the denotational semantics of STAIRS (see above).

In [62, 63] an operational semantics for UML 2.x sequence diagrams that is equivalent to the denotational semantics defined in [33] (see above) is given. This operational semantics has some similarities to the operational semantics of [60, 61]; for every execution step an event is produced and at the same time the syntactical representation of the diagram is reduced by the removal of the event produced. Contrary to [60, 61], their semantics treats sequence diagrams as complete specifications (with no inconclusive behavior). The rules are defined so that a given diagram produces a set of positive and negative traces that together exhaust the trace universe. The negative operator is replaced by a "not" operator. This operator is defined so that the sets of positive and negative traces are swapped, with the result that specifying some behavior as negative means also specifying the complement of this behavior as positive. A variant of the (positive part) of the operational semantics where each lifeline is executed separately, and an extension with channels, are given in [63].

In [64], Cavarra and Küster-Filipe present an operational semantics for UML 2.x sequence diagrams inspired by Live Sequence Charts (LSC) (see below). The semantics is formalized in pseudo-code that works on diagrams represented as locations in the diagram, but no translation from diagrams to this representation is provided. The arguments of choices have guards and there is nothing to prevent the guards of more arguments in a choice to evaluate to **true**. In this case the uppermost operand will be chosen, which means that the choices essentially are treated as nested **if-then-else** statements and may not be used for underspecification. Each lifeline is executed separately which means that synchronization at the entry of choices is necessary to ensure that all lifelines choose the same operand. They also make the same assumption about negative behavior as in LSCs, that if a negative fragment is executed, then execution aborts.

Grosu and Smolka [65] provide a semantics for UML 2.x sequence diagrams based on translating the diagrams to Büchi automata. The approach is based on composing simple sequence diagrams (no high-level operators) in high-level sequence diagrams (interaction overview diagrams), where a simple diagram may be a positive or a negative fragment of the high-level diagram it belongs to. Positive behavior is interpreted as liveness properties and negative behavior as safety properties. Hence, for a high-level diagram two Büchi automata are derived; a liveness automaton characterizing the positive behavior of the diagram and a safety automaton characterizing the negative behavior. The diagrams are composed by strict sequencing rather than weak sequencing, and hence has implicit synchronization of lifelines when entering or leaving a simple diagram. Refinement is defined as language inclusion.

Live Sequence Charts (LSC) [3, 66, 67] is a variant of MSC where diagrams may be tagged as universal or existential, and parts of diagrams as hot or cold. In addition, a diagram may have a triggering pre-chart. The semantics of LSC characterizes the execution of diagrams. It also evaluates the conditions imposed on diagrams by designating them as universal or existential, or by marking parts of diagrams as hot or cold. The semantics complies with neither the MSC nor the UML standard. Most importantly it requires synchronization between lifelines at every entry point of diagram fragments, e.g. when resolving a choice.

Harel and Maoz [68] use LSC semantics to define negative behavior of UML 2.x sequence diagrams. The operators are defined using already existing constructs of LSCs, and hence no changes or additions to the LSC semantics are needed in their approach.

In Triggered Message Sequence Charts (TMSC) [69, 70], an initial part of a diagram can be designated as a trigger diagram, with the interpretation that if the behavior described by the trigger diagram takes place, then the behavior described by the rest of the diagram must subsequently take place. Unlike the pre-charts of LSC, however, the trigger condition applies locally to each lifeline. This means that, for any given lifeline, if the events on that lifeline described by the trigger diagram take place, then the following events on that lifeline must subsequently take place. As the fulfillment of the trigger condition is determined locally on each lifeline, there is no need for synchronization between the lifelines. A refinement relation is defined, with the intuitive interpretation that a specification S_1 is refined by a specification S_2 if S_2 is more deterministic than S_1. TMSC contains two operators for choice. A delayed choice must be preserved in a refinement step. An internal choice can be resolved at any point (including at design time). In addition, an internal choice may be refined by a delayed choice.

4.4.5 Operational Semantics with Time

In [51], Letichevsky et al. claim they also have an extension to the semantics where timing concepts such as time intervals and timing of events are defined.

The operational semantics of [60, 61] has in [60] an extension with data, variables and time. Each lifeline has a set of local variables and a data state that assigns values to these variables. In order to model time a special variable *now* is introduced. Because the approach only has local variables, this variable is placed in the data state of every lifeline in a diagram. It can, however, be considered a global variable in the sense that all the local *now* variables are updated simultaneously and with equal increments, i.e. that the time of all lifelines are synchronized. Except for increments by a special tick rule, the *now* variables are read only, something that is ensured by syntactic constraints.

In [67], a time extension to LSCs is presented where a clock variable *Time* is added to the formalism. Time is then treated as data and time constraints can be expressed by means of ordinary variables.

Kosiuczenko and Wirsing [71] make a formalization of MSC-96 in a timed version of the term rewriting language Maude. Every lifeline in a diagram is translated into an object specified in Maude, and the behaviors of these objects are specified by the means of states and transition rules. This way of reducing diagrams to sets of communicating objects has the effect that all choices are made locally in the objects and the choice operator looses its semi-global nature. Hence, this formalization does not capture the intended understanding of the choice operator. With respect to time, their semantics only deals with timers, and their formalization makes restrictions on the MSC semantics.

4.4.6 Operational Semantics with Probabilities

We are not aware of any operational semantics with probabilities for sequence diagrams or similar notations.

4.5 State Machines and Similar Notations

In this section we present some of the approaches that have been taken for assigning formal semantics to models expressed in UML state machines and similar languages. UML state machines represent one of many variations that have emerged since Harel introduced the Statechart language in 1987 [4]. Over the years very much work has been dedicated to providing a satisfactory formal semantics. An extensive overview is beyond the scope of this article; our aim is to illustrate the variety of approaches that have been taken. An alternative overview from a different angle can be found in [72].

4.5.1 Denotational Semantics

Broy et al. [73, 74, 75] build a mathematical system model for UML in layers. Each layer builds an algebra consisting of a universe of elements with accompanying functions and laws for the functions. The third part, presented in [75], includes the "state machine part", which is given in terms of state transition systems. A state transition system consists of a state space (a set of states) and a state transition function. The theory of state transition systems is based on the theory of streams of FOCUS [76] for the I/O behavior, and thus inherits refinement from there. State transition systems can describe not only the behavior of a single object, but also a collaborating group of objects.

A set theoretic approach to defining a semantics is taken in [77]; object states, events, guards, and run-to-completion processing is described in set theoretic terms. The aim is to provide a compositional semantics that allows models to be subject to hierarchical and modular approaches to verification and testing.

4.5.2 Denotational Semantics with Time

In [75], the state transition systems are generalized into timed transition systems to account for time. The approach assumes a discrete global time. In each step/transition the system is provided with a finite set of input events and produces a finite set of output events; this takes a fixed amount of time corresponding to a clock tick.

Rossi et al. [78] provide a formalization of (fundamental aspects of) UML state machines in terms of a temporal logic over discrete time called LNint-e.[9] Time is represented by a discrete, linear and infinite set with a total ordering. LNint-e allows inclusion of interval expressions, and time can be treated both absolutely and relatively. The temporal primitives from which expressions can be built are instants, intervals and dates. A state machine diagram is represented by set of predicates, and the formalization can be generated automatically. States are formalized by means of expressions that can be affirmed over intervals ("hereditary interval expressions").

Hinkel, Holz and Stølen [79, 80] give a semantics for SDL specifications (whose behavioral descriptions are similar to UML state machines) based on streams and stream processing functions within the framework of FOCUS [76]. This allows properties of SDL specifications to be proved using techniques of classical higher order logic and of domain theory. Time is represented by a global clock which increases time and is accessible to all processes. Time is an orthogonal concept to system behavior, and time proceeds independently from the behavior. Timers set by processes will expire after a finite duration of time and are put in the

---

[9] Note that we have chosen to include [78] among the denotational semantics because the translation from a state machine diagram to a set of logical formulae can be viewed as a translation into a well-known domain. As there are very few approaches that give an axiomatic semantics for UML sequence diagrams and state machines, we have chosen not to have a separate category for axiomatic semantics.

input queue of the process. The refinement relations provided by FOCUS can be used also for the approach of [79, 80].

4.5.3 Denotational Semantics with Probabilities

We are not aware of any approaches that assigns a formal denotational semantics to state machines that also include probabilities.

4.5.4 Operational Semantics

In [4], Harel provides a brief discussion of how a formal semantics for statecharts could be provided, without giving definitions. The semantics is built around a function that provides the set of next possible configurations from a current configuration together with a set of conditions and a set of external simultaneous events. The set of possible next configurations represent nondeterminism. An updated and more thorough presentation of the semantics is provided in [81], which explains the executable semantics of the STATEMATE system [82].

One approach to assigning formal semantics to UML state machines is to use abstract state machines [83]. Following the description of [84], abstract state machines are transition systems whose states are multi-sorted first-order structures, i.e. sets with relations and functions. Relations can be considered as characteristic Boolean-valued functions. The transition relation is specified by rules that describe the modification of the functions from one state to the next. These update rules are of the form "if *Condition* then *Updates*", where *Updates* is a set of function updates (assigning new function values for arguments) which are simultaneously executed when *Condition* is true.

An example of an approach that uses abstract state machines is [85], which employs multi-agent abstract state machines to model the dynamic semantics of UML state machines. Their model is intended to define rigorously the UML event handling scheme so that semantic variation points become explicit, while reflecting the original structure of UML state machines. Furthermore, object interaction is formalized by combining control and data flow. This work is further extended by the authors in [86] to cover concurrent states, while [84] surveys their previous work in order to further discuss semantic variation points and unclarities of UML state machines from a formal point of view.

In [87], Jürjens extends the semantics given in [85, 86] by modeling actions, internal activities, and their operations and parameters explicitly, as well as providing message passing between different diagrams. This constitutes a further step toward formal modeling of complete UML specifications and the goal of executable UML specifications. A thorough presentation of Jürjens' work on formalization of UML is given in [88], which provides a formal semantics for UML state machines (as well as other UML languages such as sequence diagrams and static structure diagrams) in terms of so-called UML Machines and UML Machine Systems. UML Machines are inspired by abstract state machines; they are transition systems whose states are algebraic structures. In addition, UML Machines have built-in communication mechanisms similar to the corresponding mechanisms in UML. UML Machines interact by exchanging messages which are dispatched from (or received in) multi-set buffers called output queues (or input queues). Based on UML Machines, [88] defines refinement relations, as well as security properties such as integrity and authenticity, and provides proofs of preservation of security properties under refinement.

van der Beeck [89] starts with a precise textual syntax definition for UML state machines. The terms of this textual syntax is designed to closely resemble the intuitive notion of state machines. From the textual syntax a structured operational semantics is developed in two phases. First an auxiliary semantic which only deals with processing single input events is defined. Then this auxiliary semantics is used to define a semantics that also handles processing of sequences of input events. Unlike many other approaches, [89] supports the history mechanism of UML state machines, as well as entry and exit actions.

4.5.5 Operational Semantics with Time

In [81] Harel and Naamad provide two models of time: one synchronous and one asynchronous. For the synchronous model it is assumed that the system executes a single step each time unit as a reaction to the external changes that have occurred in the single time unit since the completion of the previous step. For the asynchronous model it is assumed that the system reacts whenever an external change occurs. Several external changes may occur simultaneously, and several steps may take place within a single point in time.

In [90], timed UML state machines are compiled into timed UPPAAL automata [91], which are timed automata as originally defined by Alur and Dill [92], extended with primitives for synchronization. The passage of time is represented by increasing the value of a finite number of real-valued clocks by the same amount. [90] extends the UML notation (after(t)) by allowing clocks to be explicitly declared in class diagrams. These clocks can be tested in transition guards and reset as the effect of a transition. Furthermore, clock invariants may be associated with states to model timeouts. Even though a formal semantics as such is not provided in [90], the translation of timed UML state machines has been implemented in a prototype tool called HUGO/RT. The resulting timed automata can then be analyzed by the UPPAAL model checker.

Building on ideas from timed process calculi, [93] suggests an approach to formalizing the Statechart language [4] semantics as flattened transition systems. Transition relations are defined via structured operational rules. The work is motivated by the desire to achieve a semantics that is compositional (in the sense that the semantics of a statechart can be determined from the semantics of its components), while obeying causality and synchrony. In this context, causality means the following: A statechart may respond to an event by engaging in an enabled transition, thus performing a *micro step*. This transition may generate new events which in turn may trigger additional transitions. Synchrony means that one execution step (a *macro step*) is complete as soon as this chain reaction comes to a halt. The semantics proposed in [93] represents macro steps as sequences of micro steps which begin and end with explicit global clock ticks. The flat labeled transition systems thus have two kinds of transitions: those representing the execution of a statechart transition, and those representing global clock ticks. Clock transitions are only allowed if no additional action transitions can be executed.

4.5.6 Operational Semantics with Probabilities

Jansen et al. [94, 95] define StoCharts as an extension of UML state machines to deal with quality of service (QoS) aspects. Probability is handled by allowing state transitions to select probabilistically out of different effects. In addition, the "after" operator is given a stochastic interpretation allowing the time delay to be sampled from an arbitrary probability distribution. A formal semantics is provided in the form of a mapping to Stochastic Input/Output Automata (IOSA), which is an automata model based on timed, stochastic and probabilistic (I/O-)automata extending the UML state machine semantics of [96].

Motivated by the need for quantitative dependability and performance analysis of UML behavioral models of embedded systems, [97] presents patterns for translating UML state machines with timing and stochastic information and classification of model elements (such as fault states) into Stochastic Rewards Nets (SRN). SRNs are Petri-nets that are generalized to handle rewards (various measures) and by assigning guards and distributions of the firing time to transitions. The SRN resulting from the translation gives a precise mathematical model that can be analyzed by sophisticated tools. Standard UML mechanisms are employed to achieve the required expressiveness for the UML state machines; timing and stochastic information in captured by tagged values, while classification of model elements is achieved by stereotyped states and events.

4.6 Evaluation and Comparison

The evaluation of the semantic approaches surveyed in Sects. 4.4 and 4.5 is presented in Table 4.2 and Table 4.3, respectively. In these tables we indicate with check marks whether the properties, given as evaluation criteria in Sect. 4.2, are fulfilled. It should be noted that we have been somewhat liberal in the evaluation, and the evaluation is to some degree based on the claims of the authors of the evaluated papers. With respect to refinement, we have not assessed whether or not the provided definitions of refinement correspond to our view of refinement, but checked the refinement box if any refinement relation or similar notion is defined. In the following we give further comments on the two tables.

In Table 4.2, we see that most of the approaches are evaluated to support underspecification. The general rule is that an approach providing an explicit mechanism for specifying nondeterministic choice supports underspecification, unless such choices are interpreted as *must* behavior, as in [30, 32]. The approaches evaluated as supporting trace set properties are the approaches that explicitly distinguish between underspecification and inherent nondeterminism, as for example [28], the approaches distinguishing between universal and existential behavior, as for example [3, 66, 67, 68], and the approaches distinguishing

Table 4.2. Evaluation of semantics for sequence diagrams and similar notations

	Denotational semantics?	Operational semantics?	Underspecification?	Trace set properties?	Incomplete models?	Refinement?	Real-time?	Probabilities?
Katoen, Lambert [25]	√		√					
Krüger [27]	√		√	√	√	√		
Haugen, Husa, Runde, Seehusen, Solhaug, Stølen (STAIRS) [28, 29]	√		√	√	√	√		
Störrle [30, 31, 32]	√			√	√	√	√	
Cengarle, Knapp [33]	√		√			√		
Küster-Filipe [34]	√			√	√			
Alur, Etassami, Holzmann, Peled, Yannakakis [26, 35, 36]	√		√		√		√	
Zheng, Khendek, Hélouët, Parraux [37, 38]	√		√			√	√	
Haugen, Husa, Runde, Stølen (Timed STAIRS) [39, 40]	√		√	√	√	√	√	
Faltin, Lambert, Mitchele-Thiel, Slomka (PMSC) [42, 41]	√		√	√			√	√
Refsdal, Husa, Runde, Stølen (pSTAIRS) [43, 44, 45]	√		√	√	√	√	√	√
Mauw, Reniers (MSC-92) [46, 47]		√	√					
Mauw, Reniers (MSC-96) [21, 49, 50]		√	√				√	
Letichevsky, Kapitonova, Kotlyarov, Volkov, Letichevsky Jr., Weigert [51]		√	√				√	
Alur, Etassami, Yannakakis [36, 52]		√	√					
Uchitel, Kramer, Magee [53]		√	√		√			
Graubmann et al. [54, 55, 56, 57]		√						
Jonsson, Padilla [59]		√	√					
Lund, Stølen [60, 61]		√	√	√	√	√	√	
Cengarle, Knapp, Mühlberger [62, 63]		√	√					
Cavarra, Küster-Filipe [64]		√		√	√			
Grosu, Smolka [65]		√	√		√	√		
Harel, Damm, Maoz, Marelly, Thiagarajan (LSC) [3, 66, 67, 68]		√	√	√	√		√	
Sengupta, Cleveland (TMSC) [69, 70]		√	√	√	√	√		
Kosiuczenko, Wirsing [71]		√	√				√	

Table 4.3. Evaluation of semantics for state machines and similar notations

	Denotational semantics?	Operational semantics?	Underspecification?	Trace set properties?	Incomplete models?	Refinement?	Real-time?	Probabilities?
Broy, Cengarle, Rumpe [75]	√		√			√	√	
Simons [77]	√		√					
Rossi, Enciso, de Guzmán [78]	√						√	
Hinkel, Holz, Stølen [79, 80]	√		√			√	√	
Harel, Naamad [4, 81]		√	√				√	
Börger, Cavarra, Riccobene [85, 86]		√	√					
Jürjens [87, 88]		√	√	√		√		
von der Beeck [89]		√	√					
Knapp, Merz, Rauh [90]		√	√				√	
Lüttgen, von der Beeck, Cleaveland [93]		√	√				√	
Jansen, Hermanns, Katoen [94, 95]		√	√	√			√	√
Huszerl, Kosmidis, Cin, Majzik, Pataricza [97]		√	√	√			√	√

between *must* and *may* behavior, as for example [34]. The final evaluation criteria we want to comment upon is the support for incomplete models. This is difficult to assess, as we can always *choose* to interpret a sequence diagram as an incomplete model. The evaluation was therefore based on the approaches' treatment of negative behavior, their support for existential behavior, and their definitions of refinement.

A few comments to Table 4.3 are also needed. First, we notice that none of the approaches capture incomplete models. The reason is that state machines, unlike sequence diagrams, focus on describing a single component rather than an interaction scenario. All state machine variants we are aware of describe only the behavior that the component may exhibit; behavior not explicitly described is negative in the sense that it should not occur. There is, therefore, no explicit operator for expressing negative behavior, and all behavior is either positive or negative – there is no inconclusive behavior. Second, most approaches have received a check mark under "Underspecification", but only a few under "Trace set properties". The reason is that, for approaches with only one kind of transition, we have assumed that nondeterministic choices between transitions represent underspecification, rather than explicit nondeterminism. This decision was made because a fairly standard notion of refinement is trace inclusion – the requirement that the traces of the refined specification is a subset of the traces of the original specification. Third, all approaches with probabilities have received a check mark in the "Trace set properties" column, as all alternatives with a certain (non-zero) probability are necessarily represented in a correct implementation. In this sense, probabilistic choices can be viewed as a kind of inherent nondeterminism, which means that trace set properties can be captured.

4.7 Summary and Conclusions

In this paper we have defined a set of evaluation criteria for semantics of models for embedded systems. We claim that our criteria represent an important set of the properties that semantics of models for embedded systems should support.

These evaluation criteria have been applied in an evaluation of formal semantics for models expressed in UML sequence diagrams, state machines, and similar notations. In the paper we have presented and evaluated in all more that 30 approaches, divided into four main categories: denotational semantics of sequence diagrams, operational semantics of sequence diagrams, denotational semantics of state machines and operational semantics of state machines. Our selection of approaches to evaluate is not exhaustive, but we believe that it gives a representative picture of the various approaches available.

As the evaluation reveals there is no lack of approaches to formal semantics for UML sequence diagrams and state machines, and many of these have desirable properties. We do not proclaim a winner, but we have established that formal semantics of relevant modeling languages are readily available for the developers of embedded systems. We have not evaluated to what degree the approaches presented in this paper are supported by suitable tools, nor to what degree they have been put to practical application. Still, judging from our evaluation, there should be a large potential for applying UML models supported by formal semantics in the development of embedded systems. It is up to developers to choose a suitable approach based on the nature of the system to be developed, and the background and experience of the development team.

Acknowledgements

The work on which this paper reports has partly been funded by the Research Council of Norway through the projects SARDAS (15295/431) and ENFORCE (164382/V30), and partly by the European Commission through the MODELPLEX project (Contract no. 034081) under the IST Sixth Framework Programme.

References

[1] Object Management Group: Unified Modeling Language: Superstructure, version 2.1.1 (non-change bar). OMG Document: formal/2007-02-05 (2005)
[2] International Telecommunication Union: Message Sequence Chart (MSC), ITU-T Recommendation Z.120 (1999)

[3] Damm, W., Harel, D.: LSCs: Breathing life into Message Sequence Charts. Formal Methods in System Design 19, 45–80 (2001)
[4] Harel, D.: Statecharts: A visual formalism for complex systems. Science of Computer Programming 8(3), 231–274 (1987)
[5] International Telecommunication Union: Specification and description language (SDL), ITU-T Recommendation Z.100 (2000)
[6] Labinaz, G., Bayoumi, M.M., Rudie, K.: A survey of modeling and control of hybrid systems. Annual Reviews of Control 21, 79–92 (1997)
[7] Giese, H., Henkler, S.: A survey of approaches for the visual model-driven development of next generation software-intensive systems. Journal of Visual Languages and Computing 17(6), 528–550 (2006)
[8] McLean, J.: A general theory of composition for trace sets closed under selective interleaving functions. In: Proceedings of the IEEE Symposium on Research in Security and Privacy, pp. 79–93. IEEE Computer Society, Los Alamitos (1994)
[9] Alpern, B., Schneider, F.B.: Defining liveness. Information Processing Letters 21(4), 181–185 (1985)
[10] Schneider, F.B.: Enforceable security policies. ACM Transactions on Information System Security 3(1), 30–50 (2000)
[11] Harel, D., Rumpe, B.: Meaningful modeling: What's the semantics of "semantics"? Computer 37(10), 64–72 (2004)
[12] Prinz, A.: Formal semantics of specification languages. Telektrcnikk (4), 146–155 (2000)
[13] Fecher, H., Schönborn, J., Kyas, M., de Roever, W.P.: 29 new unclarities in the semantics of UML 2.0 state machines. In: Lau, K.-K., Banach, R. (eds.) ICFEM 2005. LNCS, vol. 3785, pp. 52–65. Springer, Heidelberg (2005)
[14] Schmidt, D.A.: Denotational semantics. A methodology for language development. William C. Brown (1988)
[15] Hoare, C.A.R., Jifeng, H.: Unifying theories of programming. Prentice-Hall, Englewood Cliffs (1998)
[16] Object Management Group: Unified Modeling Language Specification, version 1.4. OMG Document: formal/2001-09-67 (2001)
[17] Facchi, C.: Formal semantics of Time Sequence Diagrams. Technical report TUM-I9540, Technische Universität München (1995)
[18] International Telecommunication Union: Information technology – Open Systems Interconnection – Basic reference model: Conventions for the definition of OSI services, ITU-T Recommendation X.210 (1993)
[19] Bræk, R., Gorman, J., Haugen, Ø., Møller-Pedersen, B., Melby, G., Sanders, R., Stålhane, T.: TIMe: The Integrated Method. Electronic Textbook v4.0. SINTEF (1999)
[20] International Telecommunication Union: Message Sequence Chart (MSC), ITU-T Recommendation Z.120 (1996)
[21] International Telecommunication Union: Message Sequence Chart (MSC), ITU-T Recommendation Z.120, Annex B: Formal semantics of Message Sequence Charts (1998)
[22] Haugen, Ø.: Comparing UML 2.0 Interactions and MSC-2000. In: Amyot, D., Williams, A.W. (eds.) SAM 2004. LNCS, vol. 3319, pp. 65–79. Springer, Heidelberg (2005)
[23] Object Management Group: UML Testing Profile, version 1.0. OMG Document: formal/2005-07-07 (2005)
[24] Object Management Group: UML Profile for Schedulability, Performance, and Time Specification, version 1.1. OMG Document: formal/2005-01-02 (2005)

[25] Katoen, J.P., Lambert, L.: Pomsets for Message Sequence Charts. In: Formale Beschreibungstechniken für Verteilte Systeme, pp. 197–208. Shaker (1998)
[26] Alur, J., Yannakakis, M.: Model checking of Message Sequence Charts. In: Baeten, J.C.M., Mauw, S. (eds.) CONCUR 1999. LNCS, vol. 1664, pp. 98–113. Springer, Heidelberg (1999)
[27] Krüger, I.H.: Distributed system design with Message Sequence Charts. PhD thesis, Technische Universität München (2000)
[28] Haugen, Ø., Husa, K.E., Runde, R.K., Stølen, K.: STAIRS towards formal design with sequence diagrams. Software and Systems Modeling 4(4), 355–367 (2005)
[29] Seehusen, F., Solhaug, B., Stølen, K.: Adherence preserving refinement of trace-set properties in STAIRS: Exemplified for information flow properties and policies. Software and Systems Modeling 8(1), 45–65 (2009)
[30] Störrle, H.: Assert, negate and refinement in UML 2 interactions. In: 2nd International Workshop on Critical Systems Development with UML (CSD-UML 2003), Technische Universität München, pp. 79–93 (2003)
[31] Störrle, H.: Semantics of interaction in UML 2.0. In: IEEE Symposium on Human Centric Computing Languages and Environments (HCC 2003), pp. 129–136. IEEE Computer Society, Los Alamitos (2003)
[32] Störrle, H.: Trace semantics of interactions in UML 2.0. Technical report TR 0403, Institut für Informatik, der Ludwig-Maximilians-Universität München (2004)
[33] Cengarle, M.V., Knapp, A.: UML 2.0 interactions: Semantics and refinement. In: 3rd International Workshop on Critical Systems Development with UML (CSD-UML 2004), Technische Universität München, pp. 85–99 (2004)
[34] Küster-Filipe, J.: Modelling concurrent interactions. In: Rattray, C., Maharaj, S., Shankland, C. (eds.) AMAST 2004. LNCS, vol. 3116, pp. 304–318. Springer, Heidelberg (2004)
[35] Alur, R., Holzmann, G.J., Peled, D.: An analyzer for Message Sequence Charts. In: Margaria, T., Steffen, B. (eds.) TACAS 1996. LNCS, vol. 1055, pp. 35–48. Springer, Heidelberg (1996)
[36] Alur, R., Etessami, K., Yannakakis, M.: Inference of Message Sequence Charts. IEEE Transactions on Software Engineering 29(7), 623–633 (2003)
[37] Zheng, T., Khendek, F., Hélouët, L.: A semantics for timed MSC. Electronic Notes in Theoretical Computer Science 65(7), 85–99 (2002)
[38] Zheng, T., Khendek, F., Parreaux, B.: Refining timed MSCs. In: Reed, R., Reed, J. (eds.) SDL 2003. LNCS, vol. 2708, pp. 234–250. Springer, Heidelberg (2003)
[39] Haugen, Ø., Husa, K.E., Runde, R.K., Stølen, K.: Why timed sequence diagrams require three-event semantics. In: Leue, S., Systä, T.J. (eds.) Scenarios: Models, Transformations and Tools. LNCS, vol. 3466, pp. 1–25. Springer, Heidelberg (2005)
[40] Runde, R.K.: STAIRS - Understanding and developing specifications expressed as UML interaction diagrams. PhD thesis, Faculty of Mathematics and Natural Sciences, University of Oslo (2007)
[41] Faltin, N., Lambert, L., Mitschele-Thiel, A., Slomka, F.: An annotational extension of Message Sequence Charts to support performance engineering. In: 8th International SDL Forum: Time for Testing, SDL, MSC and Trends (SDL 1997), pp. 307–322. Elsevier, Amsterdam (1997)
[42] Lambert, L.: PMSC for performance evaluation. In: 1st Workshop on Performance and Time in SDL/MSC, pp. 70–80 (1998)
[43] Refsdal, A., Husa, K.E., Stølen, K.: Specification and refinement of soft real-time requirements using sequence diagrams. In: Pettersson, P., Yi, W. (eds.) FORMATS 2005. LNCS, vol. 3829, pp. 32–48. Springer, Heidelberg (2005)

[44] Refsdal, A., Runde, R.K., Stølen, K.: Underspecification, inherent nondeterminism and probability in sequence diagrams. In: Gorrieri, R., Wehrheim, H. (eds.) FMOODS 2006. LNCS, vol. 4037, pp. 138–155. Springer, Heidelberg (2006)
[45] Refsdal, A.: Specifying computer systems with probabilistic sequence diagrams. PhD thesis, Faculty of Mathematics and Natural Sciences, University of Oslo (2008)
[46] Mauw, S.: The formalization of Message Sequence Charts. Computer Networks and ISDN Systems 28(1), 1643–1657 (1996)
[47] Mauw, S., Reniers, M.A.: An algebraic semantics of Basic Message Sequence Charts. The Computer Journal 37(4), 269–278 (1994)
[48] Okazaki, M., Aoki, T., Katayama, T.: Formalizing sequence diagrams and state machines using Concurrent Regular Expression. In: 2nd International Workshop on Scenarios and State Machines: Models, Algorithms, and Tools, SCESM 2003 (2003)
[49] Mauw, S., Reniers, M.A.: Operational semantics for MSC'95. Computer Networks 31(17), 1785–1799 (1999)
[50] Mauw, S., Reniers, M.A.: High-level Message Sequence Charts. In: 8th International SDL Forum: Time for Testing, SDL, MSC and Trends (SDL 1997), pp. 291–306. Elsevier, Amsterdam (1997)
[51] Letichevsky, A.A., Kapitonova, J.V., Kotlyarov, V.P., Volkov, V.A., Letichevsky Jr., A.A., Weigert, T.: Semantics of Message Sequence Charts. In: Prinz, A., Reed, R., Reed, J. (eds.) SDL 2005. LNCS, vol. 3530, pp. 117–132. Springer, Heidelberg (2005)
[52] Alur, R., Etessami, K., Yannakakis, M.: Realizability and verification of MSC graphs. Theoretical Computer Science 331(1), 97–114 (2005)
[53] Uchitel, S., Kramer, J., Magee, J.: Incremental elaboration of scenario-based specification and behavior models using implied scenarios. ACM Transactions on Software Engineering and Methodology 13(1), 37–85 (2004)
[54] Graubmann, P., Rudolph, E., Grabowski, J.: Towards a Petri net based semantics for Message Sequence Charts. In: 6th International SDL Forum: Using objects (SDL 1993), pp. 179–190. Elsevier, Amsterdam (1993)
[55] Heymer, S.: A semantics for MSC based on Petri net components. In: 4th International SDL and MSC Workshop (SAM 2000), pp. 262–275 (2000)
[56] Sgroi, M., Kondratyev, A., Watanabe, Y., Lavagno, L., Sangiovanni-Vincentelli, A.: Synthesis of Petri nets from Message Sequence Charts specifications for protocol design. In: Design, Analysis and Simulation of Distributed Systems Symposium (DASD 2004), pp. 193–199 (2004)
[57] Gunter, E.L., Muscholl, A., Peled, D.: Compositional Message Sequence Charts. International Journal on Software Tools for Technology Transfer 5(1), 78–89 (2003)
[58] Bernardi, S., Donatelli, S., Merseguer, J.: From UML sequence diagrams and statecharts to analysable Petri net models. In: 3rd International Workshop on Software and Performance (WOSP 2002), pp. 35–45. ACM Press, New York (2002)
[59] Jonsson, B., Padilla, G.: An execution semantics for MSC-2000. In: Reed, R., Reed, J. (eds.) SDL 2001. LNCS, vol. 2078, pp. 365–378. Springer, Heidelberg (2001)
[60] Lund, M.S.: Operational analysis of sequence diagram specifications. PhD thesis, Faculty of Mathematics and Natural Sciences, University of Oslo (2008)

[61] Lund, M.S., Stølen, K.: A fully general operational semantics for UML 2.0 sequence diagrams with potential and mandatory choice. In: Misra, J., Nipkow, T., Sekerinski, E. (eds.) FM 2006. LNCS, vol. 4085, pp. 380–395. Springer, Heidelberg (2006)
[62] Cengarle, M.V., Knapp, A.: Operational semantics of UML 2.0 interactions. Technical report TUM-I0505, Technische Universität München (2005)
[63] Mühlberger, H.: Eine verteile operationale Semantik für UML 2.0-Interaktionen. Diplomarbeit, Institut für Informatik, der Ludwig-Maximilians-Universität München (2007)
[64] Cavarra, A., Küster-Filipe, J.: Formalizing liveness-enriched sequence diagrams using ASMs. In: Zimmermann, W., Thalheim, B. (eds.) ASM 2004. LNCS, vol. 3052, pp. 67–77. Springer, Heidelberg (2004)
[65] Grosu, R., Smolka, S.A.: Safety-liveness semantics for UML 2.0 sequence diagrams. In: 5th International Conference on Application of Concurrency to System Design (ACSD 2005), pp. 6–14. IEEE Computer Society, Los Alamitos (2005)
[66] Harel, D., Marelly, R.: Come, let's play: Scenario-based programming using LSCs and the Play-Engine. Springer, Heidelberg (2003)
[67] Harel, D., Thiagarajan, P.S.: Message Sequence Charts. In: Lavagano, L., Martin, G., Selic, B. (eds.) UML for real. Design of embedded real-time systems, pp. 77–105. Kluwer, Dordrecht (2003)
[68] Harel, D., Maoz, S.: Assert and negate revisited: Modal semantics for UML sequence diagrams. In: 5th International Workshop on Scenarios and State Machines: Models, Algorithms, and Tools (SCESM 2006), pp. 13–19. ACM Press, New York (2006)
[69] Sengupta, B., Cleaveland, R.: Triggered Message Sequence Charts. SIGSOFT Software Engineering Notes 27(6), 167–176 (2002)
[70] Sengupta, B., Cleaveland, R.: Triggered Message Sequence Carts. IEEE Transactions on Software Engineering 32(8) (2006)
[71] Kosiuczenko, P., Wirsing, M.: Towards an integration of Message Sequence Charts and Timed Maude. Journal of Integrated Design & Process Science 5(1), 23–44 (2001)
[72] Crane, M.L., Dingel, J.: On the semantics of UML state machines: Categorization and comparison. Technical report 2005-501, School of Computing, Queens's University, Kingston (2005)
[73] Broy, M., Cengarle, M.V., Rumpe, B.: Towards a system model for UML, the structural data model. Technical report TUM-I0612, Technische Universität München (2006)
[74] Broy, M., Cengarle, M.V., Rumpe, B.: Towards a system model for UML, part 2: The control model. Technical report TUM-I0710, Technische Universität München (2007)
[75] Broy, M., Cengarle, M.V., Rumpe, B.: Towards a system model for UML, part 3: The state machine model. Technical report TUM-I0711, Technische Universität München (2007)
[76] Broy, M., Stølen, K.: Specification and development of interactive systems. In: FOCUS on streams, interface, and refinement. Springer, Heidelberg (2001)
[77] Simons, A.J.H.: On the compositional properties of UML statechart diagrams. In: Rigorous Object-Oriented Methods (ROOM 2000), Workshops in Computing, BCS (2000) (2000)
[78] Rossi, C., Enciso, M., de Guzmán, I.P.: Formalization of UML state machines using temporal logic. Software and Systems Modeling 3(1), 31–54 (2004)

[79] Hinkel, U.: Verification of SDL specifications on the basis of stream semantics. In: 1st Workshop of the SDL Forum Society on SDL and MSC (SAM 1998), pp. 241–250 (1998)
[80] Holz, E., Stølen, K.: An attempt to embed a restricted version of SDL as a target language in Focus. In: Formal Description Techniques VII (FORTE 1994), pp. 324–339. Chapman and Hall, Boca Raton (1994)
[81] Harel, D., Naamad, A.: The STATEMATE semantics of statecharts. ACM Transactions on Software Engineering and Methodology 5(4), 293–333 (1996)
[82] Harel, D., Lachover, H., Naamad, A., Pnueli, A., Politi, M., Sherman, R., Shtull-Trauring, A., Trakhtenbrot, M.: STATEMATE: A working environment for the development of complex reactive systems. IEEE Transactions on Software Engineering 16(4), 403–414 (1990)
[83] Gurevich, Y.: Evolving algebras 1993: Lipari guide. In: Specification and Validation Methods, pp. 9–36. Oxford University Press, Oxford (1995)
[84] Börger, E., Cavarra, A., Riccobene, E.: On formalizing UML state machines using ASMs. Information and Software Technology 46(5), 287–292 (2004)
[85] Börger, E., Cavarra, A., Riccobene, E.: Modeling the dynamics of UML state machines. In: International Workshop on Abstract State Machines, Theory and Applications, pp. 223–241. Springer, Heidelberg (2000)
[86] Börger, E., Cavarra, A., Riccobene, E.: Modeling the meaning of transitions from and to concurrent states in UML state machines. In: 2003 ACM Symposium on Applied Computing, pp. 1086–1091. ACM Press, New York (2003)
[87] Jürjens, J.: A UML statecharts semantics with message-passing. In: 2002 ACM Symposium on Applied Computing, pp. 1009–1013. ACM Press, New York (2002)
[88] Jürjens, J.: Secure systems development with UML. Springer, Heidelberg (2005)
[89] von der Beeck, M.: A structured operational semantics for UML-statecharts. Software and Systems Modeling 1(2), 130–141 (2002)
[90] Knapp, A., Merz, S., Rauh, C.: Model checking timed UML state machines and collaborations. In: 7th International Symposium on Formal Techniques in Real-Time and Fault-Tolerant Systems, pp. 395–416. Springer, Heidelberg (2002)
[91] Larsen, K.G., Pettersson, P., Yi, W.: Uppaal in a nutshell. International Journal on Software Tools for Technology Transfer 1, 134–152 (1997)
[92] Alur, R., Dill, D.L.: A theory of timed automata. Theoretical Computer Science 126(2), 183–235 (1994)
[93] Lüttgen, G., von der Beeck, M., Cleaveland, R.: A compositional approach to statecharts semantics. Technical report, Institute for Computer Applications in Science and Engineering (2000)
[94] Jansen, D.N., Hermanns, H., Katoen, J.P.: A QoS-oriented extension of UML statecharts. In: Stevens, P., Whittle, J., Booch, G. (eds.) UML 2003. LNCS, vol. 2863, pp. 76–91. Springer, Heidelberg (2003)
[95] Jansen, D.N., Hermanns, H.: Qos modelling and analysis with UML-statecharts: the Stocharts approach. SIGMETRICS Performance Evaluation Review 32(4), 28–33 (2005)
[96] Eshuis, R., Wieringa, R.: Requirements-level semantics for UML statecharts. In: 4th International Conference on Formal Methods for Open Object-Based Distributed Systems IV, pp. 121–140. Kluwer, Dordrecht (2000)
[97] Huszerl, G., Kosmidis, K., Cin, M.D., Majzik, I., Pataricza, A.: Quantitative analysis of UML statechart models of dependable systems. The Computer Journal 45(3), 260–277 (2002)

Part III

Modeling

5 Modeling and Simulation of TDL Applications

Stefan Resmerita, Patricia Derler, Wolfgang Pree, and Andreas Naderlinger

University of Salzburg, Austria
{stefan.resmerita,patricia.derler,wolfgang.pree,
andreas.naderlinger}@cs.uni-salzburg.at

Abstract. Most of the existing modeling tools and frameworks for embedded applications use levels of abstraction where execution and communication times of computational tasks are not captured. Thus, properties such as time and value determinism can be lost when refining the model closer to a target platform. The Logical Execution Time (LET) paradigm has been proposed to deal with this issue, by enabling specification of platform-independent execution times of periodic time-triggered computational tasks at higher levels of abstraction.

This chapter deals with modeling and simulation of embedded applications where LET requirements are specified by using the Timing Definition Language (TDL). TDL provides a programming model for time- and event-triggered components suitable for large distributed systems. We present specific TDL extensions that increase the expressiveness of the language, accommodating the needs of control applications such as minimum sensor-actuator delays. We describe simulation of TDL programs in dataflow models (using Simulink) and discrete event (DE) models (using Ptolemy II). We show how the Ptolemy II based simulation can be used to validate preservation of timing and value behaviors when mapping a DE model of an application with concurrent components into a sequential implementation platform with fixed priority preemptive scheduling.

5.1 Introduction

In complex embedded systems, execution and communication times related to computational tasks of an application can have a substantial influence on the application behavior that is unaccounted for in high level models. Consequently, the implementation of a model on a certain execution platform may violate requirements that are proved to be satisfied in the model. Explicitly considering execution times at higher levels of abstractions has been proposed as a way to achieve satisfaction of real-time properties [1]. One promising direction in this respect is the Logical Execution Time (LET) [2], which forms the foundation of several real-time programming languages [2] [3] [4]. Among these, the Timing Definition Language [4] is under active development, with commercially available support tools.

TDL is inspired from the Giotto programming model [2], which was targeted for control applications. Giotto proposed trading end-to-end latency (which must be minimal in control systems) for determinism and robustness, which have become more and more important due to increased complexity of both applications and platforms. TDL has been extended to address both requirements, in the commonly encountered case where a control task can be split into a *fast step*, used to compute controller outputs (i.e. actuator values) and a slow step, used to determine new state information. Another TDL extension, called *slot selection*, allows designation of LET values that are smaller than the invocation period for a task, providing a separation of concerns between choosing the controller sampling period, which is the job of the control engineer, and minimizing computation latency, which is the job of the software engineer. By slot selection, the designer specifies the beginning and end of a task's LET within the task's invocation period. This also enables specification of executions with fixed time offsets.

An important principle in embedded systems design is the separation between the functionality of an application and the platform where the application is implemented. This principle is adopted by modern design methodologies such as Model Driven Architecture (MDA) [5] and Platform Based Design (PBD) [6]. An MDA application consists of a platform-independent model (PIM), specifying the functionality, one or more platform-specific models(PSMs), and sets of interfaces describing the coupling between PIM and each of the PSMs. PBD proposes an iterative model-based development process where at each iteration a functionality model is mapped to a platform model. The mapped functionality becomes the functionality model for the next iteration, where it is mapped to a (usually refined) platform model. This is repeated until the final implementation is obtained for all components.

One of the main issues in the above approaches is the mapping between the functionality model and the platform model, which should be done such that the behaviors of the resultant model are in the intersection of the behavioral sets of the individual models. A commonly encountered situation is mapping a concurrent functional model into a sequential implementation platform. For a real-time application, this may lead to violation of real-time properties such as maximum sensor-actuator delays. Even if a suitable scheduling guarantees latency bounds, it may not guarantee the same value behavior as in the functionality model (a simple example of this situation will be further discussed in this paper). Using LET ensures preservation of timing and value behaviors over model refinement, by requiring that the platform model has the means to carry out the LET semantics, which refers not only to task execution (resolved by scheduling), but most importantly to data transfer (resolved by buffering). In the final implementation, LET specifications are executed by a dedicated runtime system, provided that the software components can be suitably scheduled for execution. Scheduling can be done statically for the time-triggered, periodic task executions.

Static schedulability analysis for LET-based software becomes hard to achieve, or overly conservative, in embedded control systems containing also concurrent computations triggered by environment conditions (dynamic events), where event-triggered tasks share the same execution platform as the time-triggered part, and may preempt or delay the execution of time-triggered components. Thus, a simulation platform is needed for early verification and validation of such heterogeneous systems. Existent simulation frameworks for LET-based models operate at a functional (platform independent) level, where the most influential LET benefits cannot be shown. For example, a DE model of a time-triggered application with LET-based constraints has the same behavior as a model where the LET constraints are replaced by delays. However, when mapping the functional model into a platform model, the mapped delay-based model may exhibit new behaviors, while the behaviors of the mapped LET-based model will always be included in the behavioral set of the functional LET-based model.

In the embedded systems industry, simulation is widely used for testing and validation of complex systems. It is also used for effectively demonstrating the impact of new technologies. It is therefore important to be able to simulate a model with LET specifications in order to demonstrate the benefits of the LET approach. In this paper, we consider a platform abstraction consisting of execution times and fixed priority preemptive scheduling. Thus, the mapping means assigning to each task in the functional model an execution time and a priority. We present a Ptolemy II framework which allows simulating the behavior of a LET-based application mapped to the platform. TDL is employed to specify the timing structure based on LET. We use the simulator to run an example which shows how TDL can ensure preservation of behavior over model refinement.

This paper is structured as follows. Section 5.2 describes the Timing Definition Language, including the control-specific extensions. In Section 5.3 we present the two main simulation frameworks for TDL. The relations with existing work are shown in Section 5.4, which is followed by concluding remarks in Section 5.5.

5.2 The Timing Definition Language

We describe in this section the main constructs of the core TDL, followed by extensions of the language that specifically address control applications. The ensuing presentation of TDL components is necessarily brief. The complete TDL specification can be found in [4].

5.2.1 TDL Description

The Timing Definition Language allows the specification of timing properties of hard real-time applications by employing the LET concept and the principle of separation between timing and functionality introduced in Giotto [2]. While TDL is conceptually based on Giotto, it provides extended features, a more convenient syntax, and an improved set of programming tools.

The Logical Execution Time associated with a computational unit, or task, represents a fixed logical duration between the time instant when the task becomes ready for execution and the instant when the execution finishes. A task's LET is specified at the model level, independently of the task's functionality. When deploying the model on a platform, the LET specification is satisfied if the total physical execution time of the task is included in the LET interval for every task invocation, and an appropriate runtime system ensures that task inputs are read at the beginning of the LET interval (the release time) and task outputs are made available at the end of the LET interval (the termination time). This is illustrated in Figure 5.1. Between release and termination points, the output values are those established in the previous execution; default or specified initial values are used during the first execution.

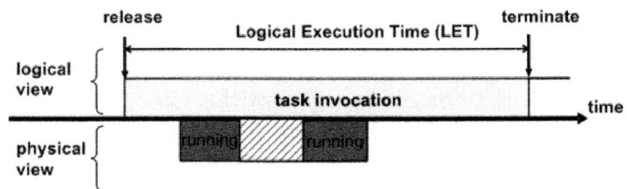

Fig. 5.1. Logical Execution Time

TDL is targeted for applications consisting of periodic software tasks designed to control a physical environment. Thus, some tasks receive information from the environment via sensors and some tasks act on the environment via actuators. A task has input ports, output ports, and state ports. State ports are employed for maintaining state information between different executions of the same task. The main structure of a task declaration in TDL is given in Figure 5.2.

```
task <task_name> {
    input <type> <list_of_input_ports>;
    ...                            //other input port declarations
    output <type> <list_of_output_ports>;
    ...                            //other output port declarations
    state <type> <list_of_state_ports>;
    ...                            //other state port declarations
    uses <external_function_call>;
}
```

Fig. 5.2. Structure of TDL task declaration

Any of the lists of ports can be empty, while exactly one external function name (possibly with arguments) must be specified after the "uses" keyword. This represents the implementation of the task functionality.

Tasks that are executed concurrently are grouped in modes. In TDL, a mode is a set of periodically executed activities: task invocations, actuator updates, and mode switches. A mode activity has a specified execution rate and may be carried out conditionally. A mode declaration is schematically shown in Figure 5.3. The frequency attribute specifies the rate of execution of the corresponding

```
mode <mode_name> [period=<time_duration>]{
  task
    [freq=<exec_rate>] <task_name>(<argument_list>);
    ...                           //other task invocations
  actuator
    [freq=<exec_rate>] <act_name>:=<task_name>.<output_port>;
    ...                           //other actuator updates
  mode
    [freq=<exec_rate>] if <condition> <name_of_target_mode>;
    ...                           //other mode switches
}
```

Fig. 5.3. Structure of TDL mode declaration

activity within one mode period. Thus, the LET of a task is expressed as the mode period divided by the frequency of task invocation. Note that the time steps of all activities in a mode period can be statically determined. Mode activities are carried out by a runtime system which performs the following operations at every time step:

(1) Update output ports of tasks whose LETs end at the current time step. At time 0, the ports are initialized rather than updated.
(2) Update actuators.
(3) Test for mode switches. If a mode switch is enabled, switch to the target mode.
(4) Update input ports of the tasks whose LETs start at the current time step.
(5) Trigger the execution of the tasks whose LETs start at the current time step.

TDL provides a top level structuring unit called a *module*, which is a logically coherent group of sensors, actuators and modes. The module concept serves multiple purposes: (1) a module provides a name space and an export/import mechanism and thereby supports decomposition of large systems, (2) modules provide parallel composition of real-time applications, (3) modules serve as units of loading, i.e. a runtime system may support dynamic loading and unloading of modules, and (4) modules are the natural choice as unit of distribution because dataflow within a module (cohesion) will most probably be much larger than dataflow across module boundaries (adhesion).

A schematic example of a TDL program is shown in Figure 5.4. Notice that a module contains declarations of sensor and actuator variables, tasks and modes. In the above example, module Sender contains a sensor variable $s1$, and an actuator variable $a1$. The value of $s1$ is updated by executing the (platform-specific) driver $getS1$ and the value of $a1$ is send to the physical actuator by using the platform specific driver $setA1$. Each module has exactly one start mode, indicated by preceding the mode declaration with the reserved word "start". The declaration of the output port of task *inc* specifies also an initial value (10). The task is invoked in mode *main* of the *Sender* module, where its input port is connected to the sensor $s1$. In the same mode, actuator $a1$ is updated with the value of the task's output port. The second module called *Receiver* imports the *Sender* module in order to connect the input of the local task *clientTask* with the output of the external task *inc*. These TDL components and their connectivity are depicted in Figure 5.5.

Let us illustrate the operations carried out by the TDL runtime system for the task *inc* during one mode period. At time 0, output ports are initialized and connected actuators are updated. Sensor s1 is read and the value is provided as input for the task, which is then released for execution. At time 5 (the end of the LET), the task's output port is updated, then actuator a1 is updated. Next, the mode switch condition in the guard function *exitMain* is evaluated. If it evaluates to true, a mode switch to the empty mode *freeze* is performed and no further actions are processed. Otherwise the mode *main* remains active and the above operations are repeated in the next mode period.

TDL enables so-called *transparent distribution* of hard real-time applications, which can be described with respect to two points of view. Firstly, at runtime a TDL application behaves exactly the same, no matter if all modules

```
module Sender {
  sensor int s1 uses getS1;
  actuator int a1 uses setA1;
  public task inc {
    input int i;
    output int o := 10;
    uses incImpl(i,o);
  }
  start mode main [period=5ms] {
    task [freq=1] inc(s1);                    //LET = 5ms (=period/freq)
    actuator [freq=1] a1 := inc.o;
    mode [freq=1] if exitMain(s1) then freeze;
  }
  mode freeze [period=1000ms] {}
}
module Receiver {
  import Sender;
  . . .
  task clientTask {
    input int i1;
    . . .
  }
  start mode main [period=10ms] {
    task [freq=1] clientTask(Sender.inc.o);   //LET = 10ms
    . . .
  }
  . . .
}
```

Fig. 5.4. Example of TDL code

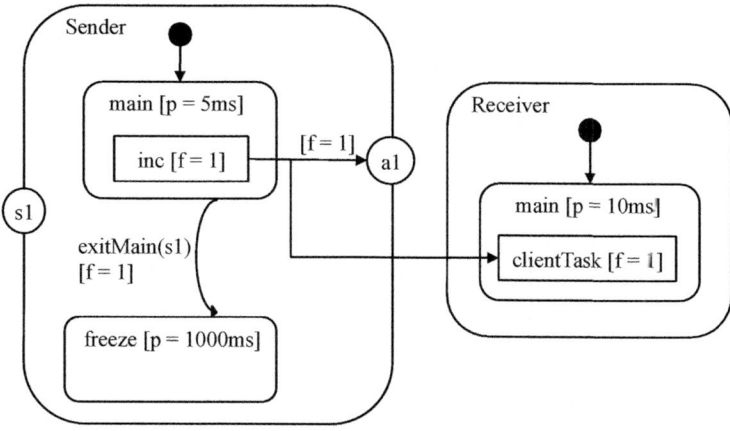

Fig. 5.5. TDL constructs defined by the code in Figure 5.4

(i.e. components) are executed on a single node or if they are distributed across multiple nodes. The logical timing is always preserved, only the physical timing, which is not observable from the outside, may be changed. Secondly, for the developer of a TDL module, it does not matter where the module itself and any imported modules are executed. The TDL tool chain and runtime system frees the developer from the burden of explicitly specifying the communication requirements of modules. It should be noted that in both aspects transparency applies not only to the functional but also to the temporal behavior of an application. The advantage of transparent distribution for a developer is that the TDL modules can be specified without having the execution on a potentially distributed platform in mind. The functional and temporal behavior of the system is independent of the mapping of modules to computation nodes, which is defined separately.

A compiler transforms TDL programs into virtual instructions called *E-Code* [7]. E-Code describes the application's reactivity, i.e. time instants to release or terminate tasks or to interact with the environment. A virtual machine, the *E-Machine* [7], interprets the instructions at runtime and ensures the correct timing behavior. According to the E-Code, the E-Machine timely hands tasks to a dispatcher and executes drivers. A driver performs communication activities, such as reading sensor values, providing input values for tasks at their release time or copy output values at their termination time.

A commercially available tool suite deals with modeling and deployment of TDL components [8]. TDL components can be written directly in textual form (TDL source code) or designed graphically by using the TDL:VisualCreator tool. The TDL:Compiler targets the TDL:E-Machine. The TDL:E-Machine exists for several different platforms, including OSEK, INtime, RTLinux, etc. The TDL:VisualDistributor can be used to assign TDL modules to a single specified computational node or a distributed system of nodes. Also, the TDL:Scheduler is employed to generate the necessary node and communication schedules. The tools also check for the schedulability of the system, based on provided worst case execution times for the tasks, under the assumption that the periodically time-triggered TDL tasks are the only significant computations competing for the platform resources.

5.2.2 TDL Extensions for Control Applications

Reducing Latency for Control Applications
The main application field for the time-triggered programming model introduced by Giotto is implementation of control systems. A control application reads environment data through sensors, and exercises control over the physical environment through actuators. In sampled data control systems, the controller is executed periodically, polling sensors and determining control actions in every period. Usually, control actions depend on the latest sensor values and on the current state of the controller, which is also updated at every period. The time delay between reading sensors and updating actuators in the same period should be as small as possible. Thus, the controller functions are organized in two steps: *update outputs* and *update state*, with the first step to be executed as soon as possible after sensor reading. To enable advance calculation of control outputs, in TDL a task's functionality code can be split in a *fast step* (corresponding to *update outputs*) and a *regular step* (corresponding to *update state*), where the fast step is executed in logical zero time at the release time of the TDL task, and the regular step is executed within the task's LET. To this end, the task declaration is modified to allow specification of *two* external function calls, the fast one being indicated by a dedicated driver annotation called "[release]", which means that the fast function has to be executed immediately when the task is released for execution (i.e. at the beginning of the task's LET). A two-step task can now be declared according to the structure shown in Figure 5.6. Syntactically, the only addition to the single-step task declaration (shown in Figure 5.2) is another *uses* line containing the *release* annotation, which is reserved for the fast step declaration. If an output port appears in the argument lists of both functions, then it acts as output of the fast function (i.e. it must be updated by the fast function) and as input to the slow function. An example is presented in Figure 5.7.

```
task <task_name> {
    input <type> <list_of_input_ports>;
    ...                         //other input port declarations
    output <type> <list_of_output_ports>;
    ...                         //other output port declarations
    state <type> <list_of_state_ports>;
    ...                         //other state port declarations
    uses [release] <fast_function_name>(<arg_list>);
    uses <slow_function_name>(<arg_list>);
}
```

Fig. 5.6. Structure of TDL declaration for a two-step task

```
task digiCon {
  input int i1,i2;
  output int o:=0;
  state double s:=0;
  uses [release] controllerOutput(i1,i2,s,o);
                              //o must be calculated here
  uses controllerUpdate(i1,i2,s,o);
                              //o is an input argument here
}
```

Fig. 5.7. Declaration example of a two-step task

The explicit declaration of a task's fast and slow steps is accompanied by the introduction of a specific mode activity, called *task sequence*, to indicate actuator updates that must take place upon execution of the task's fast step. A task sequence combines a task invocation and subsequent actuator updates. These are performed at the release time of the invoked task, if the task contains a fast step that provides the required output ports. Output ports updated in the fast step are available immediately for actuator updates if the two-step task is included in a task sequence. Figure 5.8 presents the layout of a mode declaration including task sequences. An example where the task in Figure 5.7 appears in a task sequence is shown in Figure 5.9. The effect of this code is that at every 10ms, sensors s2 and s3 are read, the function *controllerOutput* is executed and the actuator act1 is updated. Since these operations are considered as taking logical zero time, their execution times must be much smaller than the execution times of regular TDL tasks. Then the function *controllerUpdate* is executed, which may take up to 10ms. Task t0 is a regular TDL task with a LET of 50ms. Thus, at every 50ms tick, sensor s1 is read and task t0 is released for execution. The output of t0 is provided to actuator act2 at the end of the 50ms period.

```
mode <mode_name> [period = <time_duration>]{
  task
    [freq=<exec_rate>]  <task_name>(<arg_list>);
    [freq=<exec_rate>]  {<task_name>(<arg_list>);
                         <act_name>:=<task_name>.<output_port>;}
    ...                          //other task invocations
  actuator
    [freq=<exec_rate>]  <act_name>:=<task_name>.<output_port>;
    ...                          //other actuator updates
  mode
    [freq=<exec_rate>]  if <condition> <name_of_target_mode>;
    ...                          //other mode switches
}
```

Fig. 5.8. Structure of TDL mode declaration with task sequence

```
start mode main [period=100ms] {
  task [freq=2] t0(s1);
  task [freq=10] {digiCon(s2, s3); act_1:=digiCon.o;} //sequence
  actuator [freq=2] act_2 := t0.o;
}
```

Fig. 5.9. Example of task sequence

Task sequences entail a specific operational semantics. The operational steps performed by the runtime system are now as follows:

(1) Update output ports of tasks whose LETs end at the current time step. At time 0, the ports are initialized rather than updated. Exception: output ports of two-step tasks that are arguments of both functions (fast and slow) are not updated.
(2) Update actuators.
(3) Execute fast tasks. For every task sequence that occurs at the current step, update the inputs of the task, then execute the fast function, then update output ports and connected actuators as specified in the sequence.
(4) Test for mode switches. If a mode switch is enabled, switch to the target mode.
(5) Update input ports of the tasks whose LETs start at the current time step, except for those inputs already updated at step 3.
(6) Trigger the execution of the regular tasks whose LETs start at the current time step. Also, for every task sequence that occurs at the current step, trigger the execution of the slow function.

Increasing Control of Time-Triggered Activities

In TDL, the user can specify the endpoints of a task's LET within the task's invocation period. Thus, as opposed to Giotto, a task's LET may be different (i.e. smaller) than the task's period. TDL can express time-triggered executions such as the one in Figure 5.10b, which shows two tasks with the same invocation period of 8ms and a fixed offset of 3ms. TDL employs Giotto's syntax to specify a task's invocation period, by using a mode period p and a frequency f of task invocation within p. Thus, if the LET of a task equals its period of invocation, then the task's LET is p/f. TDL uses the additional feature of *slot selection* to allow the LET of any individual task invocation to be defined more explicitly as an interval that starts and ends at integer multiples of p/f. Thus, a task's LET corresponds to a *slot group*. The slots are numbered from 1 to p/f. TDL offers a compact syntax for specifying a task's slot groups within a mode period, as follows. A repeating pattern of slot groups is specified by using the character "*" after the pattern. A slot group can be optional, which means that the corresponding task execution may be skipped at runtime, if this helps in finding a feasible schedule. Some examples are:

```
slots=1*     : all slots are mandatory and LET=p/f; this is the default.
slots=~1|2*  : LET=p/f, the first slot is optional and the remaining slots
               are mandatory.
slots=1-3*   : mandatory slot groups with LET=3*p/f each.
```

Figure 5.10a shows the specification of the execution pattern depicted in Figure 5.10b.

```
start mode main [period=8ms] {
    task [freq=4,slots=2] t1();
    task [freq=8,slots=6-8] t2();
}
```

(a) TDL code with slot selection   (b) Execution pattern with offsets

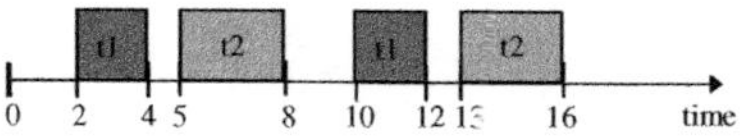

Fig. 5.10. Slot selection example

5.3 Simulation of TDL Models

Simulating TDL models means executing the operations described above on an executable model in a simulation platform rather than a physical execution platform. TDL is currently supported in two modeling and simulation frameworks: Simulink and Ptolemy II.

5.3.1 TDL Simulation in Simulink

The MATLAB extension Simulink from The MathWorks [9] is a widely used environment for modeling, simulating and analyzing dynamic and embedded systems. Simulink is based on the data flow programming paradigm and provides an interactive graphical interface. Together with automatic code generators such as the Real-Time Workshop (Embedded Coder), it has become the de-facto standard, particularly in the automotive domain.

Overview

Modeling TDL components manually with standard Simulink blocks is not feasible [10]. Typically, control systems involve multiple modes [11]. Depending on the current mode, the application executes individual tasks with different timing constraints or even changes the set of executed tasks. At the latest when mode switching logic and multiple execution rates come into play, it is all but impossible to understand or maintain the model. Instead, we use an automatic model generation approach to ensure TDL semantics in Simulink. Therefore, the TDL tool chain was extended and integrated in MATLAB/Simulink to model and simulate TDL applications and to support the code generation for particular, potentially distributed, hardware platforms.

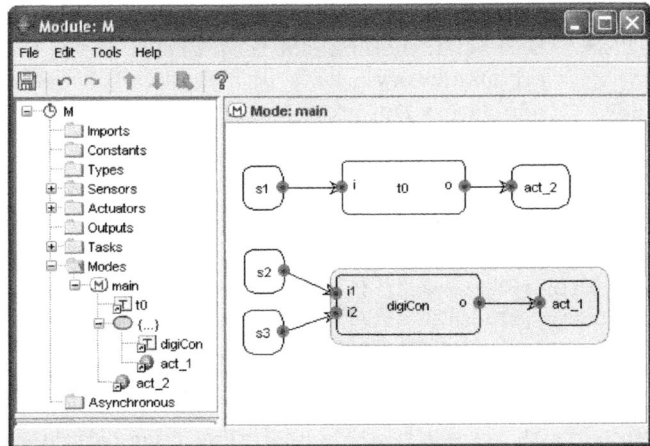

Fig. 5.11. The TDL:VisualCreator tool in Simulink

Modeling TDL in Simulink

The plant and the task respectively guard functionality is modeled with regular Simulink blocks, whereas the timing behavior, i.e. the TDL description, is specified by means of the TDL:VisualCreator tool. This graphical modeling tool is a syntax driven editor that is integrated via the *TDL Module Block* as part of the *TDL Simulink library*. Figure 5.11 shows the TDL:VisualCreator and a module M that corresponds to the mode declaration in Figure 5.7.

The activities in mode main are shown on the right, where the task sequence is indicated with the gray container that groups task digiCon and actuator act_1. Individual TDL constructs are created and managed using the tree view on the left. For each sensor (s1, s2, s3) and actuator (act_1, act_2) a corresponding Simulink Inport respectively Outport is automatically created for the module block. For each task (t0, digiCon), the tool generates a Simulink subsystem that may then be implemented by the control engineer. Again, Inport and Outport blocks are used to represent the task ports.

Simulating TDL in Simulink

For the simulation, we apply a model transformation with an E-Machine implementation for Simulink at its core. Drivers are automatically generated as *function-call subsystems* and are connected via Simulink signals. We implemented an E-Machine using the S-Function mechanism provided by Simulink to timely trigger their execution and thus to ensure TDL semantics. Therefore, the TDL:Compiler generates E-Code from the TDL description which is then interpreted by the E-Machine during the simulation.

Figure 5.12 shows the generated Simulink model for module M. The gray blocks for tasks (a) and sensors respectively actuators (b) were already generated by the TDL:VisualCreator during the modeling process. They now get linked with the rest of the newly generated model using Simulink's Goto and From blocks.

Modeling and Simulation of TDL Applications   119

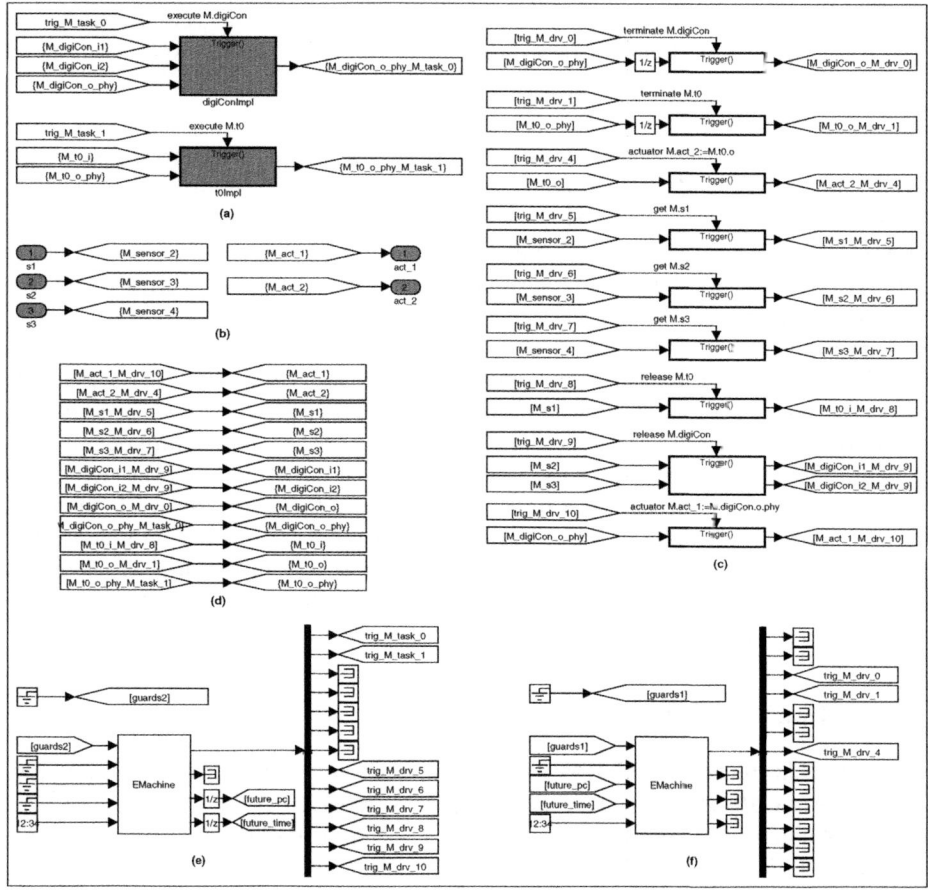

Fig. 5.12. An automatically generated TDL simulation model in Simulink

Section (c) contains the drivers (e.g. for reading sensor values or writing to actuators). The input ports of a driver are directly connected to the output ports, which corresponds with assignments in an imperative programming paradigm as soon as the system is triggered. Section (d) merges signals from drivers of different modes that write to the same port (static single assignment). As module M has only one mode, signals are simply forwarded in this example. Section (e) and (f) together implement a 2-step E-Machine architecture [12], which triggers the execution of drivers, tasks, and guards. To avoid restrictions on the set of supported blocks (e.g. for the plant) caused by Simulink's block execution strategy, we split duties of the E-Machine among two collaborating S-Functions. This allows Simulink to execute the plant or other blocks after actuators are updated and before sensors are read. Delay blocks between the release and the termination driver of a task and between the two E-Machines do not affect the timing behavior. They are, however, required to enable Simulink to resolve algebraic (feedback) loops that typically arise when simulating plants without delay or when combining LET-based with conventional controllers that are modeled as atomic (nonvirtual) subsystems [12].

Code generation
Once the simulation exhibits satisfactory behavior, one can go about generating code. Therefore, the TDL:VisualDistributor tool, which is also integrated in Simulink, may be used to define a hardware topology and to map the TDL modules to their target nodes. This also requires to specify worst-case execution times and hardware devices for sensors respectively actuators. A flexible plugin-based code generation framework generates the required C code and, in case of a distributed system, the required communication schedule. The TDL tool chain employs MathWork's Real-Time Workshop Embedded Coder [9] to generate C code for the control task functionality. For supporting the *fast step* extension, we make use of the possibility to split a Simulink task function implementation into an Output (fast step) and an Update (slow step) function. The generated code can then be compiled and linked with the platform specific E-Machine.

The main advantage of the E-Machine implementation for Simulink is that both the simulation environment and the target platform execute the same E-Code. This is a strong indicator (albeit no proof) that the simulation and the execution of TDL modules exhibit exactly the same behavior.

5.3.2 Using Ptolemy II

Ptolemy II is the software infrastructure of the Ptolemy project at the University of California at Berkeley [13]. The project studies modeling, simulation, and design of concurrent, real-time, embedded systems. Ptolemy II is an open source tool written in Java which allows modeling and simulation of systems adhering to various models of computation (MoC). Conceptually, a MoC represents a set of rules which govern the execution and interaction of model components.

Overview of Ptolemy II
The implementation of a MoC is called a *domain* in Ptolemy. Some examples of existing domains are: Discrete Event (DE), Continuous Time (CT), Finite State Machines (FSM), and Synchronous Data Flow (SDF).

Ptolemy is extensible in that it allows the implementation of new MoCs. Most MoCs in Ptolemy support actor-oriented modeling and design, where models are built from actors that can be executed and which can communicate with other actors through ports. An actor is represented by a Java class that implements the actor interface. The nature of communication between actors is defined by the enclosing domain, which is itself represented by a special actor, called the domain director. A model may define an external interface that enables it to be regarded as an actor with input and output ports. Figure 5.13 shows a sample Ptolemy model. The green block represents the local director which enforces the model of computation used in the model. The model also contains actors with input ports and output ports. Actors communicate if they are connected.

Modeling and Simulation of TDL Applications 121

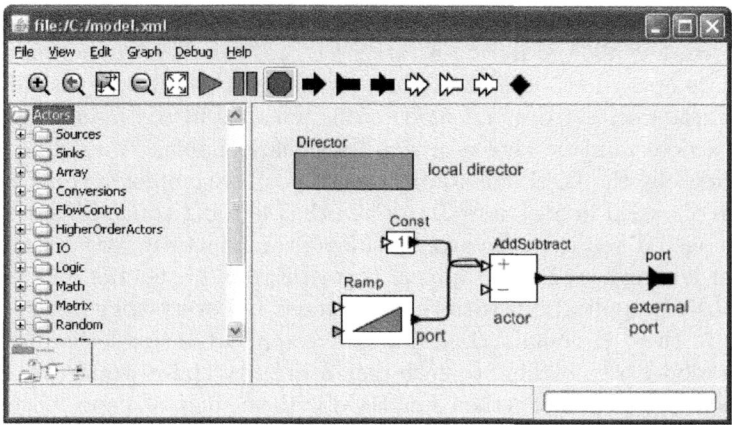

Fig. 5.13. Example of a Ptolemy model

A model can have external input and output ports and can be embedded as a composite actor in another model where it appears as an actor with local input and output ports.

Simulating a model means executing actors as defined by the top level model director. During the simulation, an actor experiences a number of iterations, where an iteration generally consists of three successive actions: *prefire*, *fire* and *postfire*. Each action is represented by a method in the actor interface. The main functionality of the actor is encoded in the *fire* method. In *prefire*, possible preconditions for execution are tested. Thus, the actor can indicate to the enclosing director that it does not wish to be fired. By convention, if the *prefire* method returns false, then the director will not call the *fire* method in the current iteration. An actor reads inputs and produces outputs in the *fire* method, which may be called multiple times in the same iteration. In *postfire*, the actor updates its persistent state and indicates to the director if the execution is complete. If *postfire* returns false, the director should perform no further iteration on the actor in the current simulation.

The TDL Domain in Ptolemy II

The implementation of TDL's modal structure is based on the modal model variant of the Finite State Machine (FSM) domain in Ptolemy, and the implementation of the LET-based semantics employs essentially a DE approach. Like modal models, TDL modules consist of modes with different behaviors, where only one mode can be active at a time. Mode switches in modal models have the same semantics as mode switches in TDL and TDL activities are conceptually regarded as discrete events that are processed in increasing time stamp order.

The TDL domain consists mainly of three specialized actors: *TDLModule*, *TDLMode*, and *TDLTask*. The *TDLModule* actor (with the associated *TDL-ModuleDirector*) restricts the basic modal model according to the TDL modal semantics. In a modal model actor, mode transitions are checked every time the actor is fired. TDL restricts the times when mode switches can be made (mode switches are not allowed during a task's LET). A similar restriction applies to port update operations. A TDL module can have guards also on task invocations and port updates, not only on mode transitions, as in the modal model. TDL requires a deterministic choice of simultaneously enabled transitions, which is not provided by the FSM domain. In this respect, we employ a convention similar to the one used in Stateflow(R), where the outgoing transitions of the active mode are tested based on the graphical layout, in clockwise order starting from the upper left corner of the graphical representation of the mode.

We consider applications with time-triggered and event-triggered components modeled in the DE domain. The functional application model is mapped to a platform model by assigning to each task a priority and a worst case execution time. The mapped model is then simulated with the help of a specialized domain controller, which is a modified DE controller. This uses an event queue and works by processing the events in the queue in increasing timestamp order. While TDL operations can be statically scheduled (they are periodic and have the highest priority), the actual moments of task executions are represented by dynamic events, as are the executions of the other event-triggered tasks.

The main difference between the implementation of the TDL-Simulink integration and the TDL domain in Ptolemy II refers to the fact that, while the former employs a Simulink implementation of the TDL:E-Machine, the latter uses no virtual machine. TDL specifications are expressed as properties of Ptolemy actors and the TDL domain uses these properties to generate an appropriate schedule of events. TDL actions are naturally represented by discrete events, and we leverage the event handling mechanism of the DE domain to achieve a correct execution of the model. In particular, this implies that any future change in the TDL semantics can be much more easily handled in the TDL Ptolemy domain, where one has to change only the event scheduling part. In contrast, in the Simulink case, changes may need to be done in the TDL compiler, in the e-code instruction set and in the TDL:E-Machine implementation. An additional advantage of the TDL-Ptolemy integration is related to the fact that mapping of a functional model to a platform model can be done much easier in Ptolemy II than in Simulink. This is due to the versatility of Ptolemy II and the availability of different models of computation. Thus, a mapped model can be obtained from the functional model by a combination of two actions: (1) Adding properties to functional actors, and (2) Choosing or defining a suitable model of computation. This enables one to simulate the (runtime) TDL operations at the platform level.

Example

In the sequel, we show how the TDL domain in Ptolemy II can be employed to demonstrate the benefits of using TDL. In the following example, a simple application with timing constraints is developed from a high-level discrete event model to an implementation on a given platform. We outline a case where, if timing constraints are expressed without TDL, the behavior of the final implementation is different than the behavior of the original model. By using TDL, the behavior of the original model remains unchanged and it is preserved in the final implementation.

Figure 5.14 shows an application modeled in Ptolemy II as a discrete event system, with one time-triggered and two event-triggered tasks. The actor TTTask is triggered by the clock signal with a period of 8 time units and it produces output with a delay of 4 units of time after being triggered. Consider a simulation of the model with two events from sensor1 at times 5 and 9, and one event from sensor2 at time 7. The execution of the task actors is shown in Figure 5.15. Notice that the time-triggered actor TTTask reads (at time 8) the value computed by the event-triggered actor ETTask1 at time 5.

This application is to be deployed on a computational platform with a fixed priority preemptive scheduling policy. Thus, code is generated from the task actors and priorities are assigned to the computational tasks. Assume that the priority of ETTask1 is higher than the priorities of both ETTask2 and TTTask, which are equal. Consider an execution of the application on the platform with the same input as in the simulation of the functional model, where the execution

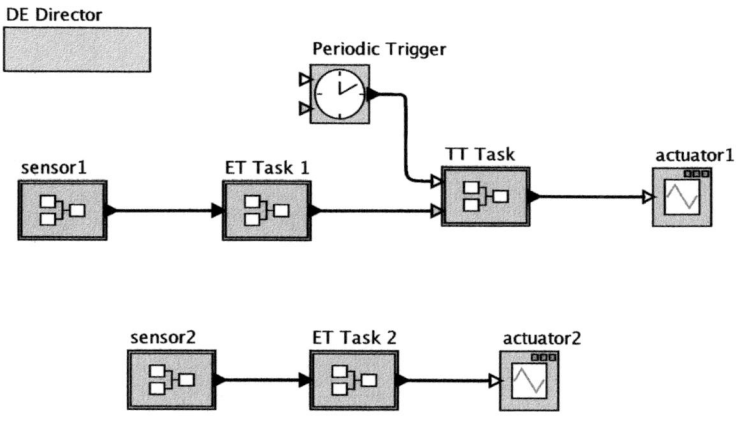

Fig. 5.14. A discrete event model

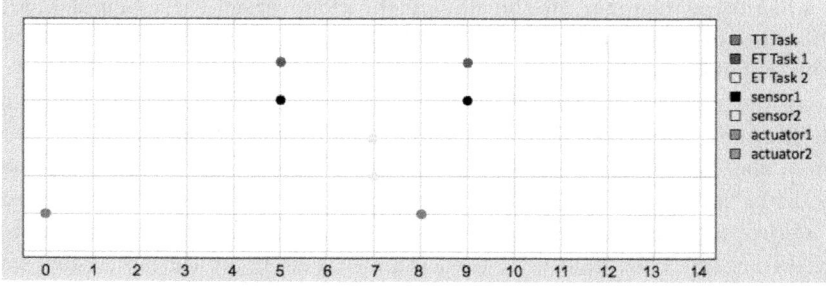

Fig. 5.15. An execution of the above model

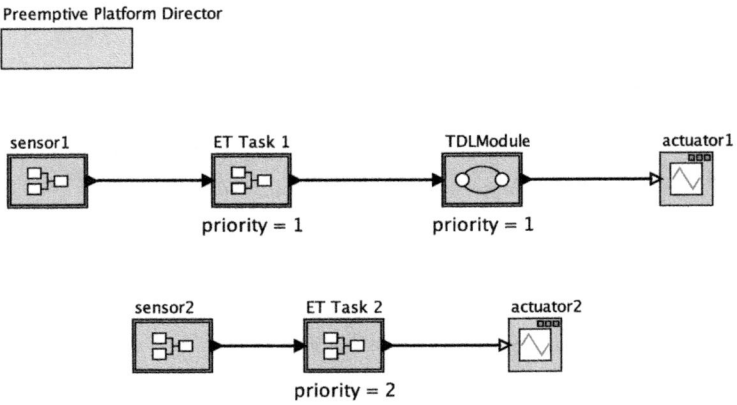

Fig. 5.16. A TDL model

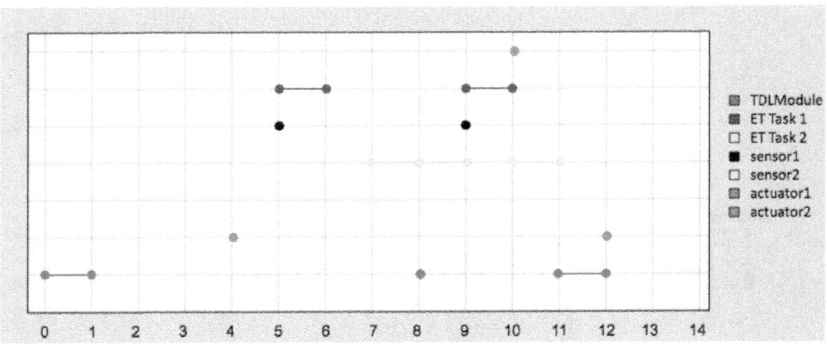

Fig. 5.17. Simulation of the TDL model

times of ETTask1, ETTask2 and TTTask are respectively 1ms, 3ms and 1ms. In this case, TTTask cannot be executed at time 8, when ETTask2 is still in execution. Also, ETTask1 preempts ETTask2 at time 9, further delaying the starting of execution of TTTask until time 11. Notice that the order of execution of TTTask and ETTask1 is changed in the implementation versus the original model. In particular, this implies that TTTask may have a different input value, hence the output behavior of the system may be changed.

Consider now a TDL model of the above application where the delay in the original time-triggered task is replaced by a logical execution time equal to 4. Let us map the TDL model into a platform model (see Figure 5.16). A specialized director (a variant of the DE director) is employed to simulate the mapped model. Figure 5.17 shows the execution of tasks under the input described above. Notice that the TDL module actor samples its input at time 8, then uses this value as input for the TDL task corresponding to the original TTTask. Thus, the mapped TDL model has the same output behavior as the TDL functional model (which has the same behavior as the functional DE model).

5.4 Related Work

TDL belongs to the family of time-triggered modeling languages and tools with roots in Giotto, such as xGiotto [14], HTL [3], and FTOS [15]. TDL stands out in this landscape due to its focus on control applications. It is, for example, the only language with a fast step feature that matches the "update outputs" part of a controller, which accommodates the need for short response times. In contrast, the Giotto software model maximizes the delay between sensor read and actuator update (placing them one LET apart), while minimizing the delay between actuator update and the next sensor read for the same task (placing them in sequence at the same time step). One important aspect in which TDL differs from Giotto is the treatment of mode switches. While Giotto allows mode switches during the LET of a task, this is not supported in TDL because it would imply a significantly more complex communication schedule generator algorithm for distributed TDL modules. Also, Giotto ensures determinism of mode switching by restricting the number of mode switch conditions that may evaluate to true to at most one. In TDL, mode switch guards are evaluated in the textual order from top to bottom and a mode switch is performed for the first condition that evaluates to true.

Among the above mentioned languages, only HTL allows flexible placement of the LET in as task's invocation period. There is also a Simulink integration of HTL [16]. In contrast to our approach, the simulation results do not match the HTL description exactly. For breaking algebraic loops, additional delay blocks are introduced which influence the observable timing. Additionally, the HTL integration in Simulink trades off accuracy for performance since it requires the sample rate of some blocks to be at least one decimal order of magnitude higher than actually required by the HTL description.

The TDL domain in Ptolemy II is related to the experimental Giotto domain in Ptolemy II[17]. The main differences between the TDL domain and the Giotto domain are as follows:

- In addition to functional models, TDL operations can be simulated also in mapped models, which contain platform specific attributes.
- The TDL domain leverages the existing DE domain while the Giotto domain is designed based on basic Ptolemy II software components.
- The implementation of the TDL domain reflects the distinction between the fundamental concepts (LET, modes) and the way these concepts are used (the operational semantics). The implementation is two-layered: the basic layer deals with scheduling LET-based tasks grouped in modes, and the operational layer corresponds to a specific time-triggered programming model. The latter extends the basic layer by specifying additional operations, as well as the order of data transfer and mode-change operations according to the programming model semantics. In principle, this enables achieving

domain controllers for other time-triggered programming models (including Giotto) by extending the basic layer.

Achieving determinism of time-triggered software is the main goal of several commercially available tools such as TTTech [18], DaVinci [19] and dSPACE [20]. A detailed comparison between TDL and each of these tools is provided in [21].

5.5 Conclusions

Timing requirements of real-time applications can be effectively achieved by using the LET approach through an established set of methodologies and tools such as the ones provided by TDL. The ability to deal with control applications was further increased by adding two extensions: (1) The *fast step*, which allows actuator update immediately upon sensor reading, and (2) The *slot selection* for flexible LET placement, which allows specification of offsets between tasks in the system.

Simulation is a powerful tool, widely used in the embedded systems industry to validate properties of complex systems. This chapter presented TDL-specific extensions of two major simulation platforms: Simulink and Ptolemy II. The TDL-Simulink integration significantly increased the accessibility of the LET-based programming model to control application developers and system integrators. The TDL tools available in Simulink make it possible to easily go through the development stages of modeling, simulation/testing, code generation and deployment to (possibly distributed) execution platforms.

The TDL domain in Ptolemy II enables, among other things, visualization of an important LET benefit: preservation of time and value determinism from high level models to lower level, platform specific, implementations. The main motivation behind its development was the observation that the influence of using LET on a system's behavior can be captured by simulation of a mapped model, even when only few platform-specific properties are considered. This could not be easily achieved by using Simulink. Ptolemy II enables experimentation and investigation of heterogeneous models of computations, where LET-based systems using Giotto and TDL can be mixed with more general, event-based systems. This can help in exploring the concept of "open" TDL models, where event-based computations can be accommodated while still guaranteeing schedulability of the system.

Acknowledgements

We thank the anonymous reviewers whose comments have been helpful in improving the presentation of this chapter.

References

[1] Stankovic, J.A.: Misconceptions about real-time computing: a serious problem for next-generation systems. Computer 21(10) (1988)
[2] Henzinger, T.A., Kirsch, C.M., Sanvido, M., Pree, W.: From control models to real-time code using giotto. IEEE Control Systems Magazine 23(1) (February 2003)
[3] Ghosal, A., Henzinger, T.A., Iercan, D., Kirsch, C.M., Sangiovanni-Vincentelli, A.: A hierarchical coordination language for interacting real-time tasks. In: Proceedings of the 6th ACM International Conference on Embedded software, Seoul, Korea. ACM, New York (October 2006)
[4] Templ, J.: TDL - Timing Definition Language 1.5 Specification. Technical report, preeTEC GmbH (2008), http://www.preetec.com
[5] Object Management Group: Model driven architecture. Technical report (2008), http://www.gigascale.org/pubs/141.html, http://www.gigascale.org/pubs/141.html
[6] Sangiovanni-Vincentelli A.: Defining platform-based design. EEDesign of EE-Times (February 2002)
[7] Henzinger, T.A., Kirsch, C.M.: The embedded machine: predictable, portable real-time code. In: PLDI 2002: Proceedings of the ACM SIGPLAN 2002 Conference on Programming language design and implementation, pp. 315–326. ACM, New York (2002)
[8] preeTEC: The TDL tool chain. Technical report, GmbH (2008), http://www.preetec.com
[9] The MathWorks (2008), http://www.mathworks.com
[10] Stieglbauer, G., Pree, W.: Visual and Interactive Development of Hard Real Time Code. In: Automotive Software Workshop San Diego, ASWSD (January 2004)
[11] Henzinger, T.A., Horowitz, B., Kirsch, C.M.: Giotto: A time-triggered language for embedded programming. Proceedings of the IEEE 91, 84–99 (2003)
[12] Naderlinger, A., Templ, J., Pree, W.: Simulating Real-Time Software Components based on Logical Execution Time. In: SCSC 2009: Proceedings of the 2009 Summer Computer Simulation Conference (2009)
[13] Brooks C., Lee E.A., Liu X., Neuendorffer S., Zhao Y., Zheng H. (eds.): Heterogeneous concurrent modeling and design in java (volume 1: Introduction to ptolemy ii). EECS Department, University of California, Berkeley UCB/EECS-2007-7 (January 2007)
[14] Ghosal, A., Henzinger, T.A., Kirsch, C.M., Sanvido, M.A.: Event-driven programming with logical execution times. In: Alur, R., Pappas, G.J. (eds.) HSCC 2004. LNCS, vol. 2993, pp. 357–371. Springer, Heidelberg (2004)
[15] Buckl, C., Regensburger, M., Knoll, A., Schrott, G.: Models for automatic generation of safety-critical real-time systems. In: Proceedings of the Second International Conference on Availability, Reliability and Security (ARES), pp. 580–587 (2007)
[16] Iercan, D., Circiu, E.: Modeling In Simulink Temporal Behavior of a Real-Time Control Application Specified in HTL. Journal of Control Engineering and Applied Informatics (CEAI) 10(4), 55–62 (2008)

[17] Brooks C., Lee E.A., Liu X., Neuendorffer S., Zhao Y., Zheng H. (eds.): Heterogeneous concurrent modeling and design in java (volume 3: Ptolemy ii domains). EECS Department, University of California, Berkeley UCB/EECS-2007-9 (January 2007)
[18] TTTech Computertechnik AG: TTP tools (2009), http://www.tttech.com/products/ttp/design-development-software
[19] Vector Informatik GmbH: DaVinci Network Designer 2.0 (2009), http://www.vector.com/vi_davinci_networkdesigner_en.html
[20] dSPACE GmbH: Real-time interface (RTI and RTI-MP) implementation guide (2009), http://www.dspace.de
[21] Farcas C., Holzmann M., Pletzer H.: The TDL advantage. Technical report, Stieglbauer G. (2004), http://cs.uni-salzburg.at/pubs/reports/T002.pdf

6 Modeling Languages for Real-Time and Embedded Systems

Requirements and Standards-Based Solutions[*]

Sébastien Gérard[1], Huascar Espinoza[2], François Terrier[1], and Bran Selic[2]

[1] CEA LIST, Laboratory of Model Driven Engineering for Embedded Systems
 (LISE), Boîte courrier 65, Gif sur Yvette Cedex, F-91191 France
 {Sebastien.Gerard,Huascar.Espinoza,Francois.Terrier}@cea.fr
[2] Malina Software Corp., Nepean, Ontario, Canada
 selic@acm.org

Abstract. Development of increasingly more sophisticated dependable real-time and embedded systems requires new paradigms since contemporary code-centric approaches are reaching their limits. Experience has shown that model-based engineering using domain-specific modeling languages is an approach that can overcome many of these limitations. This chapter first identifies the requirements for a modeling language to be used in the real-time and embedded systems domain. Second, it describes how the MARTE profile of the industry-standard UML language meets these requirements. MARTE enables precise modeling of phenomena such as time, concurrency, software and hardware platforms, as well as their quantitative characteristics.

6.1 Introduction

It is helpful to start with a clear definition of what is meant here by real-time and embedded systems (RTES). To that end we provide below a taxonomy of the different kinds of real-time and embedded systems that are of interest in this chapter. There is no generally agreed on classification of systems in the real-time and embedded domain. For our purposes, we shall use the following taxonomy (NB: this categories are not mutually exclusive) [1]:

- The *embedded* domain – This covers systems composed of both hardware and software components.
- The *reactive* domain – This sub-category covers systems that respond to discrete stimuli generated by their environment.
- The *command and control* domain – These systems are usually built to manage the running of a physical process or other systems.
- The *intensive data flow computation* domain – These systems generally deal with large amounts of physical data for applications such as signal processing, image processing, or various mobile device functions.
- The *best-effort service* domain – These are real-time systems which do not guarantee meeting all their timing and safety constraints for every individual input.

The use of abstraction as a means for coping with complexity when designing large technological systems has always been a common and effective strategy in engineering. It has proven particularly effective in the design and realization of software-intensive systems through the use of computer languages of increasing levels of abstraction, starting with assembly languages, followed by third-generation languages such as C, and on to object-oriented languages like C++ and Java. However, faced with unrelenting demands for ever more sophisticated and more dependable systems as well as for shorter time to market intervals, we seem to be approaching the limits of effectiveness achievable by using traditional code-based approaches.

Model-based design is considered by many as a suitable approach to overcoming these limits, particularly in the embedded systems domain. One of the expected advantages of this approach is the ability to exploit correct-by-construction incremental design processes, which rely on extensive use of automated transformations and synthesis, as well as formalized computer-based analyses of correctness.

Undoubtedly, much effort is required to develop the tools and methods necessary to bridge the gap between the very optimistic vision of Jacobson, who advocated that "software development is model building" [2], and the views held by more conservative software programmers, who often feel that "they don't have time to waste on modeling". In the past decade significant progress has been made in this direction, most notably the emergence of meta-modeling and practical model transformation techniques. These and related innovations are at the core of a new approach to system and software design and development often referred to as model-based engineering (MBE) or model-driven development (MDD).

The incremental nature of model-based engineering approaches is based on progressive refinement of an abstract design or system model through the gradual inclusion of more and more detail. Supported by automation-based verification and validation, this refinement is performed until the model is either (a) sufficiently detailed for relatively straightforward trouble-free implementation or (b), in case of software systems, it actually becomes the system that it was modeling. The latter in particular relies on appropriate tools that can automatically transform a model expressed using an abstract modeling language into a corresponding concrete technology-specific implementation. Thus, there are two key aspects to model-based engineering: one is the issue of selecting the right abstractions for a modeling language and the other is the matter of tool design. In this chapter, we will focus exclusively on the first aspect: the requirements for and design of modeling languages suitable for real-time and embedded systems.

For the real-time and embedded system domain, a major source of design complexity comes from the intrinsic heterogeneity of these systems. Indeed, design of modern real-time and embedded systems depends more and more on effective interplay of multiple disciplines, such as mechanical, control, electronics, and software engineering. These systems are compositions of different inter-related parts (also called components), some of which may have already been designed while others need to be designed. Given their heterogeneous nature, the parts are typically designed by different design teams, possessing different expertise, and using different tools. This is often done through vertical design chains such as, for example, in the avionics and automotive industries. A complete development chain typically involves a multitude of tools and data that are today still poorly integrated. In particular, the lack of a common modeling language to specify the overall system architecture hampers reasoning about solution trade-offs during early development phases. This results in high development costs due to long feedback cycles for issues uncovered during the integration phase.

Examples of integration needs in this area include: bridging the gap between both software and hardware models, or between software models expressed in a systems language and their implementation in terms of a target programming language. Other examples include coordinating modeling and design tools with specific engineering analysis tools (e.g., for safety or performance analysis), or connecting control engineering tools (such as Matlab/Simulink [3] or tools supporting the Modelica [4] language) with architectural design tools. Such integrations are usually complex, inefficient, and error-prone resulting in the infamous "islands of automation".

Given the importance that sharing knowledge has in embedded system development, we subscribe to the view that both system design and integration will be reduced significantly by the use of a common modeling formalism. In particular, we believe that the widespread acceptance of UML (Unified Modeling Language) [5] by both the industrial and academic communities, along with the use of UML *profiles*[1] for domain-specific purposes will considerably ease integration difficulties.

In the following section, we summarize some key requirements for modeling embedded systems. In section 6.3, we describe a standard modeling language that meets these requirements. The profile mechanism is explained first, since it is used to derive the domain-specific modeling language (DSML) out of standard UML. This is followed by an introduction to the language itself, the UML profile for modeling and analyzing real-time and embedded systems, MARTE. This profile has been adopted by the OMG as a standard technology recommendation that deals with modeling of time- and resource-constrained characteristics of systems, and includes a detailed taxonomy of relevant hardware and software patterns along with their non-functional attributes. Among other things, MARTE enables state-of-the-art quantitative analyses (e.g., performance or schedulability analysis). Section 6.4 concludes with a description of some typical scenarios that illustrate the value of MARTE in specifying real-time and embedded systems. Section 6.5 discusses contributions and shortcomings of other modeling languages for the same domain, and section 6.6 summarizes the conclusions of this chapter.

---

[1] A profile is the mechanism standardized by the OMG for creating domain-specific modeling languages by refining the concepts of an existing standard language such as UML.

6.2 Two Main Architectural Styles for Dealing with Abstraction

To cope with the complex nature of the real-time and embedded systems and their ever increasing sophistication and more stringent requirements, it is helpful to use higher levels of abstraction when specifying them. Since abstraction is one of the most powerful benefits of using models and modeling languages, we concur with the view described in [6], section 3.1.8 on page 3-13, that, when modeling a system, abstraction can be applied both vertically and horizontally.

Vertical Abstraction (Layering)
This is one of the most popular architectural patterns. It provides a graduated form of abstraction across multiple discrete levels. Two primary forms of this pattern can be identified:

Refinement layering is needed to support the iterative refinement process flows which occur during development; each layer focuses on a different level of detail. From the language point of view, what is needed is the ability to trace between corresponding model elements at two different levels of abstraction (vertical layers). From the modeling language point of view as well as from a formal perspective, it should be feasible to relate these different layer models in order to (1) enable conformance verification between them and (2) ease derivation of model elements from one level to the next. From the tooling point of view, the goal is to define and automate the process of deriving one specialized model from another (code generation is a typical example of this).

Concrete layering is used to deal with horizontal separation-of-concerns. It comes from the recurrent need in system development to model relationships between applications and their underlying software platforms (e.g., real-time operating systems or dedicated middleware) and hardware implementation platforms (e.g., SystemC and VHDL). They identify dependencies between application models and implementation choices/constraints.

Horizontal Abstraction (Slicing)
When considering abstraction from a horizontal point of view, we will use the term "slicing". Indeed, in vertical abstraction the system is divided into layers, whereas in horizontal abstraction, the system is represented as comprising different slices (i.e., partitions), with potential relationships between them. In [6], this aspect was referred to as the peer-interpretation of the client-server relationship, in contrast with its layered-interpretation that matches the layering introduced above. Again, as we did for layering, we can classify slicing into *abstract slicing* and *concrete slicing*. Slicing is intended to be used for grouping the components of a system into different sets, called slices. The rationale for grouping entities into a particular slice may vary. For example, it may be for some project-related organizational reasons or based on the need to separate distinct functional concerns. Whatever the rationale, it is important to remember that all components, regardless of which slice they belong to, share a common feature: they all coexist at the same level of abstraction!

In addition, slicing can sometimes be associated with an abstraction operation at a higher level. For example, this is typically required when it is desired to view a system from a specific perspective. Consequently, abstract slices are sometimes called *views* or *projections*. Such *abstract slicing* may contain slices that are quite different from and unrelated to the concrete slices of a system. Examples of such abstraction include task models for a schedulability analysis, or architectural models centering on system functions and scenarios models for system testing.

In contrast to a slice, a layer is a view of a complete system, but at a specific level of abstraction that is different from other layers. For example, one layer (which RM-ODP would call the "computational" layer) might show a system as a network of concurrent logical components, whereas a lower layer would represent the very same system as a set of operating system tasks (the RM-ODP "technology" layer). In that sense, layers represent different viewpoints.

6.3 Modeling Needs for Real-Time and Embedded Systems Design

In this section, we identify a set of requirements that a modeling[2] language must fulfill to support the design of real-time and embedded systems. These needs are grouped into two categories following the dichotomy introduced in the previous section.

6.3.1 Layering and Needs for RTES

First, we focus on the implications that different forms of layering have for a real-time and embedded modeling language.

Refinement
Clearly, a modeling language must support the refinement relationships between two model layers. In particular, in the real-time and embedded domain, it is necessary to be able to attach non-functional properties to such refinement relationships.

Resource
Real-time and embedded systems are computer-based systems that interact with the physical world. This means that they are not only coupled to the physical world but that they are also constrained by the physical capacities of their underlying hardware and/or software platforms. Hence, these systems are typically resource limited. It is therefore crucial that the modeling language provide facilities for a precise modeling of such resources. Specifically, this facilitates two very important capabilities.

---

[2] In this paper, the term *modeling* refers to the process of describing a system architecture and its features. *Design* refers to the activities involved in making solution-oriented decisions that satisfy given requirements and constraints for the intended product. *Analysis* is the process of verifying how well the resulting system satisfies these requirements (usually before the actual systems is fully implemented).

First, since the application software will be embedded in a specific software and/or hardware platform, the code that is generated from a model of the application must be easily interfaced with a variety of potential computing platforms, such as a real time operating systems (RTOS), middleware, micro-controllers, or specific hardware (e.g., ASIC and FPGA).

Second, achieving a balance between the need to optimize the utilization of resources for cost reasons while meeting an application's functional and non-functional (e.g., quality of service) requirements can be a very challenging design task. Consequently, real-time and embedded design generally requires facilities to perform resource optimizations.

Both of these point to a need for a modeling language that can accurately model computing platforms and other kinds of resources commonly encountered in the real-time and embedded domain.

Allocation
The design of real-time and embedded systems often follows the well-known Y-Chart scheme [7]. This approach specifies how both the application model and the resource platform model are combined to provide the full system model. This is then achieved by means of a third model, often called the mapping model, allocation model, or *deployment model*. This kind of model specifies how the elements of an application model are allocated to elements of the platform. Since an application is simply a software specification, it is the platform elements that are ultimately responsible for its physical realization. In other words, the allocation model identifies which elements of the platform are used to execute a given part of the application specification. Moreover, for our specific domain, it is very important to be able to specify the associated non-functional characteristics of the allocation. For example, when deploying the execution of a behavior on a given processor, one may need to specify its worst-case execution time.

Refinement Modeling
As explained in the section describing abstract layering, model-based development process must support abstraction refinement. This also introduces the need to trace and propagate changes up and down the layer hierarchy. Hence, a modeling language must support explicit modeling of the refinement relationships between models at different abstraction layers, and should also allow attaching non-functional properties to such relationships.

6.3.2 Slicing and Needs for RTES

As might be expected, a real-time and embedded modeling language must also support different kinds of domain-specific phenomena in a suitable manner. In particular, it needs modeling concepts dedicated to specifying quantitative characteristics such as deadlines, periods, bandwidths, processing capacities, etc. as well as qualitative features that are related to behavior, such as communication methods and concurrency. This results in a number of concrete language requirements described below.

Time

The temporal (behavioral) models of real-time and embedded systems can be grouped into three main categories as follows [1]:

(a) *Asynchronous/Causal models* are merely concerned with the proper ordering of activities (instructions, actions, so on), due to some control or data flow prescription. Some amount of scheduling may be needed if the specified flow is partial. Therefore, in such cases, time is viewed in terms of causal dependencies rather than specific quantities or durations. This model is used widely at the algorithmic software level (and in software models of hardware at the transaction level). In the presence of concurrency, the varying speeds of asynchronous components (with synchronous or asynchronous communications) generally lead to non-deterministic behavior.

(b) *Synchronous/Clocked models* add the notion of simultaneity of events and activities. Time is modeled as a discrete set of instants, and need not be connected to any physical reality, in the sense that the corresponding clock need not be regular. Henceforth we shall call this time "logical". This type of time model is used in (discrete step) simulation formalisms such as Simulink/Stateflow, in synchronous languages and Statecharts, as well as in hardware description languages at the register transaction level (e.g., VHDL, Verilog, SystemC, etc.).

(1) *Real/Continuous time models* take physical durations into account. These are important for doing various time-related analyses (e.g., deadline matches) and, in particular, for real-time scheduling as in RMA approaches [8]. They are also used for modeling the temporal characteristics of the physical environment or system with which the embedded system is interacting (usually before discretization).

A real-time and embedded modeling language needs concepts for dealing with different models of time, an, in particular, the three models of time described above, since they represent most of the common cases.

Quantitative Aspects

This concerns the use of mathematical techniques to identify or predict certain quality attributes of a system. They include, for example stress, thermal, or fluid analyses in mechanical engineering, as well as performance or reliability analyses in software engineering. One challenging problem in model-based engineering is to integrate models that are commonly used for system production or software code generation with the information that is relevant to perform these kinds of analyses. An important goal is to reduce the time required to prepare a design model for performing analysis and ensuring that the analysis model accurately represents the system. A related challenge is to hide, as much as possible, the underlying complexity of the formal mathematical model underlying the analysis methods. Both goals may be achieved by deriving the analysis model more or less directly from a suitably annotated system model using automated or semi-automated support.

To this end, it is critical to be able to capture the non-functional characteristics (e.g., performance, reliability, power consumption) in system models.

Furthermore, it should be possible to do this with precision and with maximum flexibility [9]. Thus, rather than fix in advance the set of non-functional properties that can be expressed with the language, modelers should be allowed to define the desired information in the form that is the most suitable for their specific analysis technique.. Such annotations should be interpretable by different editing or analysis tools and should not be dependent on any specific tool configuration. However, it would be impractical if modelers would have to specify all this semantic information in every design. Hence, a special requirement is a trade-off between usability and flexibility. Usability concerns favor declaring a set of fully interpretable non-functional properties for a given modeling subdomain, which are easily referred to and preserve the same meaning for every usage, whereas flexibility requires a capability for users to define their own types of non-functional properties, provided, of course, that they are semantically well-formed.

Qualitative Concerns
By qualitative concerns we refer firstly to aspects related to parallelism and related communication issues. By their very essence, real-time and embedded systems are indeed closely coupled to the real world which is inherently concurrent. Consequently, the modeling language must firstly provide the ability to specify concurrent entities capable of interacting and communicating with each other.

Underlying this preliminary concern is the more complex need to support various specific models of communication and computation. Behind this basic need is a more complex need to deal with heterogeneity of different models of communication and computation. Indeed, because of the growing complexity of systems, their development is more and more based on the possibility to consider a system as being made of a set of smaller parts. These latter can be developed using different approaches, and then different technologies, involving different models of computation and communication.. A useful modeling language must, therefore, provide a means for composing sub-systems relying on various heterogeneous models of computation and communication. This requirement may be met, for example, either by providing a means of composing different models of communication and computation, or by providing a generic model of computation and communication that can be specialized for different categories of real-time and embedded systems. In the latter case, since the model of computation and communication of the different sub-components of an application are based on the same generic model, it may be easier to compose them.

6.4 MARTE, a Standard Real-Time and Embedded Modeling Language

The Object Management Group (OMG, www.omg.org) is one of the principal international organizations promoting standards supporting the usage of model-based software and systems development. The Unified Modeling Language standard (UML, [5]) is probably the most representative of these and has had significant success in the software industry as well as in other domains such as IT and financial systems. UML is now the most widespread language used for modeling in both industry and academia. It was designed as a general-purpose modeling language as well as a foundation for generating different domain-specific languages, mainly through its profile mechanism. The latter capability allows the general concepts of UML to be specialized for a specific domain or application.

In this section, we first introduce the UML profile concept, which is a very powerful means for defining domain-specific modeling languages (DSMLs). Next, we present MARTE, which is a UML profile for modeling real-time and embedded systems and is, in effect, a domain-specific modeling language for the real-time and embedded domain.

6.4.1 UML Profiling Capabilities

Because of the diverse nature of the disciplines needed for designing real-time and embedded system, it is clear that a single modeling language will not be enough to cover all the various concerns involved in this specific area. Consequently, there has been much discussion about the suitability of UML for such domains relative to custom domain-specific modeling language designed from scratch [10]. A custom language has the obvious advantage that it can be defined in a way that is optimally suited to a specific problem. At first glance, this may seem the ideal approach, but closer examination reveals that there it can have serious drawbacks. If each individual sub-domain of a complex system uses a different modeling language, the problem will be how to interface the various sub-models into a consistent integrated whole that can be verified, tested, or simply unambiguously understood. Furthermore, there is also the problem of designing, implementing, and maintaining suitable industrial-strength tools and training for a each custom language, which can result in significant and recurring expenses.

Conversely, although UML was designed to eliminate the accidental complexity stemming from gratuitous diversity [11], it still provides a built-in mechanism, the *profile* concept, for creating domain-specific modeling languages that can take advantage of existing UML tools and widely available UML expertise. Note that we are not saying that UML profiles completely avoid DSML integration problems. However, many of the fragmentation issues[3] [12] stemming from diversity can be mitigated because all domain-specific modeling languages derived from UML share a common semantic and syntactic foundation. There is typically a lot of commonality between the various disciplines in real-time

---

[3] This is used to refer to the situation that occurs when different domain-specific languages are used to describe different aspects of a complex system. For example, one language might be used to describe the user interface function while a different one for the database management and access functions. The individual languages involved could have very different models of computation, which raises the question of how to meld the different specifications into a coherent and consistent whole.

and embedded system design. For instance, the concepts of package, composition, property and connector, which are provided by UML, are common to many disciplines, as are the basic notions of object, class, and interface.

The basic premise of profiles is that all domain-specific concepts are derived as extensions or refinements of existing UML concepts. These extensions are called *stereotypes*. A stereotype definition must be consistent with the abstract syntax and semantics of standard UML, which means that a profile-based model can be created and manipulated by any tool that supports standard UML. Moreover, because of the underlying UML foundations of a profile, it is more easily learned by anyone with a knowledge of UML.

A stereotype may have attributes and be associated with other stereotypes or existing UML concepts. From a notational viewpoint, stereotypes can also be used to adapt the concrete syntax of UML in order to provide a more domain oriented concrete syntax if needed. For instance, a class model element stereotyped as «clock» might use a picture of a clock symbol instead of the generic UML class symbol.

We can distinguish two main categories of UML profiles [13]: specification and annotation profile. *Specification profiles* are fully-fledged domain-specific modeling languages and are used to model systems from the viewpoint of a particular domain. SysML [14] is an example of this kind of profile. *Annotation profiles* are used to add supplementary information to various kinds of UML elements that can then be interpreted by specialized tools or domain experts for different purposes, such as model analyzers or code generators. Note that annotation profiles are particularly useful for defining domain-specific modeling languages that support abstract layering and slicing. As we shall describe later, some parts of the MARTE profile, namely the sub-profiles that support various analyses, are examples of this latter category.

While specification profiles is generally well understood, some explanation may be necessary to understand the second category, annotation profiles. Specifically, in case of MARTE's analysis sub-profiles, a given analysis concept may be manifested in a number of different ways in a particular model. For example, a real-time clock may be manifested as a lifeline in a UML sequence diagram or as a role in a UML collaboration diagram. From the analysis viewpoint, all of these different manifestations represent the same thing. This means that it is necessary to be able to apply the same analysis stereotype to different kinds of UML concepts, and conversely, different stereotypes (possibly from different analysis viewpoints) may be applied in the same model element.

Concepts defined in the MARTE annotation profiles that support quantitative analysis can be applied to make a standard UML model look like an analysis model (e.g., a performance model). This is achieved by tagging appropriate elements of the original UML model to represent concepts from the analysis viewpoint. These can then be used by an automated performance analysis tool to determine the fundamental performance properties of a software design. At the same time (and independently of the performance modeler) a reliability engineer might overlay a reliability-specific view on the same UML model to determine its overall reliability characteristics, and so on. Annotation profiles allow the same model to be viewed from different viewpoints (e.g., schedulability, performance, security, availability or timing). Finally, it should be noted that UML profiles have the very useful feature to be dynamically applied to a user model (e.g., to produce a domain-specific viewpoint) without changing the underlying model in any way. Subsequently, the profile can be removed to reveal the original model unchanged. As described in section 6.4.3, this feature is crucial to this type of profile usage.

6.4.2 MARTE Basics

As noted previously, UML is a general-purpose modeling language that can be specialized or extended for dealing with specific domains or concerns. The field of real-time and embedded software systems is one such domain for which extensions to UML are required to provide more precise expression of domain-specific phenomena (e.g., mutual exclusion mechanisms, concurrency, deadline specifications, and the like). The OMG had already adopted a UML profile for this purpose, called the "UML Profile for Schedulability, Performance and Time" (SPT, [6]). It provided concepts for dealing with model-based schedulability analysis, focused primarily on rate monotonic analysis, and also concepts for model-based performance analysis based on queuing theory. In addition, SPT also provided a framework for representing time and time-related mechanisms. However, practical experience with SPT revealed shortcomings within the profile in terms of its expressive power and flexibility. For example, it was necessary to support the design of both hardware and software aspects of embedded systems and more extensive support for schedulability and performance analysis, encompassing additional techniques such as hierarchical scheduling. Furthermore, when the new significantly revised version of UML, UML2, was adopted by the OMG, it became necessary to upgrade the SPT profile. Consequently, a new Request For Proposals (RFP) was issued by the OMG seeking a new UML profile for real-time and embedded systems. This profile was named MARTE (an abbreviated form of "Modeling and Analysis of Real-Time and Embedded systems," [1]). The intent was to address the above issues as well as to provide alignment with another standard OMG profile, the UML profile for Quality of Service and Fault Tolerance (QoS & FT, [15]). The latter enables specification of not only real-time constraints but also other embedded systems characteristics, such as memory capacity and power consumption. MARTE was also required to support modeling and analysis of component-based architectures, as well as a variety of different computational paradigms (asynchronous, synchronous, and timed).

In response to this request for proposal, a number of OMG member organizations collaborated on a single joint submission. This group, called the ProMARTE consortium, consisted of the following enterprises: Alcatel, ARTiSAN Software Tools, Carleton University, CEA LIST, ESEO, ENSIETA, France Telecom, International Business Machines, INRIA, INSA from Lyon, Lockheed Martin, MathWorks, Mentor Graphics Corporation, NoMagic, the Software Engineering Institute (Carnegie-Mellon University), Softeam, Telelogic AB, Thales, Tri-Pacific

Software, and Universidad de Cantabria. The resulting submission was voted on and accepted by the OMG in June 2007 [16] as a "Beta Specification".

As prescribed by the OMG's Policies and Procedures manual (P&P, [17]), following adoption, a Finalization Task Force (FTF) was instituted to prepare the new specification for its formal adoption as an official OMG technology recommendation. The working period of a finalization task force is about 18 months and its first phase (around 6 months) comprises a comments-gathering phase during which feedback from the broader user and vendor communities is collected. Of particular significance is input from commercial and other tool vendors intending to support the new specifications in their products. The second phase is then dedicated to solving the key issues identified in the initial phase resulting in a Beta 2 version of the specification. This version is first screened by the OMG's Architecture Board and, if deemed acceptable, is submitted to the OMG's Board of Directors for final approval. At that point, the resulting specification becomes formally available as version 1.0. In the case of MARTE, it is expected that this will be available by the first quarter of 2009.

6.4.3 Architecture and Some Details of MARTE

As noted, MARTE is a UML profile intended for model-based engineering of real-time and embedded systems. It consists of a set of extensions (i.e., specializations) of appropriate general UML concepts providing real-time and embedded designers and developers with first-class language constructs from their domain. Many of these extensions pertain to so-called non-functional aspects of real-time and embedded applications. Non-functional concerns of an application can be classified into two categories, quantitative and qualitative aspects. Furthermore, these extensions may be available at different levels of abstraction and, finally, they have been defined to support modeling, analysis, or both. In order to satisfy all these requirements, MARTE is structured as a hierarchy of (sub-)profiles, as depicted in the UML package diagram in figure 6.1. It has four main parts.

The topmost package, which is the foundation on which the rest of MARTE is based, consists of four basics sets of UML extensions, also called MARTE sub-profiles:

- *Non-Functional Properties Modeling (NFP)* This sub-profile provides modeling constructs for declaring, qualifying, and applying semantically well-formed non-functional aspects of UML models. The non-functional properties sub-profile supports the declaration of non-functional properties as UML data types, whereas the value specification language is used to specify the values of those types and any potential functional relationships between them. It is complemented by the Value Specification Language (VSL), which is a textual language for specifying algebraic expressions. The Value-Specification Language sub-profile is separated out in the annexes package because it was intended to be reused in other OMG profiles.
- *Time Modeling* This consists of concepts for defining time in applications, and also for manipulating the underlying time representation. The Time

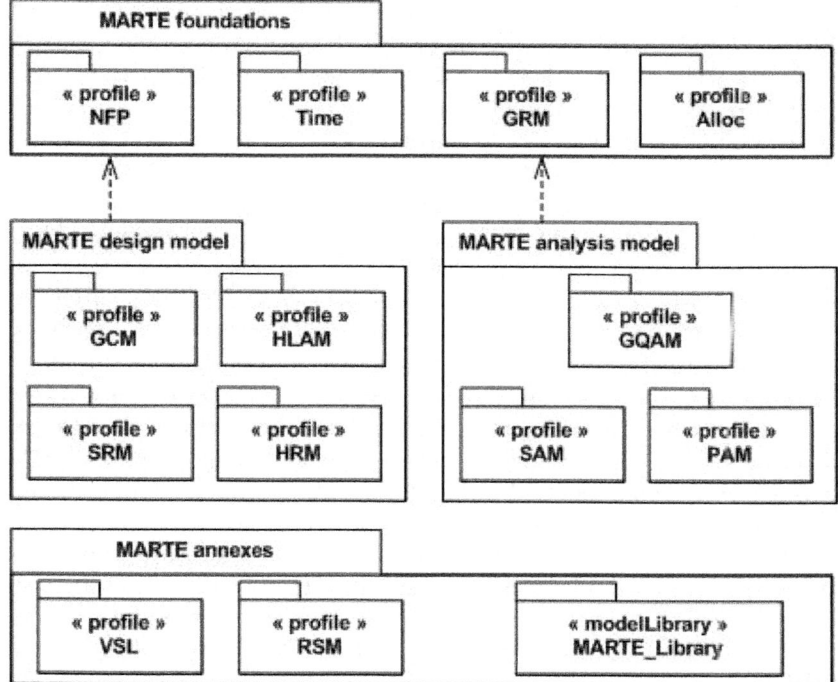

Fig. 6.1. MARTE's Architecture View

extension defined in MARTE provides support for three qualitatively different models of time: chronometric, logical, and synchronous.
- *Resource Modeling (GRM)* One important requirement with regards to real-time and embedded system modeling is to represent the set of resources underlying an application and also how the system uses them. The Generic Resource Modeling (GRM) sub-profile consists of an ontology of resources enabling modeling of common computing platforms (i.e., a set of resources on top of which an application may be allocated to be computed), and high-level concepts for specifying resource usage. The level of abstraction used here is at a general system level.
- *Allocation Modeling* This sub-profile of the foundational layer provides a set of general concepts pertaining to allocation of functionality to entities responsible for its realization. It may be either time-related allocation (i.e., scheduling) or space allocation. It also tackles the more abstract issue of refinement between models at different levels of abstraction. Note that non-functional characteristics may be attached to an allocation description (e.g., when specifying the allocation of a function to a given execution engine, it is possible to specify its worst case execution time).

Starting from these foundational concepts, MARTE is then split into two different categories of extensions: One (denoted in figure 6.1 as the "MARTE design model") is dedicated to supporting model-based design, that is to say modeling activities related to the left branch of the classical "V" development cycle[4], whereas the other, denoted by the "MARTE analysis model" package, provides support for model-based analyses of real-time and embedded applications (i.e., more devoted to validation and optimization).

Model-based design of real-time and embedded systems with MARTE proceeds mostly in a declarative way. Hence, MARTE users may annotate their models with real-time or embedded concerns using the extensions defined within the High-Level Application Modeling sub-profile (refer to the following section that illustrates this using extracts of the MARTE specification and an example). For instance, concurrent computing units with real-time features may be denoted using an extension called «rtUnit» and, by giving specific values to its properties, they can also indicate what is the model of computation for the concurrent unit. Note also that MARTE enables component-based system engineering (either software or hardware) through its specific sub-profile called the Generic Component Model (GCM). This component model supports both message- and data-based communication schemes between components. In addition, MARTE also defines very refined concepts that enable users to describe its computing platforms, either software or hardware, in a very detailed and precise manner [18, 19, 20]. These features are supported by two sub-profiles, Software Resource Modeling (SRM) and Hardware Resource Modeling (HRM). In addition, based on its Software Resource Modeling sub-profile, MARTE includes in its annexes facilities for modeling OSEK-, ARINC-, or POSIX-compliant software computing platforms. Finally, to deal effectively with the increasing degrees of parallelism available on chips, one of the MARTE annexes includes the Repetitive Structure Modeling sub-profile, which enables compact representations of multidimensional regular structures encountered in chip design.

Model-based analysis using MARTE is enabled primarily through the extensions defined either in the Generic Quantitative Analysis Modeling profile (GQAM), or using one of its two refinements, which are dedicated to schedulability analysis [21] and performance analysis [22] respectively. The annotation mechanism used in MARTE to support model-based analyses uses UML stereotypes. These typically map the UML elements of the application or platform into corresponding analysis domain concepts, including specifications of values for properties which are needed to carry out the analyses. One of the typical use cases of MARTE described in the following section provides more detail.

In summary, MARTE was designed to cover all five categories of real-time and embedded systems listed earlier. The table below summarizes how MARTE covers the requirements identified in section 6.2 of this chapter. The left-most column denotes the different parts of MARTE: a part being either a sub-profile or a specific model library.

---

[4] Note that we are not advocating the "V" development cycle as the reference process model for MARTE. We are simply using it to help orient the reader.

6.4.4 An Extract of the MARTE Specification

In this section we illustrate in practical terms some of the ideas described in preceding sections. However, due to space limitations it is not possible to provide examples covering the full specification. Therefore, we will focus on the MARTE part dedicated to high-level modeling of real-time and embedded systems design. In particular, the fragment of the MARTE profile focusing on the definition of a real-time unit and a protected passive unit.

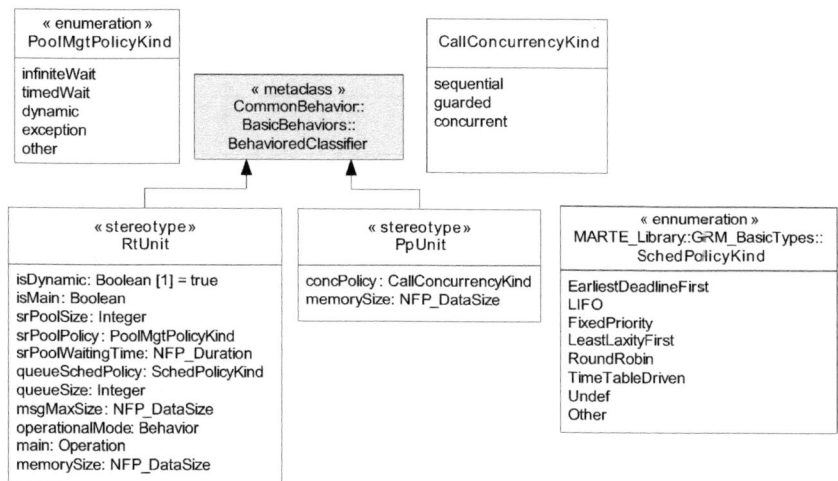

Fig. 6.2. Extract from the MARTE specification: the real-time unit and the passive protected unitmetamodels

Figure 6.2 illustrates the graphical definition of two main concepts of the MARTE specification: RtUnit and PpUnit. An RtUnit (Real-time Unit) is a unit of concurrency that encapsulates in a single concept both the object and the process paradigms. This allows concurrency control to be encapsulated within a single unit. Any real-time unit can invoke services of other real-time units, or send and receive data flows to and from those units. It owns one or more schedulable resources (i.e., threads or tasks in operating system terminology). A PpUnit (Protected passive Unit), on the other hand, is used to represent data containers that can be shared between real-time units but with some form of concurrent access protection. Therefore, a PpUnit specifies its concurrency policy, via its concPolicy attribute. It does not own any schedulable resource.

The next figure describes a UML class diagram of a very simple automotive cruise control system annotated with the two of the aforementioned stereotypes. Both classes, CruiseController and ObstacleDetector, are stereotyped as real-time units. The first of these creates dynamically schedulable resources (e.g., threads) to handle the execution of its services, while the second has a pool of ten (10) schedulable resources. Both real-time units are sharing data handled by

Table 6.1. MARTE's coverages summary in terms of RTE-specific modeling language (a box with the symbol ⊂ [23] means the MARTE part provides some support for the requirement, else it is marked as ∅)

	Slicing			Layering		
	Quantitative Concerns	Qualitative Concerns	Time	Allocation	Resource	Refinement
NFP	⊂	⊂	∅	⊂	∅	⊂
Time	⊂	⊂	⊂	∅	∅	∅
GRM	∅	∅	∅	∅	⊂	∅
Alloc	∅	∅	∅	⊂	∅	⊂
GCM	∅	⊂	∅	∅	∅	∅
HLAM	⊂	⊂	⊂	∅	∅	∅
SRM	∅	∅	∅	∅	⊂	∅
HRM	∅	∅	∅	∅	⊂	∅
GQAM	⊂	∅	∅	∅	⊂	∅
SAM	⊂	∅	∅	∅	⊂	∅
PAM	⊂	∅	∅	∅	⊂	∅
VSL	⊂	⊂	∅	∅	∅	∅
CCSL	∅	∅	⊂	∅	∅	∅
RSM	∅	∅	∅	∅	⊂	∅
MARTE Library	⊂	⊂	⊂	∅	⊂	∅

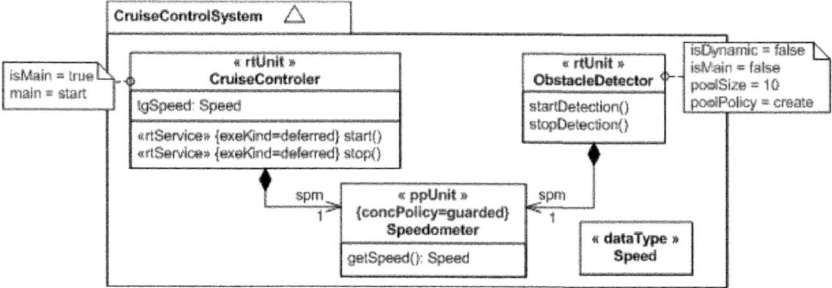

Fig. 6.3. An example of MARTE model using both stereotypes, «rtUnit» and «ppUnit»

the class Speedometer. Because real-time units execute concurrently, the access to the data encapsulated within the class Speedometer needs to be protected. To do that, the class Speedometer is tagged with the «ppUnit» stereotype. Its concPolicy property is set to guarded, meaning that only one real-time unit at a time can access a feature of Speedometer, while subsequent ones arriving later are suspended until the first user releases it.

6.4.5 Typical MARTE Usage Scenarios

The modeling capabilities of MARTE are rich enough for a wide range of design approaches. This yields the flexibility to support and integrate multiple design perspectives, but also to deal with the problem of understanding and choosing among a variety of language alternatives. In both cases, there is no standard prescriptive way of using the language constructs across the development cycle. This means that individual projects or enterprises can define their own specific modeling framework and methodology that suits their needs best. We identify below a

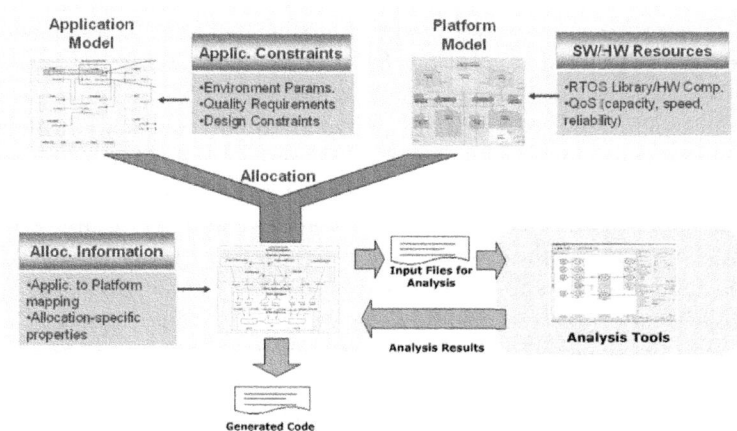

Fig. 6.4. Some typical scenarios of MARTE usage

set of representative scenarios in which using MARTE provides significant benefits. Although these scenarios certainly do not cover all possibilities, they allow us to illustrate the application of MARTE in a more focused and concrete manner.

Figure 6.4 illustrates some of example scenarios following the Y-Chart scheme [7]. For the sake of simplicity, we limit the discussion of MARTE usage here to just three simple use cases: (1) an application-oriented use case that illustrates the modeling of non-functional features of a system, (2) a platform-oriented use case that corresponds to the definition of the hardware and software resources of the system, and (3) an allocation-oriented use case that denotes how to model the deployment of an application to a platform.

Application-Oriented Use Case

The first MARTE usage scenario, depicted in the upper left-hand corner of Figure 6.4, deals with the application side and describing non-functional requirements. UML has traditionally been used to document user requirements by means of use case diagrams. Use cases follow a graphical, scenario-based approach. Although use cases may be formalized to certain degree, for example by using sequence diagrams in order to detail such usage histories, they are often criticized for a number of limitations. For instance, they are applied mainly to model functional requirements, but are not very helpful for modeling non-functional ones. One possible way of using MARTE is to add annotations that characterize non-functional constraints in use case diagrams and their corresponding sequence diagrams. This provides two important capabilities leading toward more formal requirements specification.

First, note that these non-functional requirements are specified jointly with their corresponding functional requirements. While specifying non-functional aspects is possible with UML comments, this would make their semantic relationship to the concrete functional elements hard to capture. In particular, the verification of requirements satisfaction in real-time systems is strongly dependent on the coupling between system function and timing. In MARTE, timing annotations provide semantic definitions closely related to the system behavior. For instance, one may define a jitter constraint in the arrival of an event and specify whether such an event relates either to a send, receive, or consume occurrence within a sequence diagram. Second, non-functional annotations follow a well-defined textual syntax, which is supported by the Value Specification Language of MARTE. The main advantages of this level of formalization are the ability to support automated validation, verification, and traceability, while being easily understood by stakeholders.

To model the application structure and behavior, MARTE adds key semantics to UML model elements. In particular, a common model of computation provides semantic support for the real-time object paradigm (the MARTE's High-Level Application Modeling sub-profile). This paradigm allows specifying applications at a high level of abstraction, by delegating concurrency, communication, and time-constraint aspects to a modular unit called RtUnit (see Section 6.4.4). Such units can be encapsulated in structural units (*structured components*) specifying interfaces and interactions with other structural elements. MARTE adopted the notions of port and flow from SysML [14] and extended them with the notion of message-based communications.

Platform-Oriented Use Case

MARTE can also be used to explicitly model resource patterns such as processing resources, communication buses, or power supply devices along with a set of predefined quality attributes (illustrated in the upper right-hand corner of Figure 6.4). Furthermore, the operating system (e.g., RTOS) and other software library layers can be modeled and reused for multiple application models. In this chapter we want to particularly focus on the usage of the software resource modeling capabilities of MARTE (the MARTE's Software Resource Modeling sub-profile), which deal with one of the more important open issues in Model-based Engineering: making platform models explicit.

Historically, model-driven approaches for real-time and embedded systems have focused on improving dedicated real-time and embedded platform models [6, 7] (i.e., platform models used as meta-models). At the same time, the Model-based Engineering community has proposed generic transformation languages (e.g., ATL [4] or SmartQVT [24]) which facilitate the description of dedicated bridges or transformations between meta-models. Current model-driven approaches for real-time and embedded therefore entail specific model transformations from a set of source platforms to a set of target platforms. In many case, however, the platforms are not described as explicit input models to be used in the transformation. This is a serious limitation as software platforms (e.g., RTOS, programming languages) are continuously evolving and the resulting dearth of customizable transformations hampers description of reusable generative processes.

In [25], the authors propose a model-based framework enabling explicit platform description considering the latter as an input parameter to the model transformation. The proposed approach uses the MARTE's Software Resource Modeling sub-profile for describing the specific platforms. Its principal innovation is that it avoids hard-coding the platform model in the model transformation. The main benefit is a cleaner separation of concerns within the design flow enabling easy porting to different platforms without requiring a new transformation for each case. Real-time and embedded decisions are made explicitly in the input models and not implicitly in the transformation description itself. This also improves the maintainability of the transformations. In this way, the generation process becomes more flexible, more adaptable, and more reusable for a variety of different real-time and embedded platforms. In [25], the authors also focused on the transformations dedicated to porting multitasking applications to heterogeneous computing platforms (e.g., multitasking operating systems).

Allocation-Oriented Use Case

An allocation view represents the system as a hierarchy of an application (at the top) and the software/hardware platform layers (at the bottom), as shown in the bottom of Figure 6.4. A set of MARTE stereotypes allow representing such hierarchies either using allocation/deployment or using composition relationships.

In order to generate code for a given software platform, the system model built on the real-time concepts (e.g., «rtUnit») must be allocated to a specific software platform model, as described in the preceding use case. In [26], the authors implement a real-time framework that gives meaning to the features defined in the MARTE's High-Level Application Modeling sub-profile. This is achieved by providing execution support for realizing the behavior, communications, and message management associated with this kind of objects. The code generation facility provided by this execution framework, called ACCORD, consists of a set of transformations used to apply implementation patterns on real-time concepts, and a standard C++ code generator. A methodology (ACCORD|UML) constrains the usage of MARTE concepts and parameters to a subset that are semantically meaningful in the ACCORD execution framework.

An allocation view should incorporate the non-functional annotations that results from the running of the application on a particular hardware/software platform. Some of these annotations are directly mapped to platform properties, while others require special techniques to determine their values through either computation or simulation. For instance, fine-grained timing analyzers can help to determine the worst case execution times of relevant pieces of code, which are then used in scheduling analysis to predict end-to-end response times.

In [27], MARTE is used in practical system integration scenarios (modeled as "analysis contexts"), where multiple candidate configurations are analyzed from a timing perspective, potentially by multiple techniques. In this paper, the authors use the Value-Specification Language to specify non-functional variables that are further evaluated to make efficient design decisions. One of the benefits is that multi-objective analyses can be performed and trade-off decisions can be taken on the basis of a smart binding of exposed parameters, which are used in different analysis contexts. Each analysis technique may involve specific parameters to be evaluated. Furthermore, *sensitivity analysis* can be used at the system design level to understand the degree to which the overall results are sensitive to a given parameter.

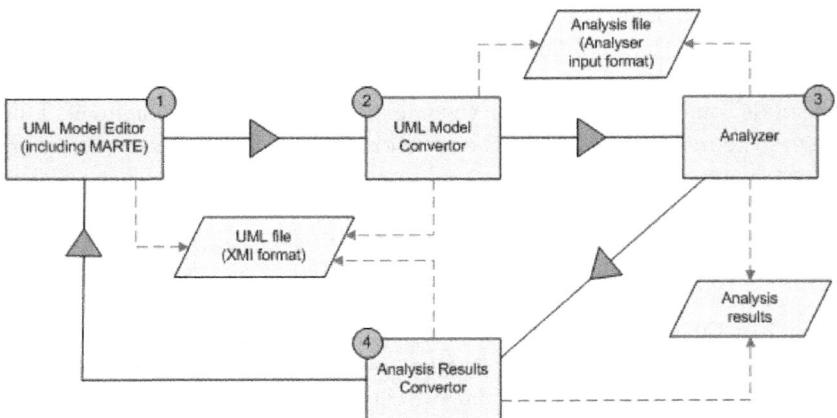

Fig. 6.5. Schema of the MARTE model-processing framework

Finally, MARTE fosters model processing in the way that it enables adding semantics to a given UML model (e.g., for code production or quantitative analysis purposes). In this context, one generic usage of MARTE enables the model-processing schema described in Figure 6.5. Note that this process can be highly automated, which, in some cases, can even eliminate the need for analysis domain experts which are often difficult to find.

6.5 Related Work

Both academia and industry have already proposed languages to support model-based development of embedded systems.

SysML [14] is an OMG standard language "for specifying, analyzing, designing, and verifying complex systems that may include hardware, software, information, personnel, procedures, and facilities". The so-called Block concept of SysML is the common conceptual entity that can represent many different kinds of system elements such as electronic or software components, mechanical parts, information units, and almost any other type of structural entity in the system under interest. Blocks articulate a set of modeling perspectives enabling separation of concerns during systems design. Among these perspectives, requirement diagrams provide constructs for specifying text-based requirements and their relationships, including requirements hierarchies, as well as derivation, satisfaction, verification, and refinement relationships between requirements. Block description diagrams and internal block diagrams enable the specification of more generic interactions and phenomena than those existing just in software systems. This includes physical flows such as liquids, energy, or electrical flows. The dimension and measurement units of the flowing physical quantities can be explicitly defined. Although most behavior constructs in SysML are similar to those of UML (interactions, state machines, activities, and use cases), SysML refines some of them for modeling continuous systems and probabilities in activity diagrams. A perspective called parametric diagram allows SysML users to describe, in a graphical manner, analytical relationships and constraints, such as those described by mathematical equations. Formally, SysML is defined as a UML profile.

MARTE and SysML are complementary in many ways [28]. The MARTE component model shares the same notions of ports and flows, and additionally extends them with the concept of client/server port. This is intended to support the request-reply and publish-consume communication paradigms. In addition, there are some actions under way at the OMG to align the semantics of data and event flows, to define a common framework to specify quantities, units, dimensions and values, and to improve some aspects such as allocation and timing modelling. This will be reflected in future versions of MARTE and SysML.

AADL (Architecture Analysis and Design Language) [29] is an architecture description language standardized by SAE (Society of Automotive Engineers). AADL has been designed for the specification, analysis, and automated integration of real-time performance-critical (timing, safety, schedulability, fault tolerant, security) distributed systems. A system modeled in AADL consists of application software components bound to execution platform components. AADL application software components are made of data, threads, and process components. AADL thread components model units of concurrent execution. A scheduler manages the execution of a thread. AADL execution platform components include processors, memory, buses and devices. Although AADL was defined as a domain-specific language from scratch, there is a MARTE rendering of AADL, as stated in Annex F of the AADL specification. This has been formalized as a subset of MARTE with some specific guidelines defined in Annex A of the MARTE specification.

In the automotive domain, AUTOSAR [30] is unquestionably the standard to specify component-based software infrastructure and it includes a standardized API. AUTOSAR's goal is to support the exchange of parts of embedded system implementation artifacts that have already been pre-designed or designed independently by different teams (e.g., by OEM's, software suppliers), without the time-consuming and costly need to re-configure, port, and re-build their code. In AUTOSAR, application models are organized in units called software components. Such components hide the implementation of the functionality and behavior they provide and simply expose well-defined connection points called ports. In particular, atomic software components are entities that support an implementation and hold behavioral entities called runnables. A runnable is an entity that can be executed and scheduled independently from other runnable entities. There is an action project funded by the European Union's Seventh Framework Program called ADAMS [31] especially dedicated to promote and show the complementarity of MARTE with other automotive and avionics standards, among which AUTOSAR is of main interest.

In addition to AUTOSAR, some of the European automotive actors have defined an architecture description language, EAST-ADL [32]. This complements AUTOSAR to cover the system level that lies outside the scope of AUTOSAR. This includes requirements modeling, feature content at the level of a vehicle description, architecture variability, functional structure of applications, middleware, plant (environment), abstract hardware architecture, and preliminary functional allocation. The ADAMS project provided some important results on the alignment of MARTE and EAST-ADL [33]. This was reflected as a set of guidelines to describe EAST-ADL-like models with a subset of MARTE concepts oriented to the design of automotive applications. Finally, note that these guidelines are part of the MARTE standard specification.

In addition to the aforementioned standards, two other non-standard approaches provide similar facilities as MARTE: MIC and Ptolemy.

Vanderbilt University's Model Integrated Computing (MIC, [34]) tool suite consists of meta-programmable model-builder (GME), model-transformation engine (UDM/GReAT), tool-integration framework (OTIF), and design space exploration tool (DESERT). This tool suite is based on specific languages for metamodeling and provides the ability to build domain-specific modeling languages. This framework is described in depth in a separate chapter of this book.

Ptolemy is a model-based tool dedicated to real-time and embedded systems [35]. This project provides support for heterogeneous modeling, simulation, and design of concurrent system. The modeling principle fostered in Ptolemy is called actor-oriented design. This relies on the concepts of models, actors, ports, parameters and channels. Actors (the core Ptolemy concept for supporting component based development) communicate with other concurrent computing actors only via their ports. Ports of two communicating actors needs to be linked via a channel. A set of communicating actors belong to a given model which may have parameters. Each model specifies a director that define its model of computation and each of the actors owned by the model will conform to the model of communication defined by the director. The key concept of actor as defined in Ptolemy was inspired by the work introduced first in 1970's by Carl Hewitt of MIT, and later formalized by Agha in [36]. The MARTE, concept of a real-time unit used for modeling real-time and embedded systems shares the same origins. More specifically, the MARTE concept was inspired by the active object concept of UML in one hand, as well as ACCORD concept of real-time object [37].

6.6 Conclusions and Perspectives

The complexity of modern real-time and embedded systems is starting to exceed the capabilities of traditional code-centered technologies. Fortunately, new model-based engineering methods have proven themselves capable of overcoming many of these limitations. These modern methods rely on intensive use of computer-based automation and take advantage of computer languages with higher-level constructs that abstract away much of the underlying implementation technology as compared to programming languages. The benefits gained from using this approach increase the closer the language is to the problem domain, which is why there is much interest in defining so-called domain-specific modeling languages (DSML). One such language is MARTE, which was specifically designed for modeling systems and phenomena in the real-time and embedded domain. It allows direct expression of domain phenomena such as time and timing mechanisms, concurrency control mechanisms, software and hardware platforms and resources, as well as precise specification of their quantitative characteristics (e.g., latency, capacity, speed and execution times).

MARTE is a profile-based language, which means that it was derived by refinement and extension of the industry-standard UML language. This allows it to reuse many existing UML tools as well as widely available UML expertise, while still retaining the expressive power and other advantages of a specialized computer language. Furthermore, MARTE itself is an industry standard, adopted and endorsed by the OMG as one of its official technology recommendations.

The domain-specific nature of MARTE enables not only more straightforward specification of real-time and embedded applications but also automated and semi-automated engineering analyses of MARTE-based models. This important capability allows candidate designs to be objectively evaluated for key performance and quality indicators early in the development cycle, before committing full development resources. Consequently, potentially expensive design flaws and shortcomings can be detected and eliminated earlier and at far less cost compared to traditional code-based methods.

At the time of this writing, MARTE is available in its version 1.0 on the OMG web site (www.omg.org). In June 2009, a revision task force was launched by the OMG. This task force is scheduled to complete its work within one year leading to a minor revision that will incorporate minor fixes for specification issues received by the OMG in the meantime.

MARTE has already been applied extensively in practice by industry and is supported by numerous tool vendors as indicated by the list of ongoing projects that identify MARTE as central to their concerns (cf. the OMG web site dedicate to MARTE, www.omgmarte.org). But MARTE, is also an interesting research subject being explored by academia and other research institutions. The expectation is that all of these research activities will lead to new proposals for using MARTE for designing and validating real-time and embedded systems based on standards. And, of course, it will also lead to proposed enhancements and extensions to the standard itself.

References

[1] Object Management Group: UML Profile for Modeling and Analysis of Real-Time and Embedded systems (MARTE) RFP (2005-02-06) (February 2005)
[2] Jacobson, I.: Object-Oriented Software Engineering: A Use Case Driven Approach. Addison-Wesley, Reading (1990)
[3] The Mathworks, http://www.mathworks.fr/
[4] Eclipse-atl, http://www.eclipse.org/m2m/atl/
[5] Object Management Group: UML Version v2.1.2 (2007-02-05) (February 2007), http://www.omg.org/spec/UML/2.1.2/
[6] Object Management Group: UML Profile for Schedulability, Performance, and Time, v1.1 (2005-01-02) (January 2005), http://www.omg.org/technology/documents/formal/schedulability.htm
[7] Chen, R., Sgroi, M., Martin, G., Lavagno, L., Sangiovanni-Vincentelli, A.L., Rabaey, J.: UML for Real: Design of Embedded Real-Time Systems. In: Selic, B., Lavagno, L., Martin, G. (eds.), pp. 189–270. Kluwer Academic Publishers, Dordrecht (2003)
[8] Klein, M., Ralya, T., Pollak, B., Obenza, R.: A Practitioner's Handbook for Real-Time Analysis: Guide to Rate Monotonic Analysis for Real-Time Systems. LNCS. Kluwer Academic Publishers, Dordrecht (1993)
[9] Espinoza, H.: An Integrated Model-Driven Framework for Specifying and Analyzing Non-Functional Properties of Real-Time Systems. Information Processing Letters (2007)
[10] Gray, J., Tolvanen, J.P., Kelly, S., Gokhale, A., Neema, S., Sprinkle, J.: Domain-Specific Modeling (in CRC Handbook of Dynamic System Modeling). CRC Press, Boca Raton (2007)

11. Selic, B.: On the semantic foundations of standard UML 2.0. In: Bernardo, M., Corradini, F. (eds.) SFM-RT 2004. LNCS, vol. 3185, pp. 181–199. Springer, Heidelberg (2004)
12. Shonle, M., Lieberherr, K., Shah, A.: XAspects: An Extensible System for Domain-Specific Aspect Languages. In: Object-Oriented Programming. LNCS. Springer, Heidelberg (2003)
13. Selic, B.: A Systematic Approach to Domain-Specific Language Design Using UML. In: ISORC (2007)
14. Object Management Group: Systems Modeling Language, Version 1 1(2008-11-01) (November 2008), http://www.omg.org/cgi-bin/doc?formal
15. Object Management Group: UML Profile for Modeling QoS and FT Characteristics and Mechanisms, v1.1 (2006-05-02) (Mai 2006), http://www.omg.org/spec/QFTP/1.1/
16. Object Management Group: UML Profile for MARTE, Beta 2 (2008-06-09) (Juni 2008), http://www.omg.org/cgi-bin/doc?ptc/
17. Object Management Group: Policies and Procedures, Version 2.7 (2008-06-01) (Juni 2008), http://www.omg.org/cgi-bin/doc?pp
18. Thomas, F., Gérard, S., Delatour, J., Terrier, F.: Software Real-Time Resource Modeling. In: Proceedings of the International Conference Forum on Specification and Design Languages (FDL). Information Processing Letters (2007)
19. Taha, S., Radermacher, A., Gerard, S., Dekeyzer, J.L.: An Open Framework for Hardware Detailed Modeling. In: IEEE Proceedings of SIES. Information Processing Letters (2007)
20. Taha, S., Radermacher, A., Gerard, S., Dekeyzer, J.L.: Marte: Uml-based hardware design from modeling to simulation. In: Proceedings of the international conference forum on specification and design languages (fdl). Information Processing Letters (2007)
21. Tawhid, R., Petriu, D.C.: Integrating Performance Analysis in the Model Driven Development of Software Product Lines. In: Czarnecki, K., Ober, I., Bruel, J.-M., Uhl, A., Völter, M. (eds.) MODELS 2008. LNCS, vol. 5301, pp. 490–504. Springer, Heidelberg (2008)
22. Espinoza, H., Medina, H.J., Dubois, H., Gerard, S., Terrier, F.: Towards a UML-based, Modeling Standard for Schedulability Analysis of Real-time Systems. In: International Workshop MARTES, MoDELS/UML 2006 (2006)
23. Selic, B.: From Model-Driven Development to Model-Driven Engineering. LNCS. Springer, Heidelberg (2007)
24. (Smartqvt), http://smartqvt.elibel.tm.fr/
25. Thomas, F., Delatour, J., Gérard, S., Terrier, F.: Toward a Framework for Explicit Platform Based Transformations. In: 11th IEEE International Symposium on Object-oriented Real-time distributed Computing. LNCS. Springer, Heidelberg (2008)
26. Mraidha, C., Tanguy, Y., Jouvray, C., Terrier, F.: Gerard: Presented in Workshop UML&AADL 2008 and Published in Proceeding of the 13th IEEE International Conference on Engineering of Complex Computer Systems. LNCS. Springer, Heidelberg (2008)
27. Espinoza, H., Servat, D., Gérard, S.: Leveraging Analysis-Aided Design Decision Knowledge in UML-Based Development of Embedded Systems. LNCS. Springer, Heidelberg (2008)

[28] Espinoza, H., Selic, B., Cancila, D., Gérard, S.: Challenges in Combining SysML and MARTE for Model-Based Design of Embedded Systems. In: ECMDA 2009, Published in Proceeding of the Conference (Model Driven Architecture- Foundations and Applications). LNCS, pp. 98–113. Springer, Heidelberg (2009)

[29] SAE: Architecture Analysis and Design Language (AADL) Annex Volume 1: Annex A: Graphical AADL Notation, Annex C: AADL Meta-Model and Interchange Formats, Annex D: Language Compliance and Application Program Interface Annex E. LNCS. Springer, Heidelberg (2006)

[30] Autosar, http://www.autosar.org/

[31] Adams-Project, http://www.adams-project.org/

[32] East-Adl, http://www.east-adl.org/

[33] Espinoza, H., Gérard, S., Lönn, H., Kolagari, R.T.: Harmonizing MARTE, EAST-ADL2, and AUTOSAR to Improve the Modelling of Automotive Systems. In: Presented in the Workshop STANDRT, Autosar (2009)

[34] (ISIS,MIC Tool Distribution), http://www.isis.vanderbilt.edu/Projects/gme/

[35] Lee, E.A.: Overview of the Ptolemy Project, Technical Memorandum No. UCB/ERL M03/25 (2003)

[36] Agha, G.: Actors: a model of concurrent computation in distributed system. MIT Press, Cambridge (1986)

[37] Terrier, F., Fouquier, G., Bras, D., Rioux, L., Vanuxeem, P., Lanusse, A.: A real time object model. In: TOOLS Europe 1996 (1996)

7 Requirements Modeling for Embedded Realtime Systems

Ingolf Krüger, Claudiu Farcas, Emilia Farcas, and Massimiliano Menarini

University of California, San Diego, USA
{ikrueger,cfarcas,efarcas,mmenarini}@ucsd.edu

Abstract. Requirements engineering is the process of defining the goals and constraints of the system and specifying the system's domain of operation. Requirements activities may span the entire life cycle of the system development, refining the system specification and ultimately leading to an implementation. This chapter presents methodologies for the entire process from identifying requirements, modeling the domain, and mapping requirements to architectures.

We detail multiple activities, approaches, and aspects of the requirements gathering process, with the ultimate goal of guiding the reader in selecting and handling the most appropriate process for the entire lifecycle of a project. Special focus is placed on the challenges posed by the embedded systems. We present several modeling approaches for requirements engineering and ways of integrating real-time extensions and quality properties into the models. From requirements models we guide the reader in deriving architectures as realizations of core requirements and present an example alongside with a formal verification approach based on the SPIN model checker.

7.1 Introduction and Overview

Requirements engineering is arguably one of the most important and least-well-understood [1] development activities. It can have a positive effect on the overall development process – systems that actually provide value to their stakeholders, i.e. systems for which there exists a good understanding of what the requirements are, as well as a match between the system's requirements and the implementation, are generally considered a success if implemented within the available resources. At the same time, it is well-known that errors made during the activities that pertain to requirements analysis and management are hard to detect and costly to fix as time progresses through the development process.

In this chapter, we discuss the challenges and opportunities of the requirements engineering process for complex embedded real-time systems (ERS) as they arise in domains such as automotive, avionics, medical, communications and entertainment systems to name but a few examples. This system class is of high economic relevance and significant technical complexity – more than 98% of processors are "embedded" [2]. In automotive systems, for instance, up to 90% of all innovations are influenced by software-enabled electronics. In some high-end vehicles more than 60 different electronic control units (ECUs), interconnected using multiple communication bus technologies and hundreds of signals, together provide thousands of externally observable functions. Heterogeneity and distribution lead to high numbers of different configurations and variants. A wide functional variety from hard real-time safety critical engine control to comfort electronics and infotainment systems, long product life cycles, demanding time-to-market, and a strong need for competitive per-piece costs compound the technical challenges. All of these aspects are directly or indirectly related to the discovery, articulation, quality assurance and continued management of a highly diverse and interdisciplinary requirements set.

In the following paragraphs we elaborate further on the challenges of requirements engineering for ERS. This account draws heavily on our experiences with automotive systems [3, 4, 5, 6, 7]; furthermore, we provide an overview of the remainder of this chapter.

7.1.1 What's in a Requirement?

Before we can gain an understanding of why requirements engineering for ERS is challenging, we first have to identify what we mean by the term "requirement". We view a requirement as a documented need of what a product or service should be or do. A requirement also identifies the necessary attributes, capabilities, characteristics, or qualities that have value or utility to a stakeholder. While sufficiently intuitive at first glance, this definition leaves open what terms such as "attribute", "quality" and "value" mean concretely. Unless these terms are precisely defined, of course, it is difficult – if not impossible – to identify whether a given requirement is well articulated, let alone whether a given system correctly implements this requirement.

Nevertheless, this definition brings forward a number of important concerns for capturing and managing requirements. First, it is important to observe that requirements are connected to stakeholders [8]. For each requirement there should be an identified party with a vested interest in seeing the requirement implemented. This implies also that the requirements, collectively, need to articulate values of the stakeholders of the system under consideration. Stakeholders include (and are not limited to) the customer who commissions and accepts the system, regulatory bodies, marketing and productization entities, suppliers, integrators, developers, architects and maintainers, and end-users. Consequently, models, techniques and tools for documenting and managing requirements necessarily need to be able to reflect the various different views [9] that each stakeholder group brings to the table. For instance, marketing representatives may articulate requirements aspects that relate functionality with cost, end-users may articulate usability requirements aspects [10, 11, 12, 13], and maintainers may articulate requirements at readability of the source code from which the software sub-system is compiled.

To address different stakeholder views, the literature distinguishes various classifications for requirements. At the highest level, there is a distinction between *business*, *product*, and *process requirements*. Here, business requirements refer to specifications of what the business wants to achieve with a specific project. An example for a business requirement is "Offer the safest car on the road today". Requirements such as this are typically directly linked to an enterprise objective, such as "Being the world leader in safe vehicles" and associated business and marketing accounts. Product requirements often encompass the functionality and operational infrastructure that is required to implement the business requirements. This includes a specification of the functions as well as the hardware/software context in which they are to be implemented. Process requirements typically refer to how the development of the system under consideration is to come about, whether and how it needs to be certified, and what methods are to be applied during development for documentation, implementation and quality assurance.

Another traditional classification distinguishes *functional* from *non-functional* requirements. Here, functional requirements include those that determine what the system is supposed to do – this amounts to a specification of the *operational capabilities* provided by the system. Non-functional requirements [14, 15] (some authors also refer to them as *quality requirements*) then are defined as constraints at the implementation level of the functional requirements. Such constraints include product requirements[1] (usability [13], performance and efficiency [16, 17, 18], reliability [19, 20] and portability [21]), organizational requirements (delivery, implementation, adherence to standards and regulations) [15], and external requirements (interoperability [14, 22], ethics, and legislation). Functional requirements address the operational capabilities of the system, non-functional requirements define the context in which these capabilities come about.

Characteristics of ERS often blur the line between the functional and non-functional requirements. Consider, as an example, the air bag controller for a modern car. The requirement "Within 10 milliseconds after impact, the airbag is to be fully inflated" identifies an operational aspect ("after impact, the airbag is to be fully inflated"), and a performance constraint ("within 10 milliseconds after impact"). Clearly, the stakeholder group "driver and passengers" would rightly argue that an airbag that misses its deadline is not a functioning product. Hence, this requirement will likely be perceived as a functional requirement, despite its performance aspect.

As we shall see below, time plays a critical role in specifying requirements for ERS. Therefore, it often becomes part of the underlying system model in relation to which requirements are expressed. In other words, timing constraints become part of the operational capabilities of the system under consideration. Then, requirements with timing constraints naturally fall into the category of functional requirements. In general, however, it often makes sense to distinguish functional and non-functional requirements aspects, and to allow specification of both aspects in a singular requirement.

---

[1] Note that this refers to a subset of what we called product requirements in the preceding paragraph.

Independently from the chosen requirements classification scheme, the way in which we are able to articulate requirements has a significant impact on how useful requirements engineering is for all other development activities. We need to be able to determine effectively and efficiently, (a) whether stakeholder values and associated constraints at the system under consideration are captured accurately in the requirements, and (b) whether the implementation is faithful to the requirements as captured. (a) and (b) are commonly referred to as the requirements validation and verification problems, respectively.

For a comprehensive approach to requirements engineering for ERS, we need to be able to address all relevant requirements aspects of the system under consideration such that they can be effectively and efficiently validated and verified. This is a hard challenge in general and specifically so for ERS, as we will elaborate in the following paragraphs.

7.1.2 Why Requirements Engineering for ERS Is Hard

We now discuss key forces that influence the difficulty of eliciting and managing requirements for ERS; many of these forces interrelate. We have already pointed out that precisely defining what a requirement is, is a challenge in and by itself. A common solution for model-based engineering approaches is to create a *system model*, which is a mathematical representation of the central phenomena exhibited by the system class into which the concrete system under development falls. We can then formally define requirements as constraints at the system model. These constraints reduce the set of all possible instances of the model to those that fulfill the requirement.

We are interested in system models that are close to the problem domain. For this purpose, system models built on top of process algebras, timed and untimed, finite and infinite automata, temporal logics, partial orders, or streams and stream, relations, and games have been devised to address a broad spectrum of properties while placing bounds on the computational complexity of validation and verification. Hence, in requirements engineering, a key challenge is to identify system models and associated specification languages that allow representation of the key domain concepts [23] within the formal model such that validation and verification are effective and efficient. In the following paragraphs we discuss the key forces that drive the selection of an appropriate system model for ERS.

Requirements do change over the lifecycle of a product [24, 25, 26] [27]. The romantic idea that requirements can be captured completely at product inception, frozen, and then implemented to satisfaction is an illusion for all but the most trivial systems. Consequently, requirements need to be managed actively throughout the product lifecycle [28, 29], from inception to retirement.

Embedding – ERS are embedded into a context by definition. This context is typically another product, such as a car, an airplane, or a medical device. The ERS often plays the role of a controller of physical processes in which the overall product engages. This means that the ERS needs to interface with the context into which it is embedded, and thus needs to have a model of its environment to appropriately react to changes in this environment.

Furthermore, each ERS has a *unit cost* associated with it. There is a natural incentive to reduce the unit cost to reduce the overall product's cost (or to increase profit margins). Consequently, ERS are equipped with just enough computational and storage resources (as in hardware) to fulfill their desired function. Especially in mass markets, such as in automotive or entertainment electronics, where comparatively tiny unit cost savings have an enormous impact on overall profitability, cost reduction at the unit level is a critical and driving business requirement. Of course, if the overall product consists of multiple ERS, this can result in many local optimizations at the expense of realizing global cost savings across ERS. As we shall see, below, this is compounded by the distributed development model for ERS, and by the absence of integrated system models that would allow articulation and optimization of cost and functionality across component ERS.

Multi-Disciplinary Stakeholder Communities. Clearly, describing the interface between the ERS and its physical environment necessitates that requirements can express the properties and constraints of the requirements that are relevant for this interface. This alone already necessitates a multi-disciplinary approach to requirements engineering for ERS. A car, for instance, provides a physical context in which mechanical engineers are key domain experts. Add to this electrical engineers for the hardware context of the ERS, as well as Human-Machine-Interface (HMI) experts for the usability aspects, to obtain an initial set of disciplines involved in ERS development besides the requirements engineers and software developers, testers and maintainers with computer science background. Each of these stakeholder groups has a different view [30] on the system under consideration, and has its own domain concepts and associated ways of articulating them. This necessitates that the chosen system model allows expression of requirements from all relevant stakeholder groups.

Two techniques allow dealing with the resulting complexity: (1) introducing mechanisms for expressing different views onto, or abstraction levels of the same system model – in the language of the respective stakeholder group, and (2) enabling the co-existence of multiple different system models, each of which is tailored to a particular stakeholder group, or view; then the challenge is how to mediate between requirements specified in the respective system models to arrive at a consistent [31, 32, 33, 34], integrated requirements specification for the overall product. This mediation can take on various forms; typically it will consist of an elaboration of the domain entities (or abstractions thereof) shared by multiple source models and the key relationships between these entities, as well as projection relations between the source models and the mediation model.

Consistency and Realizability – If there is more than one requirement needed to specify a system of interest (and for all but the most trivial ones, there is), we have to address consistency between requirements. The underlying question is whether the requirements as specified allow any system to be constructed such that all requirements are fulfilled.

For instance, if we represent requirements as predicates on the underlying system model, then we can express relationships between requirements using logical conjunction and disjunction: conjunction expresses that of two requirements both must be satisfied together; disjunction indicates alternatives among requirements. If we add negation as a logical construct, we can also express "anti-requirements", i.e. requirements that must not be fulfilled [35]. This then allows us to express conditional requirements of the form "if requirement A is fulfilled, then requirement B must be fulfilled also". The overall requirements specification can then be interpreted as a logical formula involving predicates and the mentioned logical connectives. The question then arises whether the set of tape valuations constrained by this logical formula has any elements in it.

As we have described in the preceding paragraph, we have to be able to represent different views on the set of requirements to different stakeholder groups. Consequently, we have to concern ourselves with the consistency of the resulting composite views. We have to add requirements that further restrict the set of possible models to those that are realizable.

Outsourcing and Distributed Development – ERS are, in general, developed in the interplay between an Original Equipment Manufacturer (OEM) and a supplier. The OEM is responsible for the product into which the ERS integrates. The supplier is responsible for the ERS. This necessitates precise and expressive requirements specifications [36] that elucidate the interplay between the ERS and its environment. In practice, requirements are seldom expressed precisely enough, successful projects resort to cooperative and joint development between the OEM and the supplier to ensure short feedback cycles to iteratively refine the requirements. The precision to which we are able to articulate requirements between OEM and supplier has a direct impact on the number of iteration cycles needed, and the ability for both parties to verify and validate the requirements they were able to specify.

Furthermore, the distribution of responsibility [37] between the OEM as the system integrator, and the supplier has lead to two distinguishable levels of abstraction in requirements specification, called *user* and *system requirements*, respectively. User requirements are gathered by the OEM and articulate the OEM's expectations at the outcome of the supplier's development efforts. The supplier responds to a user requirements specification with a system requirements specification that details the user requirements, and incorporates further business, product and process requirements from the supplier's point of view.

Multi-Functionality [7] means that an ERS provides not one singular, but multiple distinguishable, individually valuable functions (also called *features*) to their environment. For instance, cell phones provide calendaring, email and a wide variety of productivity and entertainment functions in addition to the all but mundane functions of placing and receiving calls.

This necessitates that requirements for ERS explicitly address the partiality of individual functions *and* precise specifications of how the individual functions integrate into the whole. In particular, the requirements need be be explicit about desirable and undesirable *feature interaction*. A desirable feature interactions emerges from the interplay between two features to the (sometimes unforeseen) benefit of a stakeholder. Undesirable feature interactions, on the other hand, reduce the value of the integrated system to a stakeholder. Again, this calls for explicit means to determine consistency among requirements for the ERS and the environment into which it is embedded.

Heterogeneity – Requirements for ERS that control physical processes often are most succinctly represented in terms of the mathematical models that are used to describe the physical processes. For instance, requirements for automotive systems need to capture the vehicle's continuous movement through space and time. The adequate mathematical model for this movement will involve differential equations. Specifically, the field of *control theory* was developed precisely to study the phenomena that arise in the interactions between the physical world and controllers that seek to influence the environment to effect a desirable condition. Any comprehensive requirements specification technique that attempts to be successful in the automotive domain, therefore, needs to be able to capture continuous behaviors (in the underlying mathematical models) to facilitate interaction with control engineers. At the same time, a car is a good example for the need to also express mixed discrete-continuous and purely discrete ERS [7].

Distribution and Integration – As mentioned above, the OEM typically acts as the integrator of a set of independently developed ERS. Consequently, the desired behavior for the integrated product emerges from the interplay of the functionality provided by the sub-systems. For the system models underlying requirements specifications this means that they need to be able to express phenomena of concurrency and synchronization. Depending on the product and the OEM, these ERS are developed by a variety of different, competing suppliers. As a consequence, the functionality valuable to the end-user is scattered across a wide variety of subsystems. This places a tremendous integration challenge on the OEM – this finds its expression today in intense and costly integration, calibration and testing activities in which the OEM engages when *all* subsystems finally are available.

Often, there is no overarching requirements specification addressing the integration challenge – the user requirements suppliers see are then underspecified in terms of these integration requirements, and the OEM has to work around the resulting implicit assumptions the suppliers make. Furthermore, as mentioned before, the absence of an overarching understanding of the integration requirements results in poor resource optimization *across* the ERS of an integrated system. Consequently, the requirements models for ERS should explicitly address the scattering of functionality and the resulting integration requirements, as well as the concerns that cut across the individual system components (see below).

Safety-Criticality – ERS in safety-critical products [38, 39, 40] such as cars, airplanes, trains, space ships, power-plant and factory control systems, heart-pacers and other medical devices are safety-critical by association. Much research has been invested into developing system models that allow the specification and verification of safety properties [41, 42, 43, 44] [45, 46, 39, 47].

A remaining research challenge is to provide *domain-specific* system models that allow articulation, validation and verification of safety requirements at the scale of thousands of integrated functions while resolving the dependencies and interactions between the requirements forces described in this section.

As a case in point we note that failure management is a critical concern for many ERS [19, 48] and specifically so in the automotive domain. Yet, none of the widely-used requirements specification techniques for automotive systems even recognizes the notion of failure as a first-class modeling entity. Of course, there are techniques such as Failure Mode and Effect Analysis (FMEA) [49] and Fault Tree Analysis (FTA) [50] – however, these techniques are rarely applied at the inception and requirements modeling phase, but rather reserved for an after-the-fact analysis, when the subsystems have already been developed. We will pick up this topic below, when we discuss cross-cutting concerns, as well as in Section 7.4, in our case study.

Multi-scale Timing, Asynchronous vs. Synchronous Communication – Time plays a critical role in ERS. Many models of physical phenomena depend on time as a variable; for instance, velocity is the derivative of position in time. ERS control properties are consequently frequently specified in the language of differential equations. However, this is typically already at the level of a solution (in the sense of specifying a particular controller) rather than at the level of a requirement. This is facilitated by the available tool support for control system development (see below), which favors the graphical specification of particular solutions rather than requirements.

In general, most system models favor a particular model of time (continuous vs. discrete), which results in awkward requirements specifications for systems with mixed discrete-continuous timing properties. Similarly, most system models favor a specific communication model (message/time-synchronous vs. message/time-asynchronous), which again can result in awkward expressions of requirements for integrated systems with mixed types of communication requirements. Of course, in concrete examples, such as cars, we find a broad range of timing constraints on scales of milliseconds (motor control) to tenths of seconds (comfort functions) to seconds (navigation). Similarly, we find a wide variety of communication mechanisms ranging from time- and message-synchronous communication within an ERS to time- and message-asynchronous communication beyond vehicle boundaries for remote operation of vehicle functions.

It is one of the key research challenges in ERS to reconcile multiple time and communication models within such that the corresponding requirements can be expressed lucidly. In practice, the timing requirements are often only informally stated on a per-component basis, following some intuitive, or implicit understanding of overall end-to-end response-time requirements, or implementation-technology-dependent constraints (processor cycle times, cache hit-rates, communication bus throughput and latencies). Using simulation and testing as the main tools, the system is then instrumented to determine worst-case execution times. The results are then matched against the per-component timing requirements.

Long Product Life Cycles – Products containing ERS typically have long product life cycles from inception to retirement. This means that over this lifecycle, many changes in the environment of any particular ERS may occur: other components may be exchanged, or the product may be placed into previously unanticipated environments. This, in turn, means that requirements specifications (and thus the system models supporting them) need to be durable and accessible throughout the product lifecycle.

Time-to-Market While product life cycles are long, especially in consumer mass-products, time-to-market is constantly under scrutiny for reduction to react more rapidly to changes in consumer, environmental or regulatory needs. A case in point is the rapidly increasing demand for hybrid vehicles on the backdrop of rapidly rising fuel prices.

This impacts requirements engineering for ERS in the sense that it needs to be able to respond to rapidly changing requirements to facilitate the agility and flexibility needs of its container. In particular, the degree to which requirements can be specified in a modular fashion will have significant impact on how rapidly the parts of an integrated system of ERS can be adapted to support changes to the product as a whole. Note that here, we are referring to the structuring of the requirements, rather than the structuring of the resulting architecture; the latter is also an important, albeit separate, topic.

Product Lines and Re-Configuration – Similarly, to amortize costs and respond to market needs, OEMs often develop platform strategies and product lines so as to be able to reuse significant parts of an integrated system, while adapting others. This leads to the challenge of managing multiple different versions of requirements sets, which correspond to multiple different configurations of the integrated system. In the automotive example, some product families have variation points amounting to hundreds of thousands of different configurations customers can order. Some of these configuration options are necessitated by regional laws and regulations, others stem from different options for feature sets of the vehicle.

Again, the major impact for requirements engineering is on the management capabilities of the associated requirements documents, so as to avoid costly rework; also, there is impact on the ability to verify consistency of the requirements configurations.

Influence of Hardware Architectures – Sometimes, products containing ERS evolve from limited feature sets to thousands of software-enabled features; again, the automotive domain is a telling example. From the beginnings of the use of electronically controlled fuel injection to today the amount and impact of software deployed in the car has grown exponentially. Now automotive engineers are faced with the challenge of integrating, and supplying power for, 30 to 80 Electronic Control Units (ECUs) per car, depending on the target market (budget vs luxury). This challenge includes calibrating the timing between the various ECUs in their attempt to communicate with other ECUs in the vehicle – largely due to the scattering of functionality across the various ECUs. To some degree, this challenge is an artifact of dominant legacy architectures that were adopted initially, and never reconsidered as more and more functions entered the vehicle – often due to long-time licensing agreements and cost savings of reuse.

The impact on requirements engineering for ERS is that such legacy architectures often become requirements constraints that also need to be articulated in the requirements model. At the same time there is a tradeoff between articulating legacy architecture requirements and writing requirements that describe more of the "what", rather than a particular "how". This tradeoff must be resolved at the level of the overall engineering process.

Deployment, Update, Diagnostics, and Maintenance – Because of their embedding, ERS are less accessible for deployment, update and maintenance tasks than desktop or laptop computers. Nevertheless, long product life cycles ultimately necessitate updates to the software or hardware components of an ERS. Especially in distributed, integrated ERS such as in cars and airplanes, however, updates are (as of this writing) difficult to deploy. As mentioned above, in such systems many ERS originate from different competing suppliers, each applying their own strategies, technologies and methods for deployment, update and maintenance (if any). This again points to the specification of dedicated requirements for these important development tasks at the integration level.

Therein lies significant research potential – today's formal requirements specification techniques have yet to broaden the range of requirements types they address. Most system models (as we shall see in Section 7.2) assume static system structures and mappings from behaviors to these system structures. This renders precise specification of deployment, update and maintenance requirements all but impossible.

The drive towards service-oriented architectures (SOAs) that has gained significant momentum in the world of business information systems is slowly gaining ground in the ERS domain as well. One of the fundamental premises of SOAs is that the location of a function is secondary to its interface due to static and dynamic advertisements, registration and binding techniques. This dynamicity today still clashes with the imperative of unit cost savings, and thus scarce resources per ERS. As attention shifts from per-unit development to integrated networks of ERS, we expect this reservation to give way to an understanding of global reuse, dependability potentials and cost-savings.

As technical solutions for these concerns are on the doorstep so should be the models for specifying the corresponding requirements – yet, there is very little support in contemporary, widely accepted requirements modeling and engineering approaches.

Quality and Cross-Cutting Concerns – Deployment, update and maintenance are good examples of cross-cutting ERS concerns that have little to no support in today's system models and corresponding systematic requirements engineering techniques. Of course, the list of cross-cutting concerns does not stop there. Other important ones are availability, fail-safety (across units in a set of networked ERS), security [51, 15], and policy/governance. Availability and fail-safety are addressed in this and other chapters explicitly, hence we focus on the other two requirements aspects here.

Security has been regarded as a secondary concern for a long time in ERS development. After all, most ERS were assumed to be inaccessible from outside of the product they were embedded into. This, of course, has changed radically with the increased networking among and beyond ERS [52] – suddenly, for instance, we find that the CLS functionality of a car is accessible via the Internet (to support remote unlocking to recover locked-in keys). Signals in car networks today are rarely encrypted, and can thus be reengineered, and reproduced in malevolent ways (for instance to gain unauthorized entry into the vehicle).

Policy and governance also come into play in networked ERS – together they address the question under which circumstances a party can (or must) perform a particular action in the system. In our automotive example this becomes important in identifying who has the authority to unlock the vehicle; another scenario is the prevention of unauthorized after-market components to participate in the exchange between the authorized ERS.

Similarly, the challenge of *diagnostics* – what and how much data to collect at what locations and time points, to identify the root cause of a failure during system operation – is an area of active research with little explicit system modeling support, and consequently no broadly accepted formal requirements specification techniques.

Furthermore, all ERS undergo a set of distinct operational modes, such as *initializing, idling, operating, resetting* and *suspending*, to name a few examples. From a modeling point of view this can be addressed with regular state-based modeling techniques such as state machines or activity diagrams. However, there is a need to explicitly provide access to all or some of these modes at the ERS-environment interface, especially in a networked ERS consisting of multiple subsystems. This need arises both for monitoring purposes and to ensure that sets of components can be steered into defined operational modes together (say, for start-up, shut-down, and testing).

Traceability – Because the requirements spectrum of ERS is vast and highly heterogeneous, traceability becomes a particularly daunting task. Success is again bound to our ability to articulate requirements at increasingly high levels of detail, and to validate and verify requirements at different levels of abstraction against each other. Furthermore, methods are needed that establish a trace between architecture specifications and implementations at various levels of abstraction and the requirements that are implemented at these levels. This challenge is again compounded by the distributed nature of the OEM-supplier relationship, and the desire to support product lines with vast numbers of possible system configurations, as well as by the tight coupling between requirements specifications and the target hardware/system platform onto which the ERS functionality is to be deployed.

At the level of system architecture specification, Model-Driven Architecture (MDA) [53] has taken a step into a more tractable direction – here, we distinguish between a Platform Independent Model (PIM) and a Platform Specific Model (PSM). The PIM can largely be regarded as a highly detailed requirements model that captures the core system entities and their interactions without specifying how these are implemented. The PSM, on the other hand, captures all aspects of the deployment architecture. Then a mapping between a PIM and multiple

PSMs can be be established to capture multiple different deployments for the same functionality set. However, further research is needed to lift the degree of abstraction from PIMs to genuine user and system requirements specifications.

Tool Landscape – A wide range of commercial and academic tools for requirements engineering and management have been developed. Few, however, cover even a small subset of the concerns we have brought forward in the preceding paragraphs to any degree of satisfaction. We attribute this largely to the absence of comprehensive system models and associated requirements specification techniques and standardized architectures that adequately capture and reduce the complexities of ERS specific to particular domains.

Of course, tools such as DOORS [54], Rational RequisitePro [55], and Cradle [56] have displayed their utility in the *management* [57] (as in organization and version control) of requirements once they are elicited. Tools such as Matlab/Simulink/Stateflow allow detailed architecture design of and even generation of efficient code for controllers for which requirements are well understood. However, the challenge of finding adequate system models and requirements specification techniques for systematic requirements discovery and refinement remain. Yes, DOORS integrates with other UML-based tools for requirements elaboration. However, for UML and its derivatives a wide variety of the challenges posed above are unsolved as of yet; we name just a few examples: consistency of description techniques and resulting requirements specifications, efficient and effective validation and verification at the model level, notations and models for system (re-)configuration, support for cross-cutting concerns in the requirements models, including failure, safety, security, and policy/governance

7.1.3 Summary and Outline

In the preceding paragraphs we have identified a broad range of challenges that render precise requirements specifications of ERS particularly difficult. We have started by identifying requirements as the expression of stakeholder values, and have established a connection between mathematical system models and requirements formalized as predicates (or constraints) over these system models. Then we have called out a number of requirements aspects that a comprehensive requirements engineering approach needs to be able to articulate and manage throughout the development process. Key challenges arise from the multi-disciplinary and heterogeneous nature of ERS requirements, their distribution, domain-specifics such as a broad range of timing specification needs, deployment, update and maintenance requirements and the associated quality, validation and verification concerns.

No currently available tool or integrated tool set addresses all of these concerns comprehensively. We conclude that this necessitates further research and development in both academia and industry – this volume is evidence of the significant research progress to date.

The remainder of this chapter is structured as follows: In Section 7.2 we review a broad range of requirements engineering techniques proposed in the literature – this provides an overview to what degree the mentioned concerns are addressed in today's models and techniques. In the absence of a formal, comprehensive requirements engineering technique, we briefly recall key best-practices of requirements engineering and how they relate to model-based development (Section 7.2) for ERS. We discuss the relationship between requirements and their traceability to architecture, and from there to implementation, in Section 7.3. In Section 7.4 we give an example for capturing safety requirements of an automotive Central Locking System using structural and behavioral modeling techniques so that these requirements can be formally verified.

7.2 Requirements Specifications and Modeling for ERS

Modeling plays an important role in all requirement engineering activities, serving as a common interface to domain analysis, requirements elicitation, specification, assessment, documentation, and evolution. Initially, domain models are created to describe the existing system for which the software should be built, covering stakeholders, human actors that interact with the system, hardware devices, and the environment in which the system will operate. In addition to behavior, domain models define "the language" of the system by capturing domain entities in a structural way [58]. Then, deficiencies in the existing system and objectives for the target system are more clearly identified. During requirements elicitation, alternative models for the target system are created, which may define different boundaries between the target system and its environment. Models can help in defining the questions for stakeholders and surfacing hidden requirements. Ultimately, the requirements have to be mapped to the precise specification of the system and the mapping should be kept up to date during the evolution of requirements or the architecture.

After requirements are specified (more or less formally), the specifications are checked for errors such as incompleteness, contradictions, ambiguities, inadequacies in respect to the real needs – which all can have disastrous effects on the system development costs and the quality of the resulted product. The choice of modeling notations is often a tradeoff between readability and powerful reasoning techniques: natural language is very flexible, useful for communicating requirements, but can not capture relationships and is often an expression of subjective reasoning [59, 60]; applied/semi-formal models (e.g., entity-relationship diagrams, UML diagrams, structured analysis) typically have a graphical representation which is very useful when communicating with stakeholders and often offers simulation and animation capabilities; and formal notations (e.g., KAOS [61, 62], RML, Telos, SCR [63, 28], process algebra, Promela/SPIN [64]) capture precise semantics, which supports rich verification techniques.

7.2.1 Requirements Models

Many challenges of requirements engineering span multiple application domains. For instance, business concerns such as conflicts from multiple viewpoints over requirements of different stakeholders are present in domains as diverse as business information systems, financial applications, avionics, car OEMs and suppliers, etc. Hence, in this subsection we first briefly present general techniques for requirements modeling that have a broader scope and can be applied on a variety of domains. As embedded systems may require dedicated techniques for some aspects such as timing, determinism, and formal verification of safety properties, we then describe particular techniques for ERS.

Business modeling Goal-based approaches such as KAOS [61, 62] and i* [65, 66] focus on modeling goal hierarchies to capture the objectives of the system, the associated tasks, and resources. The explicit modeling of goals helps in checking the requirements completeness – the requirements are complete if they are sufficient to meet the goal they are refining [67]. In KAOS [61, 62], the set of high-level goals are iteratively refined using AND/OR decomposition, obtaining a graph structure. KAOS allows to define agents and the actions they are capable of, and the goals can be operationalized into constraints assigned to individual agents. Each term is formally defined in temporal logic; therefore, a main contribution of KAOS is to prove that goal refinement is correct and complete [68], which implies proving that requirements correspond to system goals. Furthermore, [69] shows how conflicts between goals can be formally detected. The i* [65, 66] framework focuses on two models: the strategic rationale model describes the goals of the actors and the interactions between goals and tasks within each actor, whereas the strategic dependency model focuses on the relationships between actors such as dependencies on the goals or resources from other actors, or dependencies on tasks that other actors should perform. With such models, properties such as viability of an agent's plan or the fulfillment of a commitment between agents can be verified.

Another approach is to focus on business processes (workflows), business rules, and the services the system provides [70]. For this purpose, UML activity and collaboration diagrams can be used to show how actors collaborate to perform tasks. Moreover, UML class diagrams can show the roles of actors within the domain and can be used to capture business rules, although often in an implicit way through the class composition and multiplicity constraints. In UML, business rules, as well as pre- and post- conditions, can be explicitly specified in Object Constraint Language (OCL) [71].

Modeling information and behavior is an important part of the requirements specification process dealing with the structure of the system in terms of entities and their relationships; the behavior in terms of states and events that determine state transitions; and interactions in terms of communication patterns, dataflows between system components, parallelism, concurrency coordination, and dependencies – especially temporal dependencies in the case of ERS.

One way for specifying the structure of the systems is to use entity relationship (E-R) diagrams to capture domain concepts and data models. Although E-R diagrams are just notations, the concepts of objects, classes, attributes, and instances map well to domain entities and enable an easy transition to object-oriented system design. This ease of transition from requirements to design is sometimes a drawback as it becomes difficult to distinguish the real user requirements and their rationale from design decisions inferred from underspecified requirements. Also, focusing on single use cases may prevent the development of the system vision or the "big-picture". In such cases, the solution resides in operating with partial system specifications through an agile development process that iteratively refines the requirements and constructs the vision of the final system. Standards such as UML can be used to achieve consistency between models developed in different iterations.

Modeling the system behavior is generally accomplished using variants of finite state machines (FSM) [72, 73] and notations such as Dataflow Diagrams (DFD) [74]. The Structured Analysis is a data oriented approach for conceptual modeling initially intended for information systems and later adapted to ERS. It presents a development/transition path from an indicative model of the current system to an optative model of the new system. This methodology facilitates communication between stakeholders and system builder as it does not require software development expertise and can be easily used in domain terms. Abstractions and partitioning of the system into subsystems with clear boundaries make it easier to handle larger projects. However, a major drawback comes from the confusion between modeling the problem that the system is intended to solve and modeling the actual solution. Also in particular for ERS, the timing aspects are mostly invisible in the system model, making later tracing between the system behavior and its requirements a difficult task [75, 76].

Several variants of this approach exist, Structured Analysis and Design Technique (SADT) [77], Structured Analysis and System Specification (SASS) [74], Structured System Analysis (SSA) [78], Structured Requirements Definition (SRD) [79]. SASS is the closest relative of the classic structured analysis technique. SADT is a semi-formal technique supports the formalization of the declarative part of the system, but uses natural language for the requirements themselves. It provides a data model linked through consistency rules with a model for operations. It also uses activity diagrams instead of dataflow diagrams and distinguishes control data from process data. SSA uses a notation similar with [74], but adds data access diagrams to describe contents of data stores. SRD introduces the idea of building separate models for each perspective and then merging them.

Specific Requirements Models for Embedded Systems A wide range of real-time systems encountered in industrial environments, power plants, cars, airplanes, can be modeled and reasoned about as "embedded systems", because of the role of the computing system in controlling a physical process and the integration of the two aspects of "controlling" and "controlled" into a common system [80].

Modeling the requirements for embedded systems is crucial to be able to verify their behavior. Correcting requirements errors, under-/over- specifications, or similar imprecisions later in the development cycle can be extremely expensive [81, 82]. "The importance of determinism cannot be overestimated; deterministic systems are one order of magnitude simpler to specify, debug, and analyze than non-deterministic ones." [83]. Hence, formal models for specifying the requirements of ERS try to prevent costly errors [43] or that may ultimately lead to accidents.

SCR Tabular notations [84] have been used for decades to specify requirements for readability reasons. The *Software Cost Reduction* (SCR) requirements methodology [63, 28] was introduced for engineers working on the software for embedded systems. It was later refined for complete systems to incorporate both functional and nonfunctional requirements [85, 86, 87]. The method promotes a tabular notation for specifying requirements, a finite state machine model, and special constructs for expressing constraints such as modes, terms, conditions, events, inputs and outputs [63]. The method associates a table for each output, term, or mode class of the specification and enables system decomposition into smaller, more manageable parts.

Faulk's [88] initial formal foundations of this method use various classes of tables as total functions and mode classes as finite-state machines defined over events. There are *monitored* and *controlled* variables and *input* and *output* data items (provided by external devices such as sensors and actuators), where a monitored variable reflects the effect of the environment on the system behavior and a controlled variable reflects the control of the system on some environmental aspect. Events denote changes of value in the entities forming the system, where input events are trigged by the environment, whereas conditional events may also be triggered by internal system computations.

The Four-Variable Model [85] extends this method to systems by including critical aspects of timing and accuracy as mathematical relations on monitored and controlled variables. For complex systems several mode classes may operate in parallel. [86] introduces another similar abstract model. A specialized form [89, 90] of the Four Variables model is used as formal foundation for a tool suite [91, 92] consisting of a specification editor to create and maintain specifications, simulator for symbolically executing the specified system, automated consistency checker [93], and verifier for critical properties such as timing [90, 94]. These tools enable the developer to ensure proper syntax, type correctness, completeness of variable and mode class definitions, mode reachability and proper setting of initial values in all modes, disjointness (i.e., unique defined entities), coverage and acyclic dependencies.

The CoRE methodology [95] tries to address the shortcomings of its SCR ancestor, namely the lack of structuring mechanisms for variables (e.g., aggregation or generalization), models (e.g., and/or decomposition), and tables (e.g., refinement relationships). [96] proves the scalability of the approach in the context of large-scale avionics systems. [97] provides a practical comparison between SCR and CoRE within the context of a flight guidance system.

Requirements State Machine Language (RSML) [98, 99, 100] is a formal state-machine based hybrid approach using both tabular and graphical notations borrowed from Statecharts [101]. It introduces boolean tables and guards to describe state transitions in one or more high-level state machines that can communicate directly with each other. RSML tables describe transition conditions based on input events and may generate as result output events. Modes are defined explicitly as functions of input variables. The approach employs a state-based black-box model for all system components and their interfaces, which separates the specification of requirements from design aspects and enables formal analysis of the entire system its correctness and robustness [98].

[102] has a similar approach with tables and state machines but uses trace semantics for system analysis. Other specification languages such as Statemate [103], Hatley [104], Ward [105], include various models, yet not all of them are formally defined to enable automatic analysis and behavior verification. ProCos [106] provides a similar language but uses process algebra for the system model.

UML for Embedded Systems – UML can be used at different levels of the development process, especially for requirements modeling and functional design [107]. The high-level models of the system specify the requirements for behavior, domain structure, and QoS properties. The advantage of UML is its capability of modeling both system structure and behavior, specifically the structure of the problem domain and the interaction and collaboration between different agents in the system.

The profile mechanism in UML allows to define families of languages targeted to specific domains and levels of abstractions. For example, [108] presents a UML profile for a platform-based approach to embedded software development using stereotypes to represent platform services and resources that can be assembled together. Standardization activities under OMG include SysML [109] and MARTE [110], a new UML profile for modeling and analysis of Embedded Real-time Systems, in addition to the existing UML profile for Schedulability, Performance and Time [111]. UML currently supports the specification of timing and performance requirements, and could be extended to support also other QoS requirements such as for power consumption and cost.

Several embedded systems require more than one model of computation to reflect the nature of the application domain, whereas UML supports only event-based models. Therefore, several proposals have been made to extend UML: [112] introduces support for continuous-time by using stereotypes to represent continuous variables, time, and derivatives; [113] extends UML with a programming language for hybrid systems; and D-UML [114] introduces a dataflow mechanism (distinguishing between signal ports and data ports) coupled with mathematical equations in UML/Realtime.

SysML [109] customizes and re-uses a sub-set of UML concepts for systems engineering applications. It tries to be a cross-domain solution for modeling entire systems, without making domain-specific description languages obsolete. The SysML "block", which abstracts the software details in UML classes, is a significant extension in the direction of modeling complex ERS, where software is just one aspect besides electronics, mechanics, etc. Blocks can be used to decompose the system into individual parts, with dedicated ports for accessing their internals. SysML also adds requirements modeling as a key aspect of the system development process. It provides requirements diagrams, tree structures, or tables, which not only support the documenting requirements process, but also provide traceability to requirements throughout the design flow, ensuring that requirements are satisfied. SysML groups behavior, structure, analysis, and requirements in a single, integrated system model. It also supports extensions for guarding the information flow and the entities of the system. SysML is an improvement over UML in that it allows to articulate requirements concerns relevant at the system engineering level, including function networks, and requirements allocation to subsystems. However, both UML and SysML lack the binding to a concrete system model that enables formal analysis of requirements and their associated models. Also, there is still too little support for a seamless transition between requirements development and other development activities.

7.2.2 Programming Models

The observable behavior of the ERS is greatly influenced by the underlying programming model used for their construction, which plays a significant role in engineering the system requirements. High-level requirements are decomposed into requirements for individual software components according to the constraints supported by the programming model. For example, the requirement that a vehicle must stop within a given time frame since the driver pressed the brakes may translate into deadline requirements for several tasks and messages. Hence, the interaction between the ERS and its environment is governed by two different views over the notion of *time*: the stakeholders provide requirements in terms of *environment time*, whereas the system is implemented in terms of *software time*. The environment time represents the continuous time flow observable from the external environment of the ERS (i.e., wall-clock time). On the other hand, the software time is a discrete time flow of the ERS itself measured by the number of occurrences of some events such as the pulses of the CPU clock. [115] identifies three real-time programming models: synchronous, scheduled, and timed model.

The Synchronous model assumes that the ERS performs all computation and communication instantaneously [83, 116], and can always keep pace with the environment. This assumption imposes great constraints on the system requirements as an infinitely fast computer is not achievable in practice. Hence, verifying the ERS model through simulation may fail to show that, in practice, the response time of the implemented ERS may still be far from "atomic" and present output jitter.

Depending on the functional requirements of the ERS, existing synchronous languages can be classified under two categories: control-flow and data-flow oriented. The control-flow oriented languages are also imperative languages and are adequate for control-intensive applications such as communication controllers, real-time process control. Esterel [83, 116] has high-level, modular constructs that lead to a real structure of reactive programs based on the semantics of the finite-state Mealy machine. Statecharts [101] has a graphical formalism and it is not fully synchronous. Argos [117] simplifies the formalism of Statecharts and provides full synchrony. The data-flow oriented languages (also known as declarative languages) are appropriate for data-intensive applications such as digital signal processing and steady stream process-control applications. Lustre [118] is a declarative language that supports only the data-flow systems that can be implemented as bounded automata-like programs.

The synchronous approach is used in modeling tools such as Scade [119, 120], which supports the development of real-time controllers on non-distributed platforms or distributed platforms like the Timed-Triggered Architecture [121]. The Scade suite supports the design of continuous dataflows (based on Lustre) with discrete parts realized by a state-machine editor (based on Esterel). The computational models are compatible by transforming values and signals [119]. The Scade Suite is used by Airbus for the development of the critical software embedded in several aircrafts.

The Scheduled model relies on the classical scheduling theory for real-time programming. Functional requirements can be easily accommodated as the ERS may be implemented using sequential languages (e.g., C/C++), or a parallel programming language (e.g., Ada, Occam, CSP, RT-Java). Sequential languages lack concurrency and require a real-time operating system (RTOS) for inter-program communication and synchronization. Parallel languages support concurrency and communication as first-class concepts and typically have specialized run-time support systems. In this model, the software time is no longer an abstract notion equal to zero, but an unpredictable run-time variable influenced by the CPU speed, scheduler, utilization level, etc. Hence, schedulability analysis is necessary to guarantee that all computations complete in the allocated time.

There exist several UML compliant modeling tools that support code generation to C/C++, Java, Ada, different RTOS, and CORBA. UML provides a modeling framework for architecture description and behavior descriptions, but it is still work in progress to properly include real-time aspects in UML 2.0. The first step was made with the UML Profile for Schedulability, Performance, and Time [111]. Predictability analyses include the control flow analysis for sequence diagrams.

The Timed Model abstracts from ERS platform and the software time is always equal with the environment time, such that all computations and communication activities take a fixed logical amount of time, assuming that there is enough soft time to perform the computation under the real-time constraints imposed by the environment. The compiler of the specification language has to verify the time-safety of the computation and guarantee that there is enough software time to complete the computation before its deadline.

There are just a few examples of languages supporting this model, most of them based on the *Logical Execution Time* abstraction introduced by Giotto [122] – a task has a release time when it reads the inputs, and a terminate time when it provides the outputs to the environment. Within this time-span, the way the task executes on the target platform is irrelevant for the environment. Giotto is a high-level time-triggered language, which decouples the timing and functionality aspects, and abstracts from the execution platform. As a metalanguage, it describes the intended temporal behavior of a system and expects its functionality as being externally implemented in a general-purpose programming language such as C, Oberon, or Java. XGiotto [123] extends Giotto to support event-driven programming, while preserving the benefits of the timed-model with fixed response-time.

The Timing Definition Language (TDL) [124, 125] adds component support and abstracts from the distributed platforms. It provides a complete tool-chain fully integrated in the Matlab/Simulink suite. The developer can model the functional aspects of the ERS in Simulink and the timing aspects in the integrated TDL visual editor. The ERS can then be verified through simulation, which based on the timed model provides an accurate representation of the ERS behavior. The TDL compiler is extensible through plug-ins such as the bus schedule generator that enables automatic scheduling [126, 127] of the communication in a distributed system. Hence, components can be developed independently regardless of their distribution – this is the so-called transparent distribution [128] feature of the language that preserves the time and value determinism of the application regardless of how its components are deployed in a distributed solution.

7.3 Requirements Engineering Approaches: Processes and Practices

In the preceding two sections we have established why requirements engineering for ERS is inherently difficult, and have surveyed some of the techniques and tools available in the literature to address this difficulty from various angles. From this overview it becomes clear that much progress has been made and many research challenges remain in this important field. In particular, there is "no silver bullet" in sight, nor is there one to be expected. Requirements engineering demands a holistic view on the problem at hand to address the challenges we have brought forward in Section 7.1. In this section, we recall a few of the practices that have emerged in collecting requirements for ERS, with an eye on opportunities for building precise models that can be used throughout the development process. We do not attempt to give complete account of requirements engineering; we refer the reader to [129] for a comprehensive review. Instead, our aim is to draw attention to a few practices that, based on our experience, are particularly valuable for ERS projects.

7.3.1 Requirements Development and Management

It is important to recall that ERS are embedded into a container product; consequently, the requirements engineering process for ERS is embedded into and has to interact with the overall systems engineering process for the container product. This places constraints at the timelines in which requirements engineering for the ERS can occur, determines when requirements artifacts must be delivered into the overall process, and often provides a significant amount of context requirements for the interaction between the ERS and the rest of the system.

Even when not articulated explicitly, requirements play a central role throughout the development process of ERS. Following [129] we distinguish between *requirements development* and *requirements management*. Requirements development refers to all activities that lead to establishing a *requirements baseline* agreed-upon by the project's stakeholders. The baseline describes, as tightly as possible, the original understanding of all project participants about what the system to be built is. The requirements management process then starts from the baseline, and includes all activities required to respond to changes to that baseline. Its major activities include [129]:

- Define a change control process, including a Change Control Board (CCB)
- Maintain change request history
- Assess impact of change requests, and requirements volatility analysis
- Update of requirements baseline per CCB-decisions
- Establish versioning and change management tools

Explicit requirements management has the advantage that phenomena such as requirements creep (more or less sublime addition of requirements without allocation of new resources for their analysis and implementation) and requirements thrashing (a constant barrage of more or less meaningful change requests) become more transparent to all stakeholders, and can thus be addressed at the management level.

The distinction between development and management is important, because it draws explicit attention to the fact that requirements change needs to be explicitly managed throughout the development process.

Note that the core activities involved in requirements development are independent of the type of development process chosen. Any development process, be it plan-driven or agile [130], needs to find out what the system to be built is. The only difference is the value the respective process types place on formal documentation of the requirements, and how frequently the change process is triggered and executed.

A vast set of techniques has been developed and promoted to develop the requirements baseline. [129] identifies *elicitation, analysis, specification,* and *validation* as the core requirements development activities. Elicitation refers to activities that produce requirements from domain analysis and stakeholder interactions. Analysis refers to the elaboration, refinement and structuring of the requirements previously elicited with an eye towards building high-level design models that establish context for the requirements; this, again, occurs with stakeholder involvement. Specification refers to the prioritization and documentation of the analyzed requirements for transition to establish the requirements baseline. Validation refers to the inspection and testing of the specified requirements before they enter the baseline.

In practice, of course, requirements development is a highly interactive, iterative process, in which elicitation, analysis, specification and validation interleave. Depending on the overall systems engineering process of the product into which the ERS is embedded, these activities also interleave with and are informed by the activities of the overall systems engineering process.

In Section 7.1 we have seen that many requirements aspects for ERS are *cross-cutting* in the sense that they affect not only one component of the resulting system, but relate to an entire network of components. Note that this is *not* primarily a result of unnecessarily distributed architecture design (although this can be a cause as well.) Instead, this phenomenon arises from the inherent complexity of multi-functional systems, where hundreds to multiple thousands of software functions need to be offered and harnessed into a system of systems. Any decomposition of these functions into components will lead to some form of cross-cutting. Failure management is a prime example: no matter how the set of functions is sliced into logical or physical components, failures cannot be effectively managed from within these individual components – communication across components needs to occur to communicate failures, or to take remedying actions.

It is our belief that requirements engineering for ERS necessarily focuses on the *interplay* of the entities that make up the system, and the associated cross-cutting concerns. The rationale behind this is simple: because of the embedded nature of an ERS there is interaction between the ERS and its environment. Therefore, these interactions need to be understood to the maximum extent possible. Furthermore, for all but the most trivial systems, the ERS will itself decompose into a set of interacting components, each of which can be understood as an embedded component as well. Consequently the same rationale applies for the development of the ERS as a unit. In a networked system of ERS all quality properties of the system emerge from the interplay of all constituent ERS. Hence, the cross-cutting concerns that are crucial to defining the overall system's quality are naturally associated with the interplay of the constituent ERS.

Furthermore, we believe that to properly address the requirements aspects enumerated in Section 7.1, *explicit* domain models that speak to these concerns need to be constructed. As we will see in the case study of Section 7.4, creating such explicit domain models enables formal end-to-end analysis at the system of systems integration level – as opposed to the component-by-component level.

Therefore, we see the key activities in the requirements development process to bring out a sufficiently detailed domain model for ERS as follows:

(1) Identify the stakeholder group for the ERS under consideration.
(2) Identify pertinent business and process constraints for the ERS per stakeholder class.
(3) Identify the set of functions expected of the ERS by the stakeholder group.
(4) Identify the internal and external actors and data entities involved in these functions.
(5) Identify the interactions (event-, message-, control-, and data-flows) among the identified actors.
(6) Iterate over the identified functions to identify the actors and data entities needed to address the relevant cross-cutting concerns. Associate these with the interaction model built in activity 5.
(7) Identify operational infrastructure constraints, including mandated deployment contexts.
(8) Document requirements relative to the resulting models of structure and behavior.
(9) Validate requirements based on the resulting models.

Clearly, each of these activities breaks down into a variety of sub-activities and associated techniques; here, we focus on a high-level overview of these activities.

Activity 1 is critically important as per the definition of the term requirement we have given in Section 7.1. Recall that requirements are intimately linked to stakeholder values and, therefore, the set of stakeholders whose values the system is to address needs to be fully understood and modeled explicitly, sometimes via proxy elements, such as sensors.

Activity 2 serves to bring forward requirements aspects that are often neglected initially, and later turn out to be major success factors. This includes an articulation of the cost model that underlies the development process for the ERS and its container system. This is particularly important, at the integration level if the ERS is part of a system of systems. Similarly, this is the place to identify process requirements and laws or other regulations that govern the development of the ERS and its container system. This can influence the resulting domain models by creating data entities, actors or cross-cutting concerns that need to be further analyzed for functional and quality requirements.

Activity 3 is facilitated by a wide variety of techniques, such as use case [131] or user story analysis, stakeholder focus groups, flow analysis (event-, message-, control-, or workflow). For ERS we find it particularly useful to hold focused stakeholder workshops, within and across stakeholder groups to bring out not only per-component, but also across-component requirements. This is particularly critical for end-to-end and cross-cutting requirements aspects such as timing (specifically deadlines or time-budgets), failure modes and management, security, policy/governance, and deployment, update, and maintenance requirements. In these workshops we typically execute steps 4 through 9 together with the workshop participants to create initial domain model candidates on the spot.

Activities 4 through 7 amount to developing an ontology [132] and behavioral model of the core concepts that make up the domain model for the system under consideration. For an ERS this necessarily includes a model of the environment into which the ERS is placed. For the structural aspects of this domain model we favor class diagrams capturing the actor and data classes and their structural relationships. For the associated interaction model, we favor Message Sequence Charts (MSCs) and related interaction specification dialects such as Life Sequence Charts (LSC) [133], which can be augmented with constraints that reflect the cross-cutting concerns (see examples in Section 7.4).

A key observation is that the domain model should provide explicit hooks to associate the cross-cutting concerns with the interactions identified for the domain entities. This ensures (a) that the cross-cutting concerns are in the purview of the project team from the earliest stages as end-to-end aspects, rather than becoming an integration-afterthought, and (b) makes the cross-cutting concerns available for explicit validation and verification, rather than being an implicit, inaccessible aspect of the requirements model.

All entities mentioned in a textual description of a requirement should occur in the resulting model, and for each modeling entity there should be at least one requirement to which they are related.

The domain model is of such paramount importance, because we can derive a variety of other models from it, and use all models together for validation and verification. Derivative models include, for instance, a context-diagram, which shows the system entities *outside* the ERS under development, and what the structural and behavioral relationships between the two are. Furthermore, a useful domain model will capture the operational modes (high-level state transition view), major exceptions and failure modes, and the input/ouput protocols required at the interface of the ERS and its environment. It is central to the success of this exercise that it results in a model that captures the entities and relationships relevant to the problem domain and its associated stakeholder groups. This greatly facilitates validation and verification, as well as the derivation of design and implementation.

Of course, this modeling effort depends on a deep (and deepening) understanding of the problem domain. In recent years, there have been important attempts to help in building this understanding by providing catalogs of *requirements patterns* both ERS-specific, and domain-neutral. For instance, [134] have identified a catalog of ten requirements patterns that address the following concerns:

- *Controller Decompose Pattern:* decomposition of an ERS into subsystems according to responsibilities
- *Actuator-Sensor Pattern:* relationships among sensors, actuators, computational components and associated (environment) models
- *Examiner Pattern:* device monitoring and error logging
- *Fault Handler Pattern:* core entities and models for handling faults in ERS
- *Mask Pattern:* resource mediation for devices with many sensors and actuators
- *Moderator Pattern:* decoupling
- *User Interface Pattern:* reusability and flexibility for user interfaces associated with ERS
- *Channel Pattern:* communication facilitation among components
- *Monitor-Actuator Pattern:* fault management for actuators

Each of these patterns, among others, comes equipped with an explanation of the intent, motivations, constraints, applicability, entities and their structural and behavioral relationships.

Similarly, [135] presents a set of performance-related requirements patterns that are relevant for ERS. This includes patterns for response time, throughput, static and dynamic capacity (memory, computational power), and availability.

Besides these ERS-relevant requirements patterns, [135] also brings forward a rich set of more generic templates. These cover technology choices, standards compliance, inter-system interfaces, data typing and archiving, reporting, flexibility, and access control.

We call out activity 6 explicitly, because it is key to obtaining comprehensive requirements models for ERS. For each identified function of the ERS, the requirements pertaining to all quality properties [129] (availability, efficiency, flexibility, integrity, interoperability, reliability, robustness, usability, maintainability, portability, reusability, testability, security, safety, deployment, update, maintenance) should be iterated over to derive specific requirements that pertain to these qualities. Again, all of these qualities are cross-cutting in nature, and intimately linked to interactions among the identified system entities, or to the container system. Many of these qualities can thus be addressed at the integration- rather than the per-component-level. Automotive manufacturers and suppliers, for instance, have recognized this and are working together to provide a car-wide "middleware" that addresses some of these qualities *across* the components of the vehicular ERS networks.

Activity 7 serves to identify requirements that derive from the technical context of the ERS. Often, the technical infrastructure into which an ERS has to integrate is fixed long before the ERS proper is conceived. Then, this technical infrastructure injects deployment constraints into the ERS requirements set. In a clean-slate development, of course, one would seek to avoid this, or at least design the technical infrastructure after the integration requirements are sufficiently understood. In reality, however, legacy technical infrastructures exist and have to be considered. This interrelates with activity 6, of course, because some of the cross-cutting concerns may be discharged by the technical infrastructure if the latter is functionally rich enough. In any case, the capabilities of this infrastructure need to be carefully examined so as to know which of the cross-cutting concerns need to be lifted explicitly into the requirements model, and which ones are readily dealt with in the infrastructure.

Activity 8 refers to articulating the gathered requirements and their associated domain models in the form chosen by the project or mandated by a process requirement. For ERS this typically involves writing a *requirements document* that defines the scope, stakeholders, context, and all business, product, and process requirements elicited as part of the previous process activities. Discussion of an elaborate requirements document outline is beyond the scope of this text; we refer the reader to [129] for an example. However, we note that the material gathered in the previous activities typically provides a rich and authoritative source for this documentation activity.

Activity 9 can build on the models created in the preceding activities. The typical methods practiced for validation today are inspection, prototyping and simulation, automated consistency checking and verification. Each of these is facilitated greatly by detailed requirements models, as well as by broad stakeholder participation.

In reality, of course, all these activities will occur in an iterative, often interleaved fashion, rather than being executed in a prescribed sequence. The product of executing these activities, however, is a comprehensive requirements model for the ERS under development.

7.4 Example: Failure Management in Automotive Software

In the automotive domain, software has become the enabling technology for almost all safety-critical and comfort functions offered to the customer. The features supported by automotive software and electronics are increasingly dependent on the interactions of distinct components designed by different suppliers. Because of the increasing level of interaction between different components, industry standards, including OSGi [136] and AMI-C [137], introduce service based software-architectures and corresponding middleware layers as modeling and deployment abstractions. This marks a significant shift from component- to service-oriented software development in the automotive domain.

A major technological advantage of a service-based vehicle-electronics software architecture over a traditional component-based one is the ability to move the hardware-module-oriented partitioning of the vehicle system to a later point in the design cycle, allowing greater flexibility in integrating functions into hardware and potential elimination of redundant hardware across the vehicle. To exploit this advantage it is desirable to be able to model the vehicular software architecture on multiple levels, from static models of software structure to executable, time-accurate models of the actual system. This, in turn requires specifications for services that are sufficiently formal to allow tools to be built that check the integrated architecture for consistency and completeness, and to allow modeling tools to use the service-oriented specifications directly.

In the following, we illustrate the applicability of a service-oriented approach to model parts of the Central Locking System (CLS) found in typical modern cars. The CLS in the described form acts as a representative for similar problems in automotive control electronics and distributed, reactive systems in other application domains. We present aspects of requirements modeling, deriving a corresponding architecture, and performing safety-checking on the system model.

7.4.1 Central Locking System (CLS)

In modern cars, even a simple function such as locking the vehicle, i.e. central locking system (CLS), interacts with a significant number of other functions. There are not only interactions with the obvious modules, such as those controlling the individual door locks, but with less obvious systems as well, such as the vehicle speed sensor (to implement lock on drive away), the exterior lights (for remote lock acknowledgment) and the radio tuner and seat controllers (for setting driver preferences on unlock). The various interacting features in such a system are distributed across a number of different component modules, which are typically produced by different suppliers. As interactions between different subsystems increase, the features themselves become distributed across a number of components. This leads to increasing integration issues as features come to be implemented by software produced independently by a number of different suppliers.

Although we are considering a simplified version of the CLS for our study, it is evident that, given the size and distributed nature of the system, it is practically impossible to describe all the behaviors of all components involved completely. Instead, we only have a partial view on the requirements of the overall system.

7.4.2 Modeling the CLS Requirements

In the previous sections of this chapter, we have suggested a process for eliciting and managing requirements in 9 points. In this section, we present this approach using the CLS example. We demonstrate the use of a Service ADL to capture both the CLS system architecture and a set of dependability requirements along with formal verification techniques to verify the implementation of the dependability requirements.

  1. *Stakeholders Identification* represents the first step of our requirement management process and prescribes the identification of the stakeholders for the system under consideration. In our case, the groups interested in the systems are obviously "the driver and passengers" of the car that will use the given CLS. Other groups come from the car development team, for example, the "engineering team" that designs the electro mechanical actuators for locking and unlocking the car. Each supplier is also a stakeholder that will have to agree on the final integrated design and can impose constraints to other parts of the system. In addition, marketing, cost, safety and legal considerations have great influence in establishing requirements for the vehicle.

  2. *Business and process constraints per stakeholder* is the second step, which mandates to identify for each stakeholder the pertinent class of business or process constraints. In this case, we can analyze the concerns of the "safety and regulations" stakeholder and identify some critical requirements. One of such requirements is that "all the doors of a car shall be unlocked after an accident".

This type of regulation is not detailed enough to be a requirement directly. We first need to have a proper model of the car system and of the CLS to be able to articulate it further.

  3. *Identify functions expected by stakeholder* – to support the previously stated rule, the "safety and regulations" stakeholder assumes that there exists in the system a function to detect an accident and a function to unlock all doors. It is important to notice that there will also be a timing constraint on the time interval between the accident and the unlocking of the car. For example, we can assume that requirement (1) is "the system shall unlock all doors of the car within half a second from the detection of an accident". The problem with such definition is that we need to define how an accident is detected and how reliably. Moreover, during an accident there could be failures in the system, which limit the functionality of the unlocking mechanism. This requirement could then be complemented by requirement (2) stating that "even if one electronic control unit of the car completely fails in an accident requirement 1 must be fulfilled".

  4. *Identify actors and data for the functions* is the fourth step, where we analyze the requirements and functions identified so far, which leads to a number of use cases and actors. We identify the actors that participate in the services of the system under development, abstract from the concrete system elements and identify the communication roles. These roles will likely map to a variety of different component configurations depending on the concrete make and model under consideration. For instance, in a concrete implementation, the central controller (Control) and the lock management (LM) might end up on the same

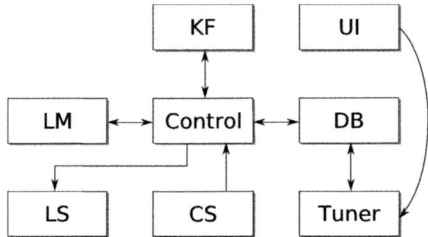

Fig. 7.1. Components and relationships in the CLS example

ECU, whereas the database (DB) and the lighting system (LS) might reside on others. Figure 7.1 depicts a simple configuration with each role being implemented by a system component. We indicate components using labeled boxes, and directed communication channels between them using labeled arrows. In a real car, most of these entities would be implemented on different ECUs (KF being a likely exception).

5. *Identify interactions among actors* – the starting point for this step is analyzing the set of relevant "use cases". In our case, we use message sequence charts to capture the identified use cases. Some of the use cases for the CLS are: locking, unlocking, lock_doors, unlock_doors, transfer_key_ID, and handle_crash. For reasons of brevity, we consider only a subset of these services here; we refer the reader to [138] for details.

transfer_key_ID is part of the unlocking process and associates seat and mirror positions, as well as tuner settings with the driver's key. handle_crash is a crosscutting service that can interrupt all others, it captures the functionality that whenever a crash signal occurs the CLS has to unlock all doors.

While both the unlocking of the car and the transfer of a key ID are triggered by the user pressing a key on the key fob, we consider these two use cases separately because there exist keys that can unlock the car (mechanically, for instance) but do not transmit key identifiers. Therefore, separating use cases and corresponding requirements enable more modularity and reuse across different models of cars.

To capture the interaction patterns defining services we use an extended version of Message Sequence Charts (MSC) [139, 140]. MSCs have proved useful as a graphical representation of key interaction protocols, originally in the telecommunications domain. They also form the basis for interaction models in the most recent rendition of the UML [141]. In our extended MSC notation, each MSC consists of a set of axes, each labeled with the name of a role (instead of a class or component name). Roles map to components in a later design step of the development process. An axis represents a certain segment of the behavior displayed by the component implementing the corresponding role. Arrows in MSCs denote communication. An arrow starts at the axis of the sender; the axis at which the head of the arrow ends designates the recipient. Intuitively, the order in which the arrows occur (from top to bottom) within an MSC defines possible sequences of interactions among the depicted roles. We also use labeled boxes in

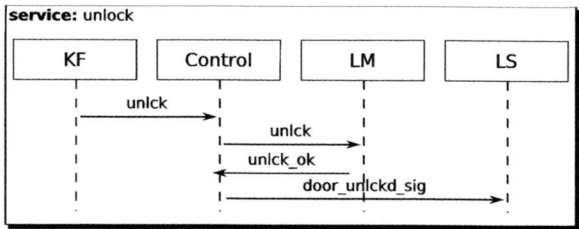

Fig. 7.2. MSC for "unlocking"

our MSCs to indicate alternatives and unbounded repetitions. High-level MSCs (HMSCs) indicate sequences of, alternatives between and repetitions of services in two-dimensional graphs - the nodes of the graph are references to MSCs, to be substituted by their respective interaction specifications. HMSCs can be translated into basic MSCs without loss of information [140].

Figure 7.2 shows an example; here we depict the interactions defining the "unlocking" service. It consists of a triggering message "unlck" from the key fob to the central controller. The latter forwards the "unlck" message to the lock management (LM). By introducing the LM role we abstract from the concrete number of locks present in the vehicle (doors front/back, trunk, moonroof, windows, security system, etc.). When the locks have been operated, LM returns an "ok" message to the control role. Upon its receipt, the control role issues a "door_unlckd_sig" message to the lighting system role, which handles the signaling of the locks' states to the driver. Clearly, this is just one course of actions that may happen during the execution of the unlocking service. The extended MSC dialect we use enables succinct specification of such alternatives [140, 142].

The next use case we turn into a service is "transfer_key_ID". Upon receipt of an unlck message the control role sends a getID message to the key fob; KF sends the id to Control, which relays it to the DB (see Figure 7.3). Again, Control switches from state LCKD to UNLD in the course of executing the service. The preceding two services are overlapping in the sense that both share references to the unlck message and states LCKD/UNLD. To compose these services into an overall service specification we have to identify the overlapping messages, and "synchronize" the execution of the services on these joint messages.

6. Identify elements to address cross-cutting concerns – along the main functional requirements for CLS we have also identified a cross-cutting requirement. In case of an accident, all doors need to be unlocked immediately. This concern comes from safety regulations that cars need to fulfill. Even if this concern is not part of the normal functions performed by CLS, it imposes a new behavior that interacts with the normal locking and unlocking behavior previously defined. Therefore, we need to identify structural elements and messages that are affected by this behavior. Moreover, we need to identify when the new behavior appears.

The handle_crash service has a particularly simple interaction pattern (see Figure 7.4): whenever the control role receives an "impact" message it responds by sending "unlck" to the lock management role, resulting in the unlocking of the

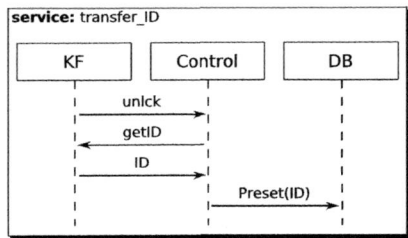

Fig. 7.3. MSC for "transfer_key_id"

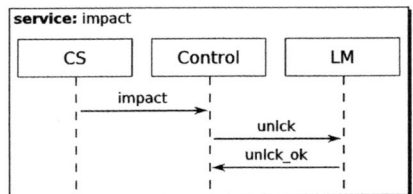

Fig. 7.4. MSC for "handle_crash"

vehicle. Methodologically this can also be handled by introducing a "preemption" concept that treats the response of the control role as the handling of a preemption triggered by the "impact" message.

7. *Identify operational constraints* – the ERS domain is characterized by tight timing constraints that can originate from several requirements. In our case study, we can consider, for example, the time constraints implied by the emergency unlocking requirement. We can capture such information in our service models using a modified MSC syntax.

Figure 7.5 shows the *unlock* function. The graphical syntax we use is derived from MSCs as described in [140, 142]. Upon receipt of the *unlck* message from KF, Control issues an *unlck* message to LM. Once LM acknowledges this with an *ok* message, Control requests signaling of the unlocking from LM by means of a *door_unld_sig* message, then returns *ok* to the keyfob.

The MSCs of Figure 7.5 is augmented with interaction deadlines, indicated by means of a labeled dashed line. The *unlock* function has a deadline of 150 ms. This means that the vehicle must be unlocked and the signaling must have occurred within 150 ms according to the interaction specification.

The deadlines we introduced in the MSC represent additional constraint that enable capturing QoS requirements directly in the service models.

8. *Document requirements relative to the models* – in our requirement elicitation process we also develop deployment models with at least a partial view of

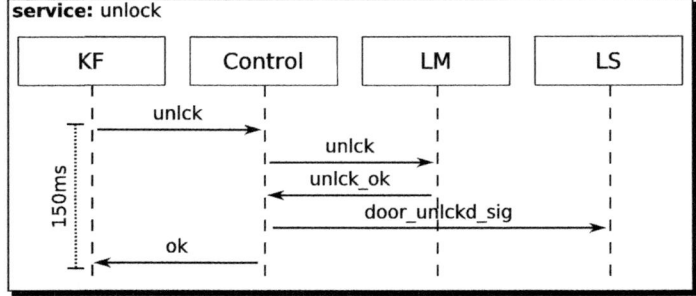

Fig. 7.5. A version of the unlock MSC with a QoS requirement added

the deployment environment. In our case study, such model is useful to identify possible failing components and ensure that the critical requirement of emergency unlocking all doors is fulfilled even when some component fails.

In our case study, we create a component model defining the ECUs that will run the CLS and the communication networks used to deliver messages to them. The behavior of each component is defined by assigning it one or more of the roles identified in our service models. This step of mapping the logical services to a concrete deployment model, makes the outcome highly specialized to the vehicle under design. On the other hand, the same functionality is often needed across vehicle platforms; this is certainly true for the CLS, which today is a standard feature across manufacturers and product lines. Therefore, the mapping process has to be repeated again to yield another specialized solution for each target platform. Because the requirements are clearly separated and we distinguish between one logical model and a deployment model, only the part of the work that deals directly with the deployment model has to be repeated while developing different car models.

The outcome of a traditional process would be eight separate component specifications; each individual component specification is complete in the sense that it has to address all the different functions the component in question might be involved in. In particular, the crosscutting nature of the functionality is lost when we look at each individual component; this results in the mentioned labor- and cost-intensive integration effort in late development stages.

We can obtain a trivial deployment domain model from the role domain model by removing the distinction between components and roles; then, each component implements precisely one role. In this state of affairs, the role domain model and the deployment domain model coincide. Another extreme case is to map all roles to a single component; this again is a trivial affair, because we simply need to treat the role domain model as a specification for the "internals" (the substructure) of one encompassing component. The most interesting and methodologically challenging case arises when we map multiple roles onto the same deployment component. All other cases (such as mapping a single role onto multiple components) can be dealt with by refactoring / refining the role domain model first, and then establishing the mapping to the deployment domain model.

In our case study, we choose to have six ECUs where we map the roles identified in our process. Figure 7.6 shows the corresponding deployment domain model. Our ADL enable us to specify communication busses (the big CAN BUS block in the middle of the figure), and electronic control unit connected to communication media (the six ECU blocks). An ECU can perform more than one role. For example in Figure 7.6 ECU1 plays the role of Control and DB, and ECU2 plays the role of UI and Tuner. On the other hand, the same role can be played by more than on ECU. This is the case of the CS role played both by ECU5 and ECU6. The reasons to replicate a role can be multiple. In the case of CS (the crash sensor role), the replication enables the detection of a crash even if one of the sensors fails.

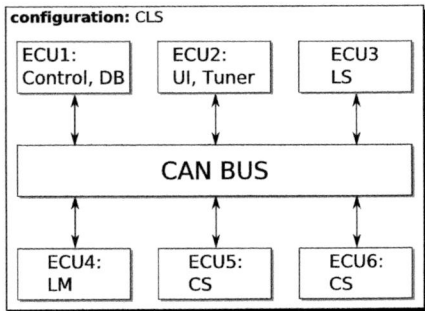

Fig. 7.6. CLS Deployment Architecture

If we work with strictly hierarchical component models such as the ones of UML2, UML-RT, or AutoFocus [143], one way to establish the mapping of multiple roles onto a single component is to take the role domain model as a staring point, and to replace the roles in question by a single component having the same input and output channels as the replaced roles taken together. Then, the entire network of replaced roles with their supporting channels becomes the hierarchical "child" of the freshly introduced component. This process can be repeated recursively into all hierarchically decomposed composites, until all role labels have been turned into component labels.

9. *Validate requirements* – failure management is particularly effective if it is performed throughout the development process[144] – rather than, as often happens, as an afterthought. For this reason, we raise awareness of failures already from the very early phases of the software and systems engineering process, during the requirements gathering phase. To this end, we have created a comprehensive taxonomy for failures and failure management entities. Failure taxonomy is a domain specific concept [144]. Our model-based failure management approach [145], leverages the interaction descriptions captured by services to identify, at run time, deviations from the specified behavior.

We enrich our standard service-oriented methodology with special services to manage failures. Hence, a key mechanism for dealing with failures is to define and decouple *Unmanaged* and *Managed* Services (see Figure 7.7). The Unmanaged Services are responsible for providing the required functionalities without considering failures, whereas the Managed Services enable the detection of failures and the implementation of mitigation strategies that avoid, or recover from, failures.

We also employ two special types of Services: *Detectors* and *Mitigators* (similar to the detector/corrector approach [146]). A Detector can detect the occurrence of a Failure based on its Effect (see Figure 7.9. This relation binds the Detector to the observable results of failures. Therefore, it is important to define what type of Effects a failure can have, and then to create appropriate Detectors.

The Detector detects the possible occurrence of a failure based on a *Detection Strategy*. One possible Detection Strategy is based on *Interactions*. In this case, a Detector compares the communication patterns captured in the service

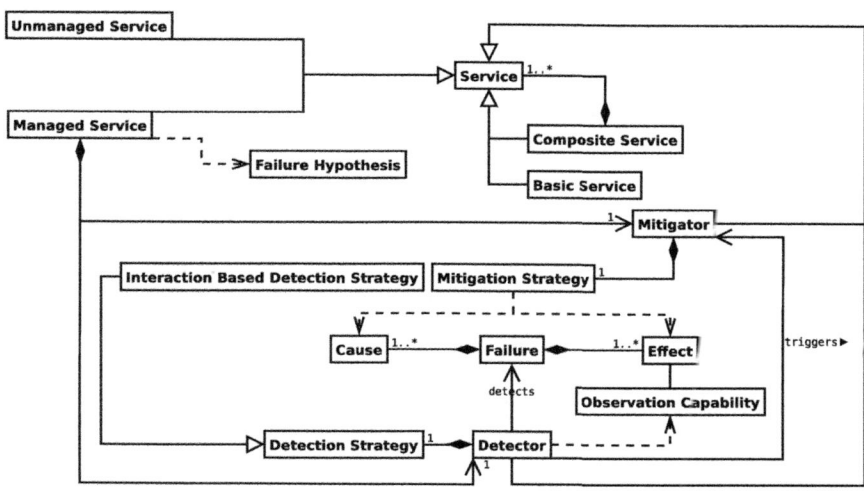

Fig. 7.7. Models of services

specification with the ones of the running system; then, it applies a mitigation strategy when behaviors don't match the specification. Mitigators are services that modify the interaction pattern of the system to recover from failure conditions.

Managed Services are a type of Services and, therefore, they can also be a component of a Composite Service. In particular, it is possible to have Managed Services that are composed of other Managed Services. Each one of them will have a Detector and a Mitigator that will address failures at its level. Using this schema, by hierarchically composing simpler services in more complex ones, and by adding Detectors and Mitigators to the various component services, it is possible to achieve a fine level of granularity in managing failures.

Each Detector is associated with a corresponding Mitigator. Upon detection of a failure, the Detector activates the corresponding Mitigator responsible for managing that specific failure. A Mitigator is another specific Service that is responsible for resolving the faulty state in order to maintain the safety of the system. A Mitigator applies its corresponding Mitigation Strategy to resolve the faulty state. Following the strategy pattern, decoupling the definition of the mitigation strategy from the entity that applies it provides flexibility to the model by allowing future changes to the strategy that is applicable to a specific failure without the need to make any additional modifications to other elements in the system.

This model allows us to compose a predefined Unmanaged Service with a Detector and its associated Mitigator in order to add failure management to it, thus, creating a Composite Managed Service. If multiple failures are supported for one Service, it will be wrapped in multiple layers of Detectors and Mitigators. This capability provides a seamless means to manage the failures that are found in further iterations of the design/development process, without redefining the existing Services. Figure 7.8 shows an example of a managed service for our case

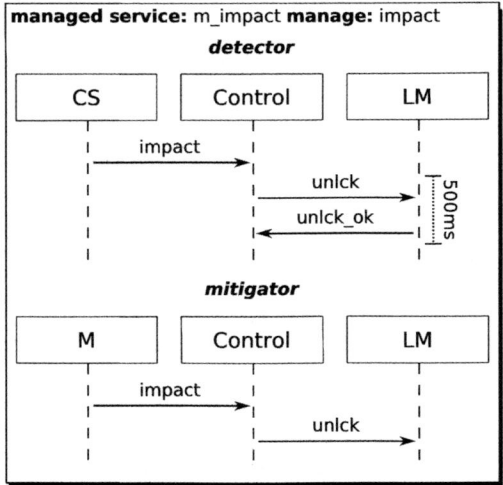

Fig. 7.8. Managed impact service

study. In the ADL fragment depicted in this figure, the detector identified if the impact service takes more than 500ms to acknowledge the unlocking of the doors after an impact, and in this case, it executes a mitigation service where an additional mitigation role (role M) repeats the unlck command.

An ontology guides the identification of failures and the activation of additional services that mitigate the effects of failures. We enrich the logical and deployment models typical of any MDA with a failure hypothesis that captures what physical and logical entities can fail in a system. It also provides a formal basis to reason about system correctness in presence of failures. Figure 7.9 shows the extended failure taxonomy using UML2 class diagram notation[147]. It captures the relationships between failures and our means for detecting and managing them. The central entity of this taxonomy is a *Failure*. A Failure has one or more *Causes* and one or more *Effects*. A failure Cause is very dependent on the application domain and could be due to either a software problem, i.e., *Software Failure*, or a hardware problem, i.e., *Hardware Failure*.

When a failure is detected, the system needs to mitigate it. This is done by following certain *Mitigation Strategies*. The Mitigation Strategy we must apply to deal with failures depends both on the associated Effects and their Causes. We identify two main strategies: *Runtime Strategy* and *Architectural Strategy*. Depending on the application domain, when a duplicated message is detected at runtime, *Ignore Message* can be a feasible Runtime Mitigation Strategy. Similarly, when a message loss is detected, *Resend Message* is a candidate Runtime Mitigation Strategy if properly supported by the interaction protocol between the exchanging parties. *Replicate Component* and *Failsafe Mode* are typical *Architectural Strategies*.

Following the outlined approach, we have lifted the management of failures to the logical architecture and started dealing with them from the early stages

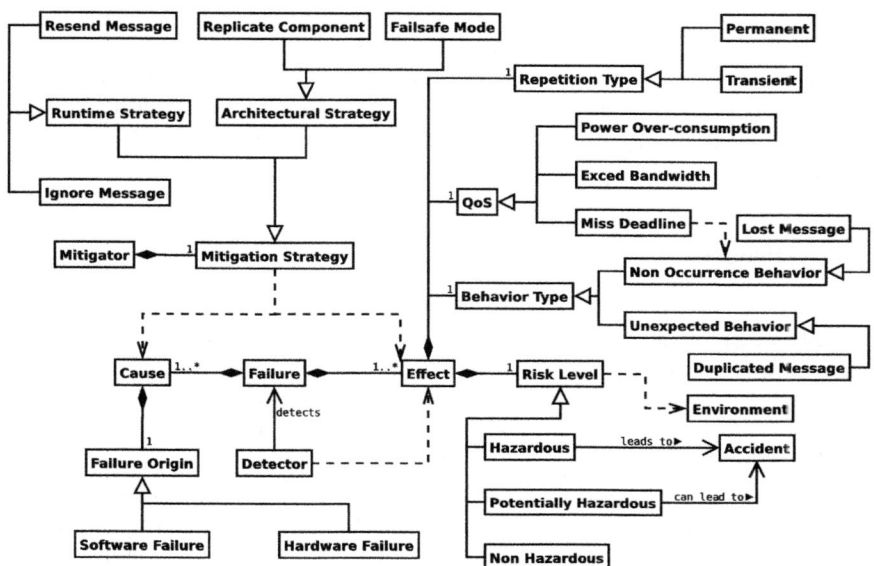

Fig. 7.9. Failure ontology

of the development process within the requirement elicitation phase. Once we have formal models of the services and a deployment architecture, along with a failure hypothesis, we can use a set of tools we developed to verify that the proposed software architecture indeed fulfills the given requirements.

The first step is to obtain an executable model form the services captured by our ADL – in [140] an algorithm to obtain state machines from MSC models is discussed. We have developed a tool [148] that can parse the service ADL and leverage the state machine synthesis algorithm to create a Promela [64] model that can be used to verify the property of a system using the SPIN model checker [149].

Each service is interpreted as a partial input/output function which defines the contributions of all participating roles to a communication pattern. The tool we have implemented uses all MSCs that define the system model to obtain a total representation of the global behavior of the system. Then it projects all messages sent or received by one role on a state machine that defines only the contribution of that role to the interactions of the system. Moreover, to cater for possible failures, the tool adds a sink state with guarded transitions from all other states. As result of this process we have one state machine for each role.

Once the tool has created all state machines for all roles, it can generate the Promela code. For each ECU in the deployment model and for each role mapped to them, the Promela code contains a concurrent process. Appropriate channel variable are used to map the communication channels of the service models to the proper ECUs. Additionally, a failure injector function, implementing the failure hypothesis is created in the Promela code. Failures are injected by killing roles (enabling the transition to the sink state) or disrupting communication channels (removing messages from channels).

Using our Service ADL for managing failures and our Promela code generator we have been able to verify the architecture of our CLS case study and ensure that the chosen architecture supports the safety requirement of unlocking all doors during an accident.

7.4.3 Discussion

In this example, we have seen that services require composition operators not generally available in component-oriented development: the concept of overlapping components is not very common. Roles, on the other hand, by definition capture a partial view on all components playing that role – to be composed with other partial views to produce the overall behavior of the component under consideration. The composition of the services as elicited above translates into a service specification. The mapping from a service specification to a set of components implementing the services in the next phase of the development process is a design step. This step entails fixing a component architecture, and an association between the components and the roles they play to support the given set of services.

In the CLS example, we could decide, for instance, to have just one component to implement the Control and LM (lock management) roles. This gives rise to a component-oriented "deployment" architecture. If the target architecture supports the definition and deployment of individual services, however, we can encapsulate the interaction protocols contained in each of the extended MSCs we have presented, and publish those as individually accessible services within service-oriented software architectures as outlined above.

We can also apply a bottom-up scheme for interaction composition. Deadlines can be applied to basic interactions. For instance, we define a deadline for a single message or a message sequence. For each composition operation, we apply defined rules that constrain the deadlines of the composite interactions. In this case, sequential composition leads to the addition of the operands' deadlines, loops to a multiplication, parallel, and join composition to the selection of the minimum deadline. All deadlines can be tightened manually. A less restrictive composition alternative (in comparison with applying the minimum constraint for join composition) would be to only consider a newly defined deadline for the composite. Doing so would allow the modeler to provide a different interpretation for the more complex composite function – it can be more than the sum of its parts. However, this may not yield a true refinement of the specification in the bottom-up sense, because the composite may not fulfill all QoS properties of the composed interactions anymore. Practical considerations would determine the concrete composition scheme used. We chose the composition variant that maintains all properties of basic interactions and allows for methodological refinement. We are aware that this is more restrictive to the modeler and requires more frequent modifications or refactorings of the specification.

In terms of methodology, we can also apply top-down refinement of deadlines, while still fulfilling all properties of bottom-up composition as described above. Starting from deadlines for entire functions, we allow the modeler to provide specific deadlines to parts of the interaction, as long as the overall deadlines are still satisfiable.

7.5 Summary and Outlook

Requirements engineering for Embedded Real-time Systems (ERS) is a tremendous challenge. In this chapter we (a) have highlighted the key aspects that render ERS requirements engineering difficult, (b) have discussed prominent approaches in the literature that tackle portions of these aspects, (c) have presented key activities that can help in model engineering for ERS across development processes, and (d) have shown how these activities play together to model and validate central failure management requirements in an automotive case study.

Clearly, there is no single technique that addresses the entire spectrum of requirements aspects from timing to distribution to failure management to local and cost drivers. Specifically, because many relevant ERS are, in fact, network-integrated systems of systems, most quality requirements are, in fact, concerns that cut across all components of the integrated system. This necessitates a modeling approach with due emphasis on the interactions among the parts to define the function of the whole system. Such an approach needs to provide models for interactions, but also for augmenting these interactions with constraints that address the cross-cutting requirements.

We have sketched the beginnings of such an approach by identifying key requirements elicitation activities for ERS, and how they can be used to produce structural and behavioral aspects of a corresponding domain model. In specifying the cross-cutting concerns we have identified interaction diagrams, such as extended UML sequence diagrams or Message Sequence Charts as a useful tool. The subsequent case study showed how to exploit this extensive domain modeling approach for the elicitation of domain-specific failure models ranging from logical to deployment architectures. The failure models capture a broad range of failures and associated detection and mitigation strategies. For a subset of these we have shown how to automatically generate [140, 148] verification models targeting Promela/SPIN to establish (or refute) fail-safety of a given architecture model. This technique can be utilized in validating requirements (does the architecture model properly reflect our understanding of fail-safety for the system under consideration?), or even the verification of proposed candidate architectures (do they fulfill the fail-safety requirement?)

This case study shows a pathway to modeling and model exploitation for ERS and can be expanded further to cover an increasingly rich set of requirements aspects. Generalizing from this example to obtain requirements engineering processes techniques and tools for a wide range of specific application domains is one promising area for future research. Another one is the seamless transition from gathered functional and (cross-cutting) quality aspects to re-configurable deployment architectures.

Acknowledgments. Much of the overview of requirements challenges is influenced by our long-time collaborations with automotive partners, as well as academic collaborators. We would like to acknowledge the influences of Manfred Broy, Alexander Pretschner, Bernhard Rumpe, Bran Selic, Wolfgang Pree, KV Prasad, Ed Nelson, Chris Salzmann, and Thomas Stauner. The authors are grateful for discussions with participants of the MBEERTS Dagstuhl Seminar, as well as for the insightful comments of the reviewers of this paper. Our work was partially supported by NSF grant CCF-0702791. Financial support came also from the California Institute for Telecommunications and Information Technology (Calit2).

References

[1] Shaw, M.: Prospects for an engineering discipline of software. IEEE Software 7(6), 15–24 (1990)
[2] Halfhill, R.T.: Embedded market breaks new ground, Embedded Processor Watch, vol. 82 (2000)
[3] Broy, M., Krüger, I.H., Meisinger, M. (eds.): ASWSD 2004. LNCS, vol. 4147. Springer, Heidelberg (2006)
[4] Broy, M., Krüger, I.H., Meisinger, M. (eds.): ASWSD 2006. LNCS, vol. 4922. Springer, Heidelberg (2008)
[5] Ahluwalia, J., Krüger, I., Meisinger, M., Phillips, W.: Model-Based Run-Time monitoring of End-to-End deadlines. In: Proc. of the Conference on Embedded Systems Software, EMSOFT (2005)
[6] Krüger, I., Nelson, E.C., Prasad, V.: Service-based software development for automotive applications. In: CONVERGENCE 2004 (2004)
[7] Pretschner, A., Broy, M., Kruger, I.H., Stauner, T.: Software engineering for automotive systems: A roadmap. In: 2007 Future of Software Engineering, pp. 55–71. IEEE Computer Society, Los Alamitos (2007)
[8] Sharp, H., Finkelstein, A., Galal, G.: Stakeholder identification in the requirements engineering process. In: Proc. Tenth Intl. Workshop on Database and Expert Systems Applications, pp. 387–391 (1999)
[9] Easterbrook, S.M., School of Cognitive, Computing Sciences, University of Sussex: Domain Modelling with Hierarchies of Alternative Viewpoints. University of Sussex, School of Cognitive and Computing Sciences (1992)
[10] Anderson, J., Fleak, F., Garrity, K., Drake, F.: Integrating usability techniques into software development. IEEE Software 18, 46–53 (2001)
[11] Bevan, N.: Usability is quality of use. Advances in Human Factors Ergonomics 20, 349 (1995)
[12] Mayhew, D.J.: The usability engineering lifecycle. In: Conference on Human Factors in Computing Systems, pp. 147–148. ACM, New York (1999)
[13] Bennett, J.L.: Managing to meet usability requirements: Establishing and meeting software development goals. Visual Display Terminals: Usability Issues and Health Concerns, 161–184 (1984)

[14] Chung, L.: Non-Functional Requirements in Software Engineering. Springer, Heidelberg (2000)
[15] Robertson, S., Robertson, J.: Mastering the requirements process. ACM Press/Addison-Wesley Publishing Co. (1999)
[16] Nixon, B.A.: Representing and using performance requirements during the development of information systems. LNCS, p. 187. Springer, Heidelberg (1994)
[17] Guinan, P.J., Cooprider, J.G., Faraj, S.: Enabling software development team performance during requirements definition: a behavioral versus technical approach. Information Systems Research 9(2), 101–125 (1998)
[18] Balsamo, S., Marco, A.D., Inverardi, P., Simeoni, M.: Model-Based performance prediction in software development: A survey. IEEE Transactions on Software Engineering, 295–310 (2004)
[19] Gobbo, D.D., Napolitano, M., Callahan, J., Cukic, B.: Experience in developing system requirements specification for a sensor failure detection and identification scheme. In: High-Assurance Systems Engineering Symposium, Proc. Third IEEE Intl., pp. 209–212 (1998)
[20] Smidts, C., Stutzke, M., Stoddard, R.W.: Software reliability modeling: an approach to early reliability prediction. IEEE Transactions on Reliability 47(3), 268–278 (1998)
[21] Mooney, J.D.: Issues in the specification and measurement of software portability. In: Poster Session at the 15th Intl. Conference on Software Engineering (May 1993)
[22] Lauesen, S.: Software Requirements: Styles and Techniques. Forlaget Samfundslitteratur (1999)
[23] Evans, E.: Domain-Driven Design: Tackling Complexity in the Heart of Software. Addison-Wesley Professional, Reading (2004)
[24] Barker, S.D.P., Eason, K.D., Dobson, J.E.: The change and evolution of requirements as a challenge to the practice of software engineering. In: Proc. of the IEEE Intl. Symposium on Requirements Engineering, San Diego, California, January 4-6. IEEE Computer Society Press, Los Alamitos (1993)
[25] Arnold, R.S.: Software Change Impact Analysis. IEEE Computer Society Press, Los Alamitos (1996)
[26] Strens, M.R., Sugden, R.C.: Change analysis: A step towards meeting the challenge of changing requirements. In: Proc. of the IEEE Symposium and Workshop on Engineering of Computer Based Systems, p. 278. IEEE Computer Society, Washington (1996)
[27] Bohner, S.A., Arnold, R.S.: Software Change Impact Analysis. Wiley-IEEE Computer Society Pr (1996)
[28] Heninger, K.: Specifying software requirements for complex systems: New techniques and their application. IEEE Transactions on Software Engineering 6(1), 2–13 (1980)
[29] Carlshamre, P., Regnell, B.: Requirements lifecycle management and release planning inmarket-driven requirements engineering processes. In: Proc. 11th Intl. Workshop on Database and Expert Systems Applications, pp. 961–965 (2000)
[30] Al-Rawas, A., Easterbrook, S.M., National Aeronautics, Space Administration, United States: Communication Problems in Requirements Engineering: A Field Study. National Aeronautics and Space Administration; National Technical Information Service, distributor (1996)
[31] Easterbrook, S.M.: Handling Conflict Between Domain Descriptions with Computer-Supported Negotiation. University of Sussex, School of Cognitive and Computing Sciences (1991)

[32] Easterbrook, S.: Resolving requirements conflicts with Computer-Supported negotiation. In: Requirements Engineering: Social and Technical Issues, pp. 41–65 (1994)
[33] Boehm, B., Bose, P., Horowitz, E., Lee, M.J.: Software requirements negotiation and renegotiation aids. In: Proc. of the 17th Intl. Conference on Software Engineering, pp. 243–253. ACM, New York (1995)
[34] Crowston, K., Kammerer, E.E.: Coordination and collective mind in software requirements development. IBM Systems Journal 37(2), 227–246 (1998)
[35] van Lamsweerde, A.: Elaborating security requirements by construction of intentional Anti-Models. In: Intl. Conference on Software Engineering: Proc. of the 26 th Intl. Conference on Software Engineering, vol. 23, pp. 148–157 (2004)
[36] Potts, C.: Requirements models in context. In: 3rd Intl. Symposium on Requirements Engineering (RE 1997), pp. 6–10 (1997)
[37] Heeks, R., Krishna, S., Nicholson, B., Sahay, S.: Synching or sinking: Global software outsourcing relationships. IEEE Software, 54–60 (2001)
[38] Lala, J.H., Harper, R.E.: Architectural principles for safety-critical real-time applications. Proc. of the IEEE 82(1), 25–40 (1994)
[39] Lutz, R.R., Helmer, G.G., Moseman, M.M., Statezni, D.E., Tockey, S.R.: Safety analysis of requirements for a product family. In: Proc. 1998 Third Intl. Conference on Requirements Engineering, pp. 24–31 (1998)
[40] Xu, J., Randell, B., Romanovsky, R.J., Stroud, R.J., Zorzo, A.F., Canver, E., von Henke, F.: Rigorous development of a safety-critical system based on-coordinated atomic actions. In: Twenty-Ninth Annual Intl. Symposium on Fault-Tolerant Computing, Digest of Papers, pp. 68–75 (1999)
[41] Leveson, N.G., Stolzy, J.L.: Safety analysis using petri nets. In: The Fifteenth Intl. Symposium on Fault-Tolerant Computing. IEEE, Los Alamitos (1985)
[42] Leveson, N.G.: Software safety in embedded computer systems. Communications of the ACM 34(2), 34–46 (1991)
[43] Lutz, R.R.: Targeting safety-related errors during software requirements analysis. ACM SIGSOFT Software Engineering Notes 18(5), 99–106 (1993)
[44] de Lemos, R., Saeed, A., Anderson, T.: Analyzing safety requirements for process-control systems. IEEE Software 12(3), 42–53 (1995)
[45] Modugno, F., Leveson, N.G., Reese, J.D., Partridge, K., Sandys, S.D.: Integrated safety analysis of requirements specifications. Requirements Engineering 2(2), 65–78 (1997)
[46] Bishop, P., Bloomfield, R.: A methodology for safety case development. In: Safety-Critical Systems Symposium, Birmingham, UK (February 1998)
[47] Hansen, K.M., Ravn, A.P., Stavridou, V.: From safety analysis to software requirements. IEEE Tran. on Software Engineering 24(7), 573–584 (1998)
[48] Napolitano, M.R., An, Y., Seanor, B.A.: A fault tolerant flight control system for sensor and actuator failures using neural networks. Aircraft Design 3(2), 103–128 (2000)
[49] United States Military Procedure: Procedure for performing a failure mode effect and criticality analysis, MIL-P-1629 (November 1949)
[50] Barlow, R.E., Chatterjee, P.: Introduction to Fault Tree Analysis (December 1973)
[51] Chung, L.: Dealing with security requirements during the development of information systems. In: Rolland, C., Bodart, F., Cauvet, C. (eds.) Proc. 5th Int. Conf. Advanced Information Systems Engineering, CAiSE, pp. 234–251. Springer, Heidelberg (1993)

[52] Landwehr, C., Heitmeyer, C., McLean, J.: A security model for military message systems: retrospective. In: Proc. 17th Annual Computer Security Applications Conference, ACSAC 2001, pp. 174–190 (2001)
[53] Frankel, D.S.: Model Driven Architecture. Wiley, New York (2003)
[54] IBM Rational DOORS (formerly Telelogic): DOORS (2009), http://www.telelogic.com/
[55] IBM: Rational RequisitePro. (2009)
[56] 3SL Cumbria, England: Cradle Requirements Management v6.0 (July 2009), http://www.threesl.com/
[57] Wiegers, K.E.: Automating requirements management. Software Development 7(7), 1–5 (1999)
[58] Jackson, M., Zave, P.: Domain descriptions. In: Proc. of IEEE Intl. Symposium on Requirements Engineering, pp. 56–64 (1993)
[59] Zave, P.: Classification of research efforts in requirements engineering. In: Proc. of the Second IEEE Intl. Symposium on Requirements Engineering, pp. 214–216 (1995)
[60] Zave, P., Jackson, M.: Four dark corners of requirements engineering. ACM Trans. Softw. Eng. Methodol. 6(1), 1–30 (1997)
[61] Dardenne, A., Fickas, S., van Lamsweerde, A.: Goal-directed concept acquisition in requirements elicitation. In: Intl. Workshop on Software Specifications & Design: Proc. of the 6 th Intl. workshop on Software specification and design, vol. 25, pp. 14–21 (1991)
[62] Dardenne, A., van Lamsweerde, A., Fickas, S.: Goal-directed requirements acquisition. In: Selected Papers of the Sixth Intl. Workshop on Software Specification and Design, pp. 3–50. Elsevier Science Publishers B.V., Amsterdam (1993)
[63] Heninger, K.L., Kallander, J.W., Parnas, D.L., Shore, J.: Software requirements for the a-7 e aircraft. Memorandum Report 3876, Naval Research Lab., Washington D.C. (November 1978)
[64] Holzmann, G.J.: The model checker SPIN. IEEE Transactions on Software Engineering 23(5), 279–295 (1997)
[65] Yu, E.S.: Modelling strategic relationships for process reengineering. PhD thesis, University of Toronto (1995)
[66] Yu, E.S.: Towards modelling and reasoning support for early-phase requirements engineering. In: Proc. of the Third IEEE Intl. Symposium on Requirements Engineering, pp. 226–235 (1997)
[67] Yue, K.: What does it mean to say that a specification is complete? In: Proc. IWSSD-4, Fourth Intl. Workshop on Software Specification and Design, Monterey (1987)
[68] Darimont, R., van Lamsweerde, A.: Formal refinement patterns for goal-driven requirements elaboration. In: Proc. of the 4th ACM SIGSOFT symposium on Foundations of software engineering, San Francisco, California, United States, pp. 179–190. ACM, New York (1996)
[69] van Lamsweerde, A., Darimont, R., Letier, E.: Managing conflicts in goal-driven requirements engineering. IEEE Transactions on Software Engineering 24(11), 908–926 (1998)
[70] Greenspan, S., Feblowitz, M.: Requirements engineering using the SOS paradigm. In: Proc. of IEEE Intl. Symposium on Requirements Engineering, pp. 260–263 (1993)
[71] Warmer, J., Kleppe, A.: The object constraint language: precise modeling with UML. Addison-Wesley Longman Publishing Co., Inc., Boston (1998)

[72] Gill, A.: Introduction to the Theory of Finite-state Machines. McGraw-Hill, New York (1962)
[73] Hennie, F.C.: Finite-state Models for Logical Machines. Wiley, Chichester (1968)
[74] DeMarco, T.: Structured analysis and system specification, pp. 409–424. Yourdon Press, New York (1979)
[75] Gotel, O.C.Z., Finkelstein, C.W.: An analysis of the requirements traceability problem. In: Proc. of the First Intl. Conference on Requirements Engineering, pp. 94–101 (1994)
[76] Ramesh, B., Jarke, M.: Toward reference models for requirements traceability. In: IEEE Transactions on Software Engineering, 58–93 (2001)
[77] Ross, D.T., Schoman, J.K.E.: Structured analysis for requirements definition, pp. 363–386. Yourdon Press, New York (1979)
[78] Gane, C.P., Sarson, T.: Structured Systems Analysis: Tools and Techniques. Prentice Hall Professional Technical Reference (1979)
[79] Orr, K.: Structured requirements definition. K. Orr, Topeka, Kan. (1981)
[80] Burns, A., Wellings, A.: Real-time Systems and Programming Languages, 3rd edn. Addison Wesley, London (2001)
[81] Boehm, B.W.: Software Engineering Economics. Prentice Hall PTR, Englewood Cliffs (1981)
[82] Fairley, R.: Software engineering concepts. McGraw-Hill, Inc., New York (1985)
[83] Berry, G., Gonthier, G.: The Esterel Synchronous Programming Language: Design, Semantics, Implementation. Institut National de Recherche en, Informatique et en Automatique (1992)
[84] Barnes, B.H.: Decision Table Languages and Systems. In: Metzner, J.R. (ed.) Academic Press, Inc., London (1977)
[85] Parnas, D.L., Madey, J.: Functional Documentation for Computer Systems Engineering. Queen's University at Kingston, Dept. of Computing & Information Science (1990)
[86] Schouwen, J.V.: The A-7 requirements model: re-examination for real-time systems and an application to monitoring systems. National Library of Canada (1991)
[87] van Schouwen, A., Parnas, D., Madey, J.: Documentation of requirements for computer systems. In: Proc. of IEEE Intl. Symposium on Requirements Engineering, pp. 198–207 (1993)
[88] Faulk, S.R.: State determination in hard-embedded systems. PhD thesis, The University of North Carolina at Chapel Hill (1989)
[89] Heitmeyer, C., Labaw, B., Kiskis, D.: Consistency checking of SCR-style requirements specifications. In: Proc. of the Second IEEE Intl. Symposium on Requirements Engineering, pp. 56–63 (1995)
[90] Heitmeyer, C., Mandrioli, D.: Formal Methods for Real-Time Computing. John Wiley & Son Ltd., Chichester (1996)
[91] Heitmeyer, C., Bull, A., Gasarch, C., Labaw, B.: SCR*: a toolset for specifying and analyzing requirements. In: Reggio, G., Astesiano, E., Tarlecki, A. (eds.) Abstract Data Types 1994 and COMPASS 1994. LNCS, vol. 906, pp. 109–122. Springer, Heidelberg (1995)
[92] Heitmeyer, C., Kirby, J., Labaw, B.: The SCR method for formally specifying, verifying, and validating requirements: Tool support. In: Proc. of the 1997 (19th) Intl. Conference on Software Engineering, pp. 610–611 (1997)
[93] Heitmeyer, C.L., Jeffords, R.D., Labaw, B.G.: Automated consistency checking of requirements specifications. ACM Trans. Softw. Eng. Methodol. 5(3), 231–261 (1996)

[94] Landwehr, C.E., Heitmeyer, C.L., McLean, J.: A security model for military message systems. ACM Trans. Comput. Syst. 2(3), 198–222 (1984)
[95] Faulk, S., Brackett, J., Ward, P., Kirby, J.: The core method for real-time requirements. IEEE Software 9(5), 22–33 (1992)
[96] Faulk, S., Finneran, L., Kirby, J., Shah, S., Sutton, J.: Experience applying the CoRE method to the lockheed C-130J software requirements. In: Reggio, G., Astesiano, E., Tarlecki, A. (eds.) Abstract Data Types 1994 and COMPASS 1994. LNCS, vol. 906, pp. 3–8. Springer, Heidelberg (1995)
[97] Miller, S.P.: Specifying the mode logic of a flight guidance system in CoRE and SCR. In: Proc. of the second workshop on Formal methods in software practice, Clearwater Beach, Florida, United States, pp. 44–53. ACM, New York (1998)
[98] Jaffe, M.S., Leveson, N.G., Heimdahl, M.P.E., Melhart, B.E.: Software requirements analysis for real-time process-control systems. IEEE Transactions on Software Engineering 17(3), 241–258 (1991)
[99] Leveson, N.G., Heimdahl, M.P.E., Hildreth, H., Reese, J.D.: Requirements specification for process-control systems. IEEE Transactions on Software Engineering 20(9), 684–707 (1994)
[100] Heimdahl, M., Leveson, N.: Completeness and consistency in hierarchical state-based requirements. IEEE Transactions on Software Engineering 22(6), 363–377 (1996)
[101] Harel, D.: Statecharts: A visual formalism for complex systems. Science of Computer Programming 8(3), 231–274 (1987)
[102] Parnas, D.L., Wang, Y.: The trace assertion method of module interface specification. Queen's University, Dept. of Computing & Information Science, Kingston, Ont., Canada (1989)
[103] Harel, D., Lachover, H., Naamad, A., Pnueli, A., Politi, M., Sherman, R., Shtull-Trauring, A., Trakhtenbrot, M.: STATEMATE: a working environment for the development of complex reactive systems. IEEE Transactions on Software Engineering 16(4), 403–414 (1990)
[104] Hatley, D.J., Pirbhai, I.A.: Strategies for real-time system specification. Dorset House Publishing Co., Inc., New York (1987)
[105] Ward, P.T., Mellor, S.J.: Structured Development for Real-Time Systems. Prentice Hall Professional Technical Reference (1991)
[106] Ravn, A.P., Rischel, H.: Requirements capture for embedded real-time systems. Proc. of IMACS-MCTS 91, 1147–1152 (1991)
[107] Douglass, B.P.: Doing hard time: developing real-time systems with UML, objects, frameworks, and patterns. Addison-Wesley Longman Publishing Co., Inc., Amsterdam (1999)
[108] Chen, R., Sgroi, M., Lavagno, L., Martin, G., Sangiovanni-Vincentelli, A., Rabaey, J.: UML and platform-based design, pp. 107–126. Kluwer Academic Publishers, Dordrecht (2003)
[109] Object Management Group: SysML Specification Version 1.0 (2006-05-03) (August 2006), http://www.omg.org/docs/ptc/06-05-04.pdf
[110] Rioux, L., Saunier, T., Gerard, S., Radermacher, A., de Simone, R., Gautier, T., Sorel, Y., Forget, J., Dekeyser, J.L., Cuccuru, A.: MARTE: a new profile RFP for the modeling and analysis of real-time embedded systems. In: UML for SoC Design Workshop at DAC 2005, UML-SoC 2005 (2005)
[111] Object Management Group: UML profile for schedulability, performance, and time (September 2003)
[112] Axelsson, J.: Real-world modeling in UML. In: Proc. 13th Intl. Conference on Software and Systems Engineering and their Applications (2000)

[113] Berkenkötter, K., Bisanz, S., Hannemann, U., Peleska, J.: The HybridUML profile for UML 2.0. Intl. Journal on Software Tools for Technology Transfer (STTT) 8(2), 167–176 (2006)
[114] Bichler, L., Radermacher, A., Schürr, A.: Integrating data flow equations with UML/Realtime. Real-Time Syst. 26(1), 107–125 (2004)
[115] Kirsch, C.: Principles of real-time programming. In: Sangiovanni-Vincentelli, A.L., Sifakis, J. (eds.) EMSOFT 2002. LNCS, vol. 2491, pp. 61–75. Springer, Heidelberg (2002)
[116] Berry, G.: The foundations of Esterel. In: Stirling, C., Plotkin, G., Tofte, M. (eds.) Proof, Language and Interaction: Essays in Honour of Robin Milner. MIT Press, Cambridge (2000)
[117] Maraninchi, F.: The Argos language: Graphical representation of automata and description of reactive systems. In: IEEE Workshop on Visual Languages, Kobe, Japan (October 1991)
[118] Halbwachs, N., Caspi, P., Raymond, P., Pilaud, D.: The synchronous data-flow programming language Lustre. Proc. of the IEEE 79(9), 1305–1320 (1991)
[119] Camus, J.L., Dion, B.: Efficient Development of Airborne Software with Scade Suite. Esterel Technologies (2003)
[120] Caspi, P., Raymond, P.: From control system design to embedded code: the synchronous data-flow approach. In: 40th IEEE Conference on Decision and Control, CDC 2001 (December 2001)
[121] Kopetz, H., Bauer, G.: The Time Triggered Architecture. In: Proc. of the IEEE Special Issue on Modeling and Design of Embedded Software (2002)
[122] Henzinger, T., Horowitz, B., Kirsch, C.: Giotto: A time-triggered language for embedded programming. Proc. of the IEEE 91(1), 84–99 (2003)
[123] Ghosal, A., Henzinger, T.A., Kirsch, C.M., Sanvido, M.A.A.: Event-driven programming with logical execution times. In: Alur, R., Pappas, G.J. (eds.) HSCC 2004. LNCS, vol. 2993, pp. 357–371. Springer, Heidelberg (2004)
[124] Templ, J.: TDL Specification and Report. Technical report, Department of Computer Science, University of Salzburg, Austria (March 2004)
[125] Farcas, C.: Towards Portable Real-Time Software Components. PhD thesis, University of Salzburg (2006)
[126] Farcas, E.: Scheduling Multi-Mode Real-Time Distributed Components. PhD thesis, University of Salzburg (2006)
[127] Farcas, E., Pree, W., Templ, J.: Bus scheduling for TDL components. In: Reussner, R., Stafford, J.A., Szyperski, C. (eds.) Architecting Systems with Trustworthy Components. LNCS, vol. 3938, pp. 71–83. Springer, Heidelberg (2006)
[128] Farcas, E., Farcas, C., Pree, W., Templ, J.: Transparent distribution of real-time components based on logical execution time. ACM Press, Chicago (2005)
[129] Wiegers, K.E.: Software Requirements: Practical Techniques for Gathering and Managing Requirements Throughout the Product Development Cycle. Microsoft Press, Redmond (2003)
[130] Boehm, B., Turner, R.: Balancing Agility and Discipline: A Guide for the Perplexed. Addison-Wesley Professional, Reading (August 2003)
[131] Holbrook, I.H.: A scenario-based methodology for conducting requirements elicitation. SIGSOFT Software Engineering Notes 15(1), 95–104 (1990)
[132] Gruber, T.: A translation approach to portable ontology specifications. Knowledge Acquisition 5, 199 (1993)
[133] Damm, W., Harel, D.: Lscs: Breathing life into message sequence charts. In: Formal Methods in System Design, pp. 293–312. Kluwer Academic Publishers, Dordrecht (1998)

[134] Konrad, S., Cheng, B.: Requirements patterns for embedded systems. In: Proc. IEEE Joint Intl. Conference on Requirements Engineering, pp. 127–136 (2002)
[135] Withall, S.: Software Requirement Patterns. Microsoft Press, Redmond (2007)
[136] OSGi: OSGi Alliance Specifications (2007), http://www.osgi.org/
[137] Automotive Multimedia Interface Collaboration: AMI-C Software API Specifications – Core API (2003), http://www.ami-c.org/
[138] Krüger, I.H., Ahluwalia, J., Gupta, D., Mathew, R., Moorthy, P., Phillips, W., Rittmann, S.: Towards a process and Tool-Chain for Service-Oriented automotive software engineering. In: Proc. of the ICSE 2004 Workshop on Software Engineering for Automotive Systems, SEAS (2004)
[139] ITU-T Geneva: ITU-T Recommendation Z.120 – Message Sequence Chart (MSC 1996) (1996)
[140] Krüger, I.H.: Distributed System Design with Message Sequence Charts. PhD thesis, Technische Universität München (2000)
[141] Object Management Group (UML 2.0), http://www.omg.org/uml/
[142] Krüger, I.H.: Capturing overlapping, triggered, and preemptive collaborations using MSCs. In: Pezzé, M. (ed.) FASE 2003. LNCS, vol. 2621, pp. 387–402. Springer, Heidelberg (2003)
[143] Munich University of Technology: AutoFocus (1996-2002), http://autofocus.informatik.tu-muenchen.de/index-e.html
[144] Leveson, N.G.: Safeware: system safety and computers. ACM Press, New York (1995)
[145] Ermagan, V., Krüger, I., Menarini, M., Mizutani, J.I., Oguchi, K., Weir, D.: Towards Model-Based Failure-Management for Automotive Software. In: Proc. of the ICSE 2007 Workshop on Software Engineering for Automotive Systems, SEAS (2007)
[146] Arora, A., Kulkarni, S.S.: Component based design of multitolerant systems. IEEE Transactions on Software Engineering 24, 63–78 (1998)
[147] Object Management Group: UML 2.1.1 Superstructure Specification (2007)
[148] Ermagan, V., Farcas, C., Farcas, E., Krüger, I.H., Menarini, M.: A service-oriented approach to failure management. In: Proc. of the Dagstuhl Workshop on Model-Based Development of Embedded Systems, MBEES (April 2008)
[149] Holzmann, G.J.: The Spin Model Checker: Primer and Reference Manual. Addison Wesley, Reading (2003)

8 UML for Software Safety and Certification
Model-Based Development of Safety-Critical Software-Intensive Systems

Michaela Huhn[1] and Hardi Hungar[2]

[1] Institute for Software Systems Engineering, Technische Universität Braunschweig,
Mühlenpfordtstr. 23, 38106 Braunschweig, Germany
m.huhn@tu-bs.de
http://www.cs.tu-bs.de/sse
[2] OFFIS eV, Escherweg 2, 26121 Oldenburg, Germany
hungar@offis.de
http://ses.informatik.uni-oldenburg.de

Abstract. With the proliferation of UML in the development of embedded real-time systems, the interest in methods and techniques integrating safety aspects into a UML-based software and system development process has increased. This chapter provides a survey on relevant UML profiles and dialects as well as on design and verification methods and process issues supporting a safety assessment. These subjects are discussed in the light of norms and standards on software development for safety-critical systems.

8.1 Introduction

Nowadays, software has become an integral part of safety-critical systems in nearly all technical domains, from aeronautics or power generation, to traffic control or medical devices. Due to advances in mechatronics and communication the role which software plays is expected even to grow in future. In addition, the complexity of control to be implemented increases permanently. The adaptation of the well established model-based software engineering paradigm to the specific needs of safety engineering is an obvious and frequently proposed approach to systematically cope with the challenges of developing software components in safety-critical systems.

As stated by N. Leveson [1] and others, safety is an issue to be solved on the system and not the component level. Since software is immaterial, it differs from physical entities: Software by itself will not harm persons, property or the environment. But as an integral part controlling the behavior of physical components, its correct functioning contributes to safe operation or hazardous situations [2], as any other component of a safety-critical system. Software failures are mostly considered as systematic, having their cause in the safety analysis or software development process, whereas physical components may also fail at random.

It is the purpose of the discipline of *software safety engineering* to prevent software failures to occur. According to [3], software safety engineering has three major sub-processes: (1) *Software safety analysis* extends system safety analysis to software components in that hazards particularly relevant for software and software/hardware interaction are identified. The software safety analysis sub-process results in software safety requirements and safety design strategies aiming at elimination or mitigation of the identified hazards. (2) In *software safety design*, the software is designed and implemented according to the requirements and safety strategies. Safety design activities take the needs of safety assurance for traceability, documentation and safety argumentation into account. (3) *Software safety assurance* is concerned with all activities that provide evidence that the software meets its safety objectives. Verification and validation (V&V) activities are essential constituents of this sub-process. They are by themselves not sufficient, but their results have to be incorporated into an overall safety argumentation that integrates them into the system safety process.

In this paper, we focus on UML-based approaches to the sub-processes of software safety design and assurance. These two are also considered together in standards like the CENELEC standards for railway applications [4], the RTCA-DO-178B for airworthy software [5], or the IEC 61508-3 [6] on software requirements for the functional safety of electrical / electronic / programmable electronic controlled systems. Model-based techniques for the preceding system and software safety analysis like failure modes, effects and criticality analysis (FMECA), hazard and operability analysis (HAZOP), or event tree analysis (ETA) [7] are beyond the scope of this paper. They are partially addressed in Chapter 10 "Model-based Analysis and Development of Dependable Systems". An approach to a further aspect not covered here, namely UML-based dependability analysis, can be found in [8, 9].

As a consequence, we assume the software safety requirements to be provided. Another important input, coming from safety analysis, is the criticality level that classifies the software's contribution to system safety. The criticality level determines a so-called *software integrity level* (SIL) in the IEC 61508-3 [6], and the CENELEC standard [4][1]. Each SIL is equipped with requirements and recommendations on processes, activities and roles, and on software engineering design and V&V techniques. For the higher SILs, formal models are highly recommended for requirements analysis, design and verification. However, the relation of formalisms, which are mentioned in present standards, to artifacts of the development process remains vague. Thus, model-based software development and in particular the integration of model-based design and V&V techniques is a lively research field and major challenge in safety-critical systems engineering [10].

We start our presentation in Section 8.2 by briefly recapitulating the essentials of software development for safety-critical systems conforming to existing norms and standards, and deriving from that a categorization of usages of software models in software safety processes. In Section 8.3 we survey safety-related extensions of UML and classify them according to their purposes and usages. Section 8.4 sketches a seamless certification-oriented process based on UML. The perspectives of model-based V&V techniques and tool support are discussed in Section 8.5. Section 8.6 concludes.

---

[1] A similar concept in RTCA-DO-178B [5] are *development assurance levels* (DAL).

8.2 Development of Certifiable Software

The standards [4, 5, 6] do not prescribe a specific process model, but they require clearly distinguished development phases or activities with predefined input and output documents. The classical V-model (see Fig. 8.1) is well-accepted for software development for high assurance systems, in particular in the context of the CENELEC standards [4] and IEC 61508 [6] that both refer to it. The standards regulate key objectives to be addressed by the activities and in the documents. The development has to assure quality criteria like conciseness, completeness, traceability or testability. Safety-related requirements and constraints have to be distinguished and traced throughout the development phases. They are the major subject of the recommended verification and validation techniques. Moreover, for achieving safety, programming strategies and mechanisms like defensive programming or cyclic self-tests [4] are to be applied.

As stated in the introduction, we concentrate on UML-based approaches employed in software safety design and assurance of critical systems. Safety analysis, which we mention briefly in Sec. 8.3.4, is a mandatory preceding sub-process in a safety-critical system's life cycle. For the phases of safety design and assurance, two results of the safety analysis are of major interest: (1) The product-specific requirements for functional safety, i.e., goals to be achieved constructively in order to eliminate or mitigate the identified hazards. (2) The association of a SIL classifying the risk resulting from a failure of a software component. A SIL 0^2 classification means that the component is not related to system safety functions, whereas a SIL 3 and SIL 4 classification is assigned if a component failure may cause a severe or even catastrophic accident.

The standards associate with each SIL a set of process requirements or objectives to be met, concepts to be employed and techniques to be applied in order to achieve an acceptable level of confidence that systematic flaws in software development are eliminated. For software developed under SIL 3 or 4, specific formal and semi-formal model-based techniques are highly recommended for software specification, for software verification, and to complement software validation (see table A.2, A.5, and A.8 in [4]). However, today's standards [4, 6], do not state clearly which software engineering techniques should or may be used to achieve the required software quality characteristics.

An advanced view is taken by the Committee Draft for Voting (CDV) of IEC 61508-3 [11]: It suggests an explicit semi-quantitative *quality model* to relate particular usages of software engineering techniques to detailed quality characteristics of development artifacts. This relation is expressed in terms of a degree of rigour by which a certain software engineering technique can achieve a quality characteristic.

Another issue that hampers the proliferation of model-based development methodologies in software safety design is the fact that traditional programming is assumed for module design and implementation in the standards (see for

---

[2] In RTCA-DO-178B, the corresponding classification ranges from A to E in reversed order.

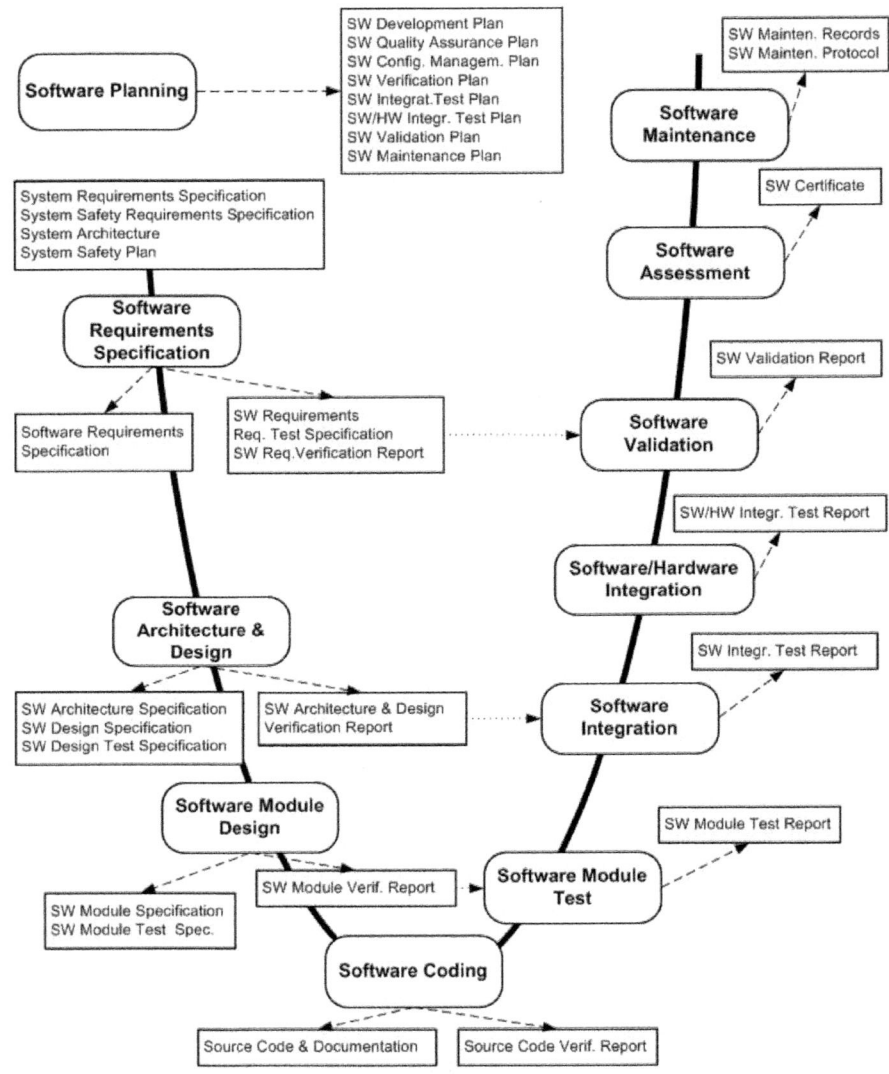

Fig. 8.1. V-Model according to EN 50128

instance the V-model according to EN 50128 [4] in Fig. 8.1). Restrictions to a safe subset of programming languages and approved compilers according to the SIL are highly recommended. How to establish a corresponding notion of safe model-based programming and code generation is discussed in Sec. 8.4.

All tools which are employed within the development process of safety-critical software have to be qualified. In general, tools that facilitate design automation – as in particular model-driven approaches incorporating code generation - are requested to be qualified with the same rigour as the safety-critical software itself. Whereas tools that are intended for use in the safety assurance process – i.e. that support testing, validation or verification - can be qualified by a more light-weight assessment process. An example for a certified code generator is the SCADE Software Factory [12].

We summarize this elaboration in the following observation: Any scientific approach to customize UML as a modeling notation for the use with safety-critical systems has not only to fulfil the intentions but also the practical certification-oriented requirements set by the standards. Only then, it will come into operation in industrial safety-critical system development. Practical certification-oriented requirements are in particular: (1) A conclusive argument for the usage of a UML-based method for specific design or V&V activities has to be provided to prove the method's adequacy for the quality characteristics required for a certain artifact. This part will benefit from progress in standards: E.g., the upcoming CDV of IEC 61508-3 is less concrete with respect to the referenced modeling formalisms than [4] and [6], and more explicit with respect to the quality characteristics to be achieved by a technique. (2) As explained in the previous paragraph, a tool supporting a UML-based method has to be qualified according to its usage in the development process. Even if an approach does not strive for *design automation* (see Sec. 8.5), the standards' requirements are satisfied only by very few UML tools today.

In the view of our recapitulation of the standards, we identify the following six categories where software models can be used in the development and certification of safety critical systems:

Usage 1 - Precise specification of safety and software requirements:
As a starting point for software safety design, the domain and safety engineers identify a set of safety requirements and constraints that have to hold on the system under development or evolution. Subsets of these are allocated to software components in the architectural steps of decomposition and partitioning. Requirements models are used to assist a common and detailed understanding of all safety-related issues between the software engineers and the safety and system experts. This usage scenario aims at enhancing the communication processes on safety-related requirements at the interface between system level and software component view.

Usage 2 - Software design and evolution: The next step after requirements specification is software design. In a model-based approach, the software architects and engineers describe the design or evolution task in terms of models representing various views. In case of embedded safety-critical software, hardware-dependent runtime properties like real-time behaviour, power consumption or resource utilization are an integral part of the functional safety requirements, and thus these properties should be addressed in the models. Additionally, a number of specific software safety strategies and techniques (e.g., defensive programming or multiple version dissimilar software, watchdogs or voters) are recommended for architectural design. Hence, this usage scenario describes the process of designing and thereafter implementing software components of safety-critical systems.

Usage 3 - (Partial) code generation: In a model-driven approach, automated model transformations and code generation are employed to obtain target-specific, executable models from design models. This scenario extends the model-based design scenario described before: Code generation moves efforts from manual implementation and extensive testing from the code level to model analysis. But a prerequisite is a qualified development environment that assures that the semantics of the models within the modeling environment corresponds to that of the generated code as it is executed by the runtime environment on a specific target. In addition, from the safety engineering viewpoint this usage of models faces a number of difficulties as discussed in Sec. 8.5.

Usage 4 - Verification and testing: At all stages of software development, the software engineers have to show that the outcomes meet the specifications and constraints induced by the previous stage. Verification and testing are part of the safety assurance process. Formal modeling of the software and system behaviour and its specification plays a prominent role here: The standards, e.g. [4], recommend a number of formal modeling notations that were considered potentially useful at the time the standards were published. However, the standards remain unspecific in which technique should be employed for which kind of safety requirement: Software safety requirements that are derived from well established safety analysis techniques like SHARD (Software Hazard Analysis and Resolution in Design) [13] cover a broad spectrum of software failures like omitted or untimely reactions, or unexpected or missing parameters. In difference to that, the referenced techniques like HOL [14] or CCS [15] focus on subsets like functional correctness and correct interaction behaviour. Moreover, questions of model validation, i.e. showing evidence that the formal model truly represents the relevant behaviours of the real system, are not addressed explicitly though they are of course highly important.

In practice, this phase is dominated by testing applied either to code or to executable models. In research, the usage of formal models for different verification techniques is considered at least as relevant as model-based testing.

Usage 5 - Software validation: Validation is the process of establishing conclusive, documented evidence that a system satisfies its requirements. In early phases, validation can be supported by animating, resp., simulating executable models. In later phases, software models may be part of the informations available to the validator, in particular for systems developed under SIL 3 or 4. For these, the standards require that the validation and design tasks have to be performed by independent teams. To transfer this principle to model-based approaches, independence between the models that are used for design and those to derive tests from has to be guaranteed.

Model validation is again, as already mentioned in Usage 4, an inevitable prerequisite for accepting results from model analyses as evidence for requirements compliance of the software.

Usage 6 - Software certification: In the certification process, assessors from a certification authority will examine whether the system will operate adequately safe. Therefore, the manufacturer delivers a so-called safety case. In the safety case, the safety claims for the system in its operating environment are identified and linked by structured, sufficient and comprehensible arguments. These address the documentation on the development process, design artifacts, and verification and test results that provide evidence that the claims are valid. Traceability of the requirements through the whole process as they are realised step by step in the design, and an underlying rationale are major prerequisites for certification.

In practice, maturity of processes, techniques and tools is also mandatory. In addition, particular software engineering techniques that are well accepted in safety engineering, like the restriction of programming constructs to a safe subset in the implementation phase or Modified Condition Decision Coverage (MC/DC) as a testing technique oriented towards code coverage, are an integral part of the documented evidence of conformity to the standards. Formal models may be part of the design documentation (usage 1 or 2) or the basis of analyses that support the evidence of the safety arguments.

It has to be pointed out that UML does not belong to the notations explicitly referenced in safety standards. Hence, employing UML models in the software safety design and assurance process requires conclusive safety case arguments on several aspects:

(1) The use of UML models in activities and through the development process has to be clarified as for any other artifact. It has to be demonstrated how and to which confidence level the requested safety objectives and quality characteristics can be achieved by UML-based techniques.
(2) The standards recommend a rich portfolio of safety strategies ranging from defensive programming, design diversity or restriction of programming languages to elements that are statically verifiable to formal V&V techniques and testing. Those strategies are widely accepted in safety engineering and they should be supported by a modeling approach.

8.3 Safety-Related Extensions of UML

In this section we survey UML profiles and dialects dedicated to safety-critical software development. From the numerous works, we have selected a subset of approaches aiming at a seamless and tool-supported model-based software safety and assurance process. As the software safety process is complex and multi-faceted, the approaches differ significantly in their aims and methodologies:

UML Profile for Developing Airworthiness-Compliant Safety-Critical Software [16] aims at a tight linkage of a UML-based software design with the safety argumentation in the context of RTCA DO-178B (see Sec. 8.3.1).
rtUML and the OMEGA-RT Profile [17] focusses on seamless integration of UML design models and a rich collection of formally founded, tool-supported V&V techniques (see Sec. 8.3.2).

Safe-UML [18] tailors UML for certifiable software safety design in the railway domain. Besides a formal foundation, also best practices to achieve quality characteristics for the design and safety-directed issues in model-based programming are considered. (see Sec. 8.3.3).

UML Profile for Modeling and Analysis of Real-Time Embedded Systems [19] explicitly addresses resource allocation and SW/HW integration. It supports the specification and analysis of real-time and performance properties (see Sec. 8.3.4).

Railway Control System Domain Profile [20] targets seamless support for formally founded design, code generation and verification of interlocking functionality in the railway domain (see Sec. 8.3.5).

A concise comparison of the particular strengths of these UML profiles is given in Table 8.1.

SysML [21] is not discussed here, because it has a more general objective of extending UML from software to system development and safety is not addressed by particular modeling elements.

EAST-ADL [22] is an architecture description language for the automotive domain defined upon UML 2.0. EAST-ADL offers notational elements for the Goal Structuring Notation [23] to model arguments of a safety case in context of the upcoming automotive standard ISO 26262 [24]. However, its support for safety remains rudimentary compared to other UML profiles.

8.3.1 The UML Profile for Developing Airworthiness-Compliant (RTCA DO-178B) Safety-Critical Software

In [16], Zoughbi, Briand, and Labiche address the explicit representation of safety information within UML models that constitute the requirements, the design, the deployment, or the finally installed configuration of a software system. The authors aim at a better understanding of safety issues during development and certification. They want to improve the communication between safety engineers, software developers, and assessors from the certification authorities (usage 1, 2 and 6).

The authors identified 65 safety-related concepts in the airworthiness standard [5] that are relevant for software models. The concepts are grouped into eight categories: *safety, reliability, integrity, concurrency, performance, certification, design, and configuration*. However, all concepts contribute (at least indirectly) to software safety. The relationship between the concepts is formalized in a conceptual meta-model. The meta-model is the basis for the definition of stereotypes, tagged values and constraints of the UML profile, and it reflects the key idea of integrating the safety argument into the UML models for software design.

A central concept is `Safety Critical` which is used to stereotype entities with direct impact on system safety. By the tagged values `Criticality Level` and `Confidence Level` the developer may declare the criticality level of a safety critical component determined in a safety analysis[3] and his/her confidence that the requested criticality level will be reached. Thus, a direct link is established between the design elements in the UML model and the safety argumentation according to a standard. The link between the safety terms from the standard and the UML model is strengthened by two major groups of concepts in the meta-model that are connected via the Safety Critical concept. The first group supports argumentations on design by offering concepts like (safety) Requirements, Rationale, Strategy, or Deviation. They describe design decisions, architecture rationale and modifications of approved plans that the developers make when they transform the original requirements into a design. The second group enables to explicitly represent technical safety engineering expertise in the UML model. For instance, design elements to detect and handle any kind of safety-related event are uniformly structured and classified by the stereotypes Monitor, Handler, Event, or Reaction. In a similar way, a group of concepts related to Replication Group allows to characterize the safety strategy of a replicated group of components whose elements are connected to a voter. Among others, concepts like Style are provided to describe the kind of a selected solution in common software safety terminology on the level of detailed design and implementation.

Since the proposed UML extension is defined as a UML profile, integration in existing UML modeling tools is possible. The authors propose an integration into frameworks like Rhapsody by IBM [25] or the Eclipse Modeling Framework (EMF)[26]. Thereby, the designer and certifier are supported in searching UML designs for occurrences of specific stereotypes or tagged values either by using a proprietary API or the Object Constraint Language (OCL). Thus, certain information - like listings of all COTS components used or all hardware-software interfaces - that are required for certification in the context of RTCA DO 178B [5] can be generated automatically. Additionally, traceability can be achieved if the model is fully elaborated according to the methodology suggested by the authors. Therefore, not only the UML model must contain different views on the software architecture and the design. The requirements linked to the design rationale and safety considerations leading to that design have to be represented in the model, too. Then the designer may traverse the model guided by the stereotypes provided by airworthiness profile to comprehend the safety argumentation.

To summarize, the airworthiness profile by Zoughbi, Briand. and Labiche is tailored for UML-based development and certification of safety-critical software according to the RTCA DO-178B. Safety information supporting the communication, and reasoning for safety cases is integrated into UML design models. With this focus on incorporating the safety argumentation *into* software design models, the airworthiness profile can be understood as a standard-specific alternative to approaches that provide the safety argumentations *externally* like the *Goal Structuring Notation* by Kelly [23] or *Assurance Based Development* by Knight et al. [27].

---

[3] e.g. "A" to "E" if the component is developed according to RTCA DO-178B.

8.3.2 *rtUML* and the OMEGA-RT Profile

rtUML and the OMEGA-RT Profile were defined in the context of the EU funded project *Correct Development of Real-Time Embedded Systems* OMEGA[4] as an extension of UML 1.4. The OMEGA approach integrates functional views and extra-functional properties, mainly timing, into functional views on specification, architecture and detailed design. A main goal is a formal foundation enabling tool-supported formal verification and validation techniques [17] (usages 1,2, 4, and 5).

rtUML comprises those functional concepts from UML that are considered most relevant in the embedded domain: For a structural design view, object-oriented concepts like polymorphism, inheritance, aggregation as well as various kinds of associations can be used in class diagrams. Active, passive and reactive objects are distinguished. To model the behaviour of a class or object resp., hierarchical state machines with a rich action language are included in *rtUML*. Interaction between so-called activity groups can be modelled as synchronous or asynchronous inter-object communication. *rtUML* can be pre-compiled to a kernel language *krtUML* containing only basic concepts from class diagrams and flat state machines. An operational, discrete time semantics in terms of *Symbolic Transition Systems* (STS) for *krtUML* was defined by Damm, Josko, Pnueli and Votintseva in [28].

rtUML is extended for requirement specification, architectural descriptions and in particular for the specification and verification of real-time aspects: Requirements can be specified scenario-based as Live Sequence Charts (LSCs) [29] or by temporal logic formulae. On the architectural level, a component-connector view comprises required and provided interfaces, protocol state machines, and OCL constraints. The OMEGA-RT Profile distinguishes different kinds of internal events like the send and accept of signals or state enter and exit events. Additionally, matching clauses and filters can be used to specify constraints on the duration between two event occurrences. The model can be extended by classes stereotyped as observers to express more involved timing requirements. Observers emulate timed automata in the UML modeling setting.

A rich portfolio of verification and validation techniques and tools supports software development with *rtUML* and the OMEGA-RT Profile [17]: Live Sequence Chart specifications can be animated with the Play Engine tool for requirements validation. Formal verification on finite state design models can be performed in two ways: Either a model checker specifically optimized for *rtUML* models can be used to prove specifications in terms of LSCs or temporal logic formulae. Alternatively, a model transformation to the IF framework [30] can be applied, enabling discrete and continuous time verification. Automated time and data abstraction mechanisms are offered for state space reduction as a preparatory step for model checking. To enable formal verification for infinite state models, a model transformation from *rtUML* to PVS (Prototype Verification System) [31] is provided. Using the interactive theorem prover PVS, infinite

---

[4] www-omega.imag.fr

value domains or unbounded message queues can be handled. As the transformation includes type information and OCL constraints on the model, these can be checked in PVS as well.

8.3.3 Restricting UML for Specification and Programming in a Certification Context

Motivated by the wish to be able to use UML in a way compatible with the railway norms (mainly EN 50128), Safe-UML has been designed as a restriction of (a part of) general UML. It is intended to address the functional viewpoint, expressed in class diagrams and statecharts [18]. To adequately cover an application range from documentation over specification (artifacts in the early phases of the design process) to actually UML-based programming, the definition has been organized in two levels:

Safe-UML (S): (*S* for *Superstructure*) applies to the OMG standard [32] for superstructures. It takes the definitions of state machines and class diagrams of UML and eliminates all semantic ambiguities, sources of underspecification, unclarity and unboundedness of system resources. In particular, it considers the parallelism (and its potentially sequentialized implementation).

Safe-UML (P): (*P* for *Programming*) applies to IBM's Rhapsody in Cpp as an instance of a UML implementation which enables programming in UML via the Cpp code generation. Safe-UML (P) gives directions on how to achieve conformance of the generated code with coding guidelines. Together with the rules from Safe-UML (S) it defines a set of restrictions which turn UML with Cpp annotations into a programming language suitable for the development of safety-critical systems.

Though originally designed for the rail domain – for instance, Part 42730 of the Mü 8004 [33] was taken for the definition for an admissable subset of Cpp – it is applicable also in other domains, particularly if the IEC 61508 is the source for the standard to be adhered to.

Safe-UML (S) and the Principles Guiding its Definition

In the following, we will give a short overview of essential features of Safe-UML, grouped as instances of four main principles which guided the definition of the language. The cross-cutting issue of parallelism and communication is treated separately.

Unambiguity: Every construct used must have a clearly (unambiguously) defined semantics. General UML, for instance, explicitly includes "semantic variation points" such as the handling of incoming events. In such cases, Safe-UML restricts to a particular interpretation, such as a bounded FIFO queue.

Determinacy: Usually, UML behavior specifications are nondeterministic. This is, for instance, the case if there are conflicting transitions leaving the same state, or if behavior is executed in orthogonal regions of a state machine. Safe-UML (S) tackles these problems by adding constraints to (a) prevent these situations to occur (e.g. guards of conflicting transitions must be exclusive, if they are triggered by the same event), or, if this is not possible, (b) ensure that the outcome is the same for each possible execution order, so that the internally nondeterministic behavior cannot be observed externally.

Clarity: Clarity addresses the question of accessibility and understandability of a specification or program. As an example, the state machines may be influenced severely by the context in which they are used (e.g. a transition triggered by an event may never fire, because the event is deferred in an enclosing state machine). Such effects are targeted by adding constraints which try to reduce context influence to a minimum (e.g. a constraint that events should not be deferred).

Boundedness: Consumption of time and space are particularly important aspects of a safety-critical system. I.e., system reactions shall come in time, the system must never deadlock nor run out of memory, etc. So, among other things, unbounded multiplicities are forbidden in class diagrams, and transition loops are ruled out in state machines.

Multiple Threads and Communication: To capture this major source of problems, Safe-UML requires a conservative system structure which is closely related to the one assumed by *rtUML* – in fact it also bases on [28] and its semantic definition. A major feature is the requirement of a finite, static structure where all active objects are organized in *active groups*, each featuring one active object with a common set of queues for events, timers, calls and completion events (one set for the active group). Problems related to sequentialization within queues, potential queue overflow, deadlocks due to multiple calls are in general hard to avoid. Safe-UML forbids some constructs and, for the rest, refers the developer to proven patterns of communication and scheduling, resp., to methods establishing correctness (like an abstraction to a decidable Petri-net property).

Safe-UML (P) — Safely Programming in UML

The objective in the definition of Safe-UML (P) is to turn state machines with Cpp annotations and class diagrams into a graphical programming language which by itself adheres to principles underlying the definition of coding guidelines ([33, 34]), and, taking the code generator from Rhapsody, translates into a fragment of Cpp meeting these restrictions.

First, of course one must restrict the Cpp annotations to the UML constructs accordingly. Part of the remaining answer is given by importing the Safe-UML (S) restrictions, which essentially restrict the (mostly) graphical UML constructs in a way one would restrict a programming language for safety – see the four principles exemplified above. And last but not least, the implementation dependent (Rhapsody-specific) code generation has to be considered.

The generator translates the UML constructs to a Cpp program using a library, the so-called framework, which essentially provides all necessary objects and methods to execute them, i.e., the equivalent of a runtime system. The code generator, if parametrized properly [35], produces rather well-structured code, so that only minor issues do arise. This analysis has been performed on a large set of examples systematically covering the graphical constructs and annotations.[5] The framework, which is part of the resulting Cpp, is itself not programmed according to strict safety guidelines. It can freely be modified by the Rhapsody user, so that one can remedy the defects identified in [36]. Framework modifications may also be employed to complement the restrictions on the graphical UML level by adding safety features to, e.g., event communication. Such an approach has already be used successfully in a signaling application which has been certified by the German Railway Authority.

Summarizing, Safe-UML defines a way to rigorously specify and safely program using UML in the rail domain and similarly regulated contexts (usage 2,3 and 6). It is, however, not yet integrated into design environments, and its (P) level is geared towards a particular implementation.

8.3.4 The UML Profile for Modeling and Analysis of Real-Time Embedded Systems (MARTE)

In 2008, the OMG published the Beta Specification for a UML Profile for "Modeling and Analysis of Real-Time Embedded Systems" (MARTE) [19] that shall replace the existing UML Profile for "Schedulability, Performance and Time" (SPT Profile) [37]. As stated in the title, the primary concern of the MARTE profile is real-time in embedded (RTE) systems, and not safety. However, the correct timing is part of functional correctness. With its modeling extensions, the MARTE profile supports detailed design and verification of safety-critical RTE systems (usages 2 and 4). Since MARTE is already realized as a plug-in of Papyrus for UML [38], tool support is available.

The MARTE foundations offer elements for modeling logical and physical time, resources and the spatial and temporal allocation of functional application entities onto them. The MARTE design model contains a so-called "RTE Model of Computation and Communication" to characterize the concurrency and synchronization behavior. A generalized, UML-conformant description of standardized APIs of real-time operating system like POSIX, QNX, or OSEK is supported. The extensions of the MARTE analysis model aim at the integration of state-of-the-art techniques for schedulability and performance analysis at the level of detailed design. Techniques like SymTA/S [39] or Modular Performance Analysis [40] offer tool supported analyses for various, common scheduling strategies and communication protocols. Their use is twofold - either predictive or verifying: For *predictive* use, a design model is enriched with estimated values on execution times and communication loads and with a specification of the planned scheduling situation. The analysis result is predictive and can (only) increase the

---

[5] This approach parallels a widely used practice of compiler validation.

confidence that the system design will fulfil its requirements on response times (worst, average or best case) or path latencies. Moreover, the analysis can be used to optimize the real-time dimensioning of a design [41]. In the case that values from the implementation are available, the analysis formally *verifies* whether real-time requirements are met in all possible situations. To pave the way for this kind of real-time analysis in the MARTE context, the concepts from the real-time analysis models are included in the MARTE profile. Thus, the definition of model transformations into the analysis framework is straightforward.

Additionally, the MARTE profile provides a package for the declaration of *non-functional properties* and an associated value specification language. Thereby the developer may annotate the model with further information relevant for safety-critical systems. In particular, reliability and availability issues addressed in the UML profile for "Modeling Quality of Service and Fault Tolerance Characteristics and Mechanisms" [42] can be integrated seamlessly by these means [43].

The MARTE profile has been extended towards dependability [44] analysis by several authors: Pataricza [45] introduced the concept of error propagation[6] to the General Resource Model of the MARTE predecessor, the SPT profile [37], to enable efficient system level diagnosis based on partial diagnostic information. He used quality of service parameters to characterize errors and the error behavior is modeled explicitly. Recently, Bernardi, Merseguer and Petriu [8] proposed a more general extension of the MARTE profile for analyzing and modeling dependability. Their "Dependability Analysis Model" addresses reliability, availability, maintainability, and safety, so-called RAMS properties, as major attributes of dependability. Among others, a "Threat Model" is introduced to describe either errors and failures when reasoning on reliability and availability or hazards that are relevant for safety. In that, the Dependability Analysis Model in [8] mainly focuses on the analysis of RAMS properties which is a usage of UML that precedes the software design and assurance process.

As shown by Thomas, Delatour, Terrier, and Gérard [46], the rich set of concepts for resource allocation provided by MARTE permits to clearly separate the model of the application from an explicit model of the real-time execution platform. The explicit platform model is taken as input to govern the model transformations to different target platforms in an MDA approach. In that, the Software Resource Modeling sub-profile has the potential to support deployment and code generation (usage 3) that goes beyond existing approaches that address the RTE characteristics only implicitly.

An alternative approach was chosen in the OMEGA-RT profile by Graf, Ober, and Ober [47] where a specific RTE platform model is explicitly addressed in the formal semantics that fosters automatic, correct code generation.

---

[6] According to [44], we call an event, at which a violation of the specified behavior becomes observable at the system boundary, a *failure*. An *error* describes the occurrence of a deviation from the intended behavior that may be internally compensated. If the error is propagated to the system's interfaces, a failure occurs.

8.3.5 The Railway Control System Domain Profile (RCSD)

Berkenkötter and Hannemann [20] conservatively extend UML 2.0 by a domain specific profile for railway and tram control systems. The RCSD profile supports the precise specification of railway networks with the aim to automatically generate code for a specific interlocking functionality. It shall foster the unambiguous communication between railway experts and embedded software designers and lay a foundation for the automated generation of verified controller software (usages 2, 3 and 4 for a specific rail application).

The RCSD profile offers basic entities to model railway tracks, namely track segments, points, and crossings. Additionally, there are elements for signals indicating the driving instruction for the following track segment, specific sensors for detecting whether a track element is occupied by a train, and automatic train runnings, which enforce braking if a train does not obey the signaling. The states of these elements are described by attributes for which specific datatypes are introduced. The topology of a railway network is modeled through the neighboring relation given by specific associations of sensors to track elements. Additionally, a set of top level constraints is included to ensure global consistency and completeness.

A class diagram models a restricted pattern or sub-problem of the RCSD domain, employing for instance further constraints on some entities. An object diagram can then be used to describe the track layout of a concrete network as an instance of the sub-problem of the corresponding class diagram. On both modeling levels, the static semantics is precisely defined by an elaborated set of OCL constraints that can be evaluated automatically on class and object diagrams annotated according to the RCSD profile [48].

The dynamic semantics is defined as a *Timed State Transition System*. Timed transitions are defined locally for the RCSD elements. The behavior of a network is the parallel composition of its component behaviors. To ensure safe train passage through the network, a *controller* realizing the interlocking functionality has to be added to the network model. Haxthausen, Peleska et al. [49] have shown how to generate the controller automatically from sets of generic transitions patterns that are instantiated according to the concrete network and the set of pre-defined routes when synthesizing the controller.

In addition, formal verification is supported on the level of the configured network by employing bounded model checking and inductive reasoning. A set of generic functional safety requirements is provided that covers the specific interlocking problem addressed by RCSD. Thereby all those states of the network are characterized - in terms of train locations, moving directions, and point positions - that are considered hazardous. The safety predicates on the configured network to be enforced by the controller are derived automatically by instantiation. The formal system description is transformed into SystemC that serves for both, verification input and executable code. To validate software/hardware integration the authors propose automated hardware-in-the-loop testing.

The RCSD profile provides proprietary prototype tool support for the development of a specific class of controllers in the railway domain on a non-standard compliant, formal semantics. The profile's application domain is clearly restricted for the benefit of rigorous formalization of the specification and design and an intertwined set of verification techniques covering several process phases.

8.4 Using UML in Certification-Oriented Processes

8.4.1 Questions to Be Addressed by a Certification-Oriented Process

The central idea of model-based software development is to employ models as key design elements, expressing design aspects in a tangible way. For safety criticality or even certification, it is desirable to extend the role the models play: We want to integrate them into the documentation of the development entering the safety case. Thus, not only the detailed usage of UML models within each activity or process phase has to be explained, but also quality characteristics requested for artifacts have to be substantiated, and a way to achieve this quality has to be delineated. Common quality characteristics are completeness and correctness wrt. the requirement specification, traceability, simplicity and understandability, behavioral determinacy, testable and (statically) verifiable design, fault tolerance, and last but not least, linkage to the safety analysis sub-process in both parts, design and assurance. If UML models are employed in more than one activity, their relation and consistency becomes an issue, too. Wrt. achieving the quality of models, one may note that models which document the finished design usually are not finalized in an early phase, but have to change over the course of the development. This is an issue to be reflected in the process definition. In this section, we sketch a process framework which emphasizes the aspect of iterative model development in a way compatible with standard requirements. It is a framework of a process in that it will have to be instantiated to the concrete development context and to the project requirements.

8.4.2 Purpose and Scope of the Proposed Process

The process outlined here has been defined to be compliant with the EN 50128 for developing safety critical software for the railway domain. The sketch is based on results of the OpRail[7] project [50], and we will call it the OpRail process in the following. Its primary goal is to delineate a way to harmonize the use of UML for expressing design artifacts and tools related to UML development with the strict requirements of the EN 50128. As this norm has been derived from the more widely applicable IEC 61508, the sketched process outline is useful beyond the railway domain.

The main motivation for the definition of the process was to introduce expressive modeling features from UML into the development, to enhance precision and

---

[7] This project has been funded by the German Ministry of Education and Research (BMBF) under grant No. 01|SC26A. The process has been mostly developed by the project partner Berner& Mattner.

Table 8.1. Safety-related UML profiles and their usages

	Usage of UML Models						Tool Support
	1 Specification	2 Design	3 Code Generation	4 Verification & Testing	5 Validation	6 Certification	
Airworthin. UML (aerospace)	annotations for SW safety requirements	built-in SW safety design patterns & strategies				traceability of RTCA DO-178B safety concepts through UML models	(extendable UML tools) + prop. support for search & review
rtUML & OMEGA-RT	use cases, scenarios, constraints, observers	components & interfaces: structure, behavior, timing		model checking & theorem prov., refinem. & RT verification	LSC animation, consistency checks, property deduction		tool coupling: IF 2.0 toolbox, Play Engine, PVS
Safe-UML (railway)	use cases, scenarios, temp. logic formulae	component structure & behavior, module behavior	restricted Cpp code generation from classes & state machines	model checking, test generation: sequ. diagr., coverage criteria (MC/DC)		model-based process definit., adaptation of UML due to EN 50128	tool integration: Rhapsody + ATG + prop. extensions
MARTE	extra-functional properties, resource modeling, timing	SW&HW comp., allocation, RT constraints, dependability extensions	PIM → PSM: operating system & protocol configuration	schedulability & performance analysis			Papyrus for UML + interfaces to sched. analysis tools
RCSD (railway sub-domain)	constraints, generic temporal logic safety properties	domain model for track topology & railway control	controller synthesis & SystemC code generation	model checking, induct. reasoning, autom. HIL testing	consistency checks		coupling of prop. tools

communicability of design artifacts, to explicitly represent the iterative nature of development activities – for instance the way in which early prototyping is employed – and to profit from the wide offering of tool support available for UML. It is mainly the intention to permit iterations and early prototyping which made it necessary to deviate from the V shape as depicted in Fig. 8.1. Despite the V model being quasi mandatory (as we mentioned above), norm compliance can be achieved by mapping the components of the V process to the new one.

The OpRail process covers the software development only, with interfaces to other development activities, including the integration of legacy code. Also sketched is the role tools could play and the requirements they would have to satisfy. In this short presentation, we only hint at these latter aspects.

8.4.3 Terms and Definitions

Process. A process defines who is doing what, when and how. A sketch of a process model like the one given in Fig. 8.1 is an illustration with a focus on the temporal aspect. A process may be composed of a set of sub-processes and is divided into different phases.

Sub-Process. A sub-process is a part of a process with is either focused on a particular aspect (like *Safety Management*) or the collection of actions to perform a logical step like the components in Fig. 8.1. A sub-process may span several phases of the process.

Phase. A phase is the period in a project begun and ended by major project milestones. A phase may encompass several sub-phases that may be repeated multiple times (iterations). Within a phase, a well-defined set of objectives is to be met and certain artifacts are to be produced.

Step. Within a phase, a (sub-)process can be divided into a number of more elementary steps.

Milestone. A milestone is an important event (completion of specified products) during the course of a project which can be scheduled and monitored and may be used for evaluating the progress of the project. The decision to move a project to the next phase is taken at a milestone. If the decision is negative, the milestone must be rescheduled and repeated.

Artifact. An artifact is an outcome of a sub-process or phase. It may be a required result of the process or some other piece of information that facilitates and supports the process. Artifacts may be grouped to artifact sets that are assigned to different sub-processes. For example, an artifact set can be composed of documentation, models, software modules etc. Artifacts shall be clearly specified by a given version number.

Activity. An activity is the execution of a step of the process.

Iteration. An iteration is a repetition of an activity, with the purpose of improving the end result, usually until a certain condition is met. Iterations are introduced to capture explicitly that many activities are often performed in this way, and to reflect agile development styles.

8.4.4 Phases and Sub-processes

Since we do not consider maintenance activities in the definition of our process, there are eleven sub-processes (of the twelve from Fig. 8.1) to be mapped.

Software Planning, Software Requirements Specification, Software Architecture & Design, Software Module Design, Software Coding, Software Module Test, Software Integration, Software/Hardware Integration, Software Validation and Software Assessment.

The OpRail process organizes these into (only) four phases, where these phases are not executed sequentially but overlap each other. Accordingly, there are five milestones, M0 to M4: M0 starts the first phase of the project, M1 to M3 mark transitions between phases, and M4 ends the project.

The artifacts which are tied to the sub-processes are transferred to the phases. If a sub-process starts within a phase, this phase produces versions of the artifacts which are defined as an outcome of the phase, continuing a sub-process means revising the artifact, and finishing a sub-process finalizes the artifact. Accordingly, an artifact may be produced in stages *draft*, *revised* and *final*, where of course the version *final* corresponds to the artifact as defined in the standard. That is, the OpRail process produces a documentation of the development of the system as if it was carried out according to the V process, thus presenting a familiar view suitable for certification.

In short, the phases are defined as follows.

Concept Phase (M0 to M1). Typically the concept phase consists of one to two iterations. In practice, the first iteration can be interpreted as the offer phase. Firstly, the main focus is to analyze and understand the problem. All input documents shall be reviewed. All SW related requirements, architectures and plans shall be proposed in draft.

Definition Phase (M1 to M2). In this phase the SW requirements, architecture and detailed design shall be fixed and reviewed. The design shall be simulated and tested early to figure out cost-intensive design flaws. This phase shall ascertain that the proposed system can be realized as specified. Afterwards the requirements set can be approved by e.g. the EBA (Eisenbahn-Bundesamt) or assessed by an ISA (Independent Safety Assessor). After approval, it is not allowed to change the requirements. Further changes have to be realized as changes within the change management workflow.

Realization Phase (M2 to M3). The main focus of this phase is to realize the solution and construct a product. First of all, this means implementation of the functionality fixed in the System and SW Requirement Specifications. The implementation is accompanied by unit and SW integration and SW/HW integration tests. This phase ends with approved SW and SW/HW integration testing.

Qualification Phase (M3 to M4). This phase includes validation tests. Furthermore it includes tasks for the assessment and certification of the system. This phase should end with extensive field tests which include the approval of the customer.

The full description (resp., a full instantiation) of the process contains a mapping of the stages of all design artifacts to the phases. Fig. 8.2 illustrates such a mapping.

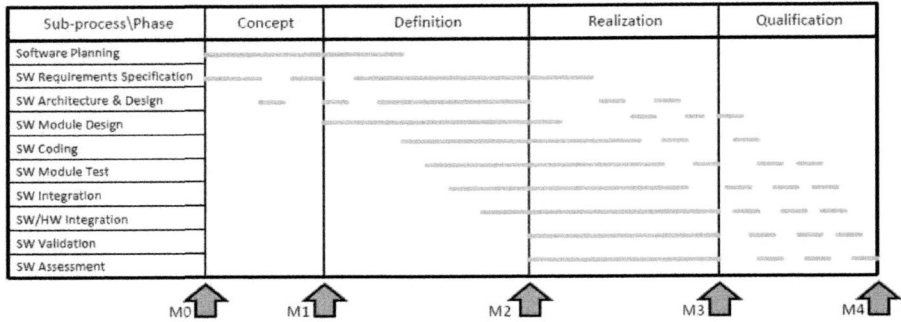

Fig. 8.2. Subprocesses and Phases in the OpRail-Process

8.4.5 The Use of UML in the Process

In general, UML diagrams can be used as parts of design artifacts, or even to replace some textual artifact completely. We indicate an elaboration which employs the most common diagram types to the various purposes.

Requirements. For the formulation of requirements (System Requirements Specification, SW requirements Specification etc.), the most suitable UML diagrams are *use case diagrams* to provide an overview and *sequence diagrams* to visualize behaviors. These can complement textual use cases (see e.g. [51]), as well as other, more traditional representation formats.

Architecture. An architecture may be visualized in a *structure diagram*, e.g. as a *component diagram* in the SW Architecture Specification, or as a *class diagram* on the level of module specifications. Depending on the nature of the design object and the level of abstraction, *sequence diagrams* illustrating the interaction of components may be useful.

Design/Module Specification. Detailed behavioral aspects can be specified with *state machines* or *activity diagrams*, while *class diagrams* define the software structure.

Test Specifications. Tests can be specified in *sequence* or *timing diagrams*.

The mapping above does not yet reflect the specific nature of a safety-critical design process. To be in line with the requirements coming from the safety criticality, one must observe that the diagrams are

- equipped with an unambiguous semantics, like for instance as given for Safe-UML, see Sec. 8.3.3,
- embedded in a context completing the usually partial specification given by the annotated graphics,

- adequate in their level of detail to the development stage. For instance, it is difficult to formulate a global requirement in a state machine without referring already to a component of a particular implementation.

What makes the process specifically suitable for UML is its flexible phase structure which permits gradually refining models, switching between specification and prototypically implementing components or aspects for explorative purposes, and thus gaining a much clearer picture of characteristics of the remaining design space than with most other development approaches. In short, it gives a well-defined, norm-compliant elaboration for an agile software development style, taking full advantage of the expressiveness of models for rich specifications at different levels of abstraction, in particular in early stages.

8.4.6 Realization

There are several additional aspects to be taken into account when implementing the OpRail employing UML and UML tools.

Models as Documents. Traceability of requirements and accessibility of artifacts require specific measures to be taken to integrate diagrams into the development process. Often, model-based development environments come equipped with model management mechanisms. But these are, usually, not sufficient, so that integration with other tools (e.g., for requirement management) and further measures become necessary which may go to the point where all models used in the design are printed on paper.

Models as Specifications. While for ordinary development projects it makes sense to have different views of one and the same object at the same or at different stages of the development, which even need not be consistent, this is not acceptable in a safety-critical process. All inconsistencies and semantic ambiguities have to be ruled out. As already indicated, using a restriction defined for such purposes like Safe-UML from Sec. 8.3.3 is one possible approach for functional and communication aspects, while, for instance, OMEGA-RT from Sec. 8.3.2 is useful to address the real-time aspect.

Models and Code. If code is generated from models, this too has nontrivial ramifications in the context of safety-criticality. As one aspect, the relation between model and code has to be clarified. If the semantics of the model itself is given by such a translation, all arguments relying on the model must be rooted in the code it represents, which may be awkward. Otherwise, formal relations between the model and the code semantics must be established. One facet of this problem can be addressed by certifying the code generator as it has been done for SCADE. Another facet comes with the execution environment, where the development environment (e.g., a Windows PC) and the target system (e.g., a real-time operating system running on a small processor in a simple architecture) may differ considerably. And if the generated code is modified later, be it for reasons of efficiency or platform compatibility, this must be reflected in the model (e.g., via roundtrip engineering) or addressed in the respective artifacts, the SW Module Verification Report, to name an example.

Models and Formal Verification. Complete models with an operational semantics permit the application of formal verification tools like theorem provers and model checkers. These, with their promise of assuring complete coverage of the model behavior seem attractive for high SILs. Indeed formal verification in the long run may further increase the usefulness of model-based design. However, currently the tools and methods are rarely mature enough for using them in industrial practice. Computational complexity and tool qualification are common obstacles. Nevertheless, formal verification techniques may already today help the designer in exploring models at abstract design stages, or specific components with great scrutiny, without making use of the result in safety cases. A more complete treatment of these issues and the following point can be found in Sec. 8.5.

Models and Tests. Test generation from models or models as components in test scenarios constitute another possible benefit of employing the model-based design paradigm. Mature techniques are available which can be put to use in the OpRail process. Generating tests covering models can partly automate the construction of the Requirement Test Specification or the Software Module Test Specification.

Taking into account all these issues, we conclude that it is possible to move from traditional design processes to one which profits from the use of models at different steps and levels of abstraction. The scheme provided by the OpRail process has been favorably evaluated by the TÜV Süd Rail as being suitable for an instantiation to a real-life process apt for the development of safety-critical rail applications.

8.5 Verification and Validation Techniques

Almost all safety-related UML profiles come along with a number of formally founded V&V techniques. It is far beyond the scope of this paper to present their technical background to a satisfactory level of detail. Instead, we discuss issues on the integration of formal verification techniques in the software safety assurance process for certifiable systems. As it is the case for the use of models in design, scope and objectives to be achieved by employing models in V&V activities have to be made explicit for safety assessment in certification processes. A verifiable, mathematical proof of a theorem on a formal model can serve as a piece of evidence for a safety claim only, if conclusive arguments are provided how the mathematical statement relates to the real-world system.

8.5.1 General Remarks on Verification and Validation Techniques in Model-Based Development of Certifiable Software

In accordance with the standards, we perceive a technique as *verification method* if it is adequate to evaluate whether or not the system or a component complies to a specification or constraints imposed at a preceding development phase. A *formal analysis* technique is based on a mathematical model of the system and the requirements and uses mathematical deduction for reasoning. In many cases the reasoning can be mechanized. In case, the primary and immediate goal of the technique is to prove that the (sub-)system satisfies the specification, we call it a *formal verification technique* in the narrow sense; if it directly aims at disproving the conformance of the system or component to the requirements it is called an *error detection technique*. In this article, we use *testing* for V&V techniques that execute the (sub-)system with selected inputs and compare its outcomes with the expected ones. A *validation* technique increases our confidence that the system accomplishes its intended requirements.

Following the terminology of Dwyer [52], we call a technique *sound* if a positive result of the evaluation constitutes conclusive evidence that the stated claim holds. Thus, a sound method does not generate false positives[8]. In this sense, exhaustive state exploration is a sound formal verification technique on finite state system models and testing usually is sound for error detection, but not for verification because exhaustive testing is impossible in most cases. Pure bounded model checking is sound for error detection only, as the state space is explored to a limited depth. But enhanced with inductive reasoning it may be extended to a verification technique. In particular in the realm of safety-critical systems, the limitations of a verification technique have to be clarified carefully as the evidence provided by the reported analysis results can only have relevance in a safety argumentation if the technique is applied in a sound manner.

Furthermore, software verification and validation – whether model-based or not – do not prove that software will not contribute to serious hazards under any circumstances. The best what can be achieved is to demonstrate that the software accomplishes its functional and safety requirements that have been derived from the aggregated knowledge on the system, its environment, and the foreseen hazards.

With models as design artifacts new V&V techniques can be applied. If design models are executable, simulation of the functionality provides additional validation facilities already in design phases. In case models are formally founded, model checking, abstract interpretation and theorem proving offer a powerful formal verification tool kit that can be further enhanced with various abstraction heuristics or compositional reasoning. If safety-related, extra-functional characteristics like reliability or the error behaviour, real-time or performance are explicitly represented in the model, then these can be subject of the analysis, too. However, reliable data for extra-functional runtime characteristics are most often only available when software/hardware integration is finished. Hence, analyses performed at earlier design stages on the basis of estimated values have to be repeated to approve the results.

Additional V&V techniques are not only a possibility, but are also a necessity in a model-based development process for several reasons: First of all, manual

---

[8] While false positives principally compromise the value of a verification technique, false negatives may cause additional effort, but do not put the technique in question.

review and inspection as traditionally performed on text documents have to be significantly adjusted for models. Without denying the well-known deficiencies of textual documents like incompleteness, inconsistency, poor structure, and the lack of traceability, these problems are at most disarranged but not solved without effort in a model-based integrated development environment:

Comprehensibility of a model-based design can be negatively affected by aspects of the method, the modeling language and the tooling: The design is usually scattered over several views and kinds of diagrams. Moreover, UML is a rich notation that often offers a set of alternative modelings to express the same issue. The developer may not oversee all semantic interdependencies between complex issues like object creation or deletion, event handling, transition selection, or run-to-completion-steps, even if the semantics is precisely defined. Tools often hide the details in the top view on a diagram, like, e.g., attribute or method declaration in class diagrams or inner hierarchy levels in statecharts. Moreover, specific settings severely influencing the semantics are often accessible only via nested preference lists or attribute tables within internal model browsers. Another open issue is the accessibility of different versions of a model stemming from earlier design phases or abstraction levels within one model repository.

"Collateral validation", as the implicit team validation has been called by Heimdahl [10], is lost, if model-based development comes along with large scale automation: Traditional development processes of safety-critical systems involve a plurality of experts whose expertise covers a broad field ranging from domain knowledge and software architecture to detailed questions of process and communication integration and hardware drivers. In the V&V phases, test experts and validators have a good chance to identify the tender points in a design due to their experience. Model design encompasses tasks from the whole field and is performed by fewer developers who may not always distinguish all consequences.

Also from a more technical perspective, several issues have to be considered to provide a conclusive safety argument for a model-based development approach.

Model paradigms: Software design and verification models are based on a model of communication and computation (MoCC) defining an abstraction from physical time, the granularity of steps, a concurrency paradigm etc. These abstractions may be adequate on a certain level of abstraction and in certain contexts. On the level of implementation, the safety-critical software applications are going to be executed in an environment of real-time operating systems (RTOS) and communication protocols like IMA [53] or time-triggered protocols [54]. Only if these support safe abstractions to analyzable MoCCs – which is not always the case – one may develop the software applications independently from the RTOS. For applications themselves, an answer to the model abstraction problem is given by approaches that establish a direct correspondence between the formal model and the code like Safe-UML(P) (see Sec. 8.3.3 [18] or SystemC models in RCSD [49]).

Model content: It has to be justified by thorough model-based validation that the formalized description of the requirements in a model-based specification and their implementation in a design model meet the informal, intended requirements. Only then a formal modeling framework can benefit from the enormous pool of techniques on model-based verification. Supplementary vacuity checks can assure the specification in fact covers the relevant behavior of the model (see Heimdahl for an overview [10]). Another caveat is the impact of simplifications and omissions: For scalability reasons or due to an early design stage, sub-systems or parts of the functionality are modelled very coarsely or omitted at all. Obviously, verification results have to be proven robust against such simplifications.

Backend questions: The more behavioral abstraction a modeling notation provides, the more is added in a code generation step that can only roughly be configured by the designer. In particular, extra-functional run-time properties like execution times and storage consumption may heavily depend on a prudent choice and combination of modeling elements.

Software-intensive technical systems are mostly assembled with proprietary hardware and operating systems for good reasons. But code generation of commercial modeling tools is optimized and approved for usage on standard processors. Thus, the code generator and linker have to be customized, which is a delicate task for specialists with joint expertise on the tool and the target system.

Tool qualification: The fundamental soundness requirements on tools offering early simulation, code generation, or formal verification are the *coincidence of the simulation, the verification, and the execution semantics* and *sound reasoning mechanisms*. If the execution semantics diverges from any of the other, or the deduction mechanism is corrupted, V&V results achieved on the basis of the simulation or verification semantics become worthless.

In contrast to many papers advertising verification techniques, successful industrial applications of formal model-based techniques mainly address detailed component design, not only for scalability reasons, but also for the validation needs and the caveats mentioned above. To benefit from formal verification and early simulation, a model must be precise and detailed with respect to all aspects that are the subject of verification. This can usually be carried out in the detailed design phase at the earliest.

8.5.2 Testing

Testing is the predominant V&V method applied in practice. For safety-critical systems, the standards explicitly recommend testing. Major test purposes are (1) to explore the functional specification in appropriate detail, (2) to execute the code to a sufficient degree of completeness, and (3) to ensure that the software is running properly on the target hardware. Therefore, a number of testing techniques are listed in the standards like testing based on equivalence partitioning of inputs, boundary value analysis or structural coverage criteria referring to data and control flow. Additionally, prototyping and animation for design validation, stress testing and exploratory or risk based testing are advised.

Executable design models pave the way to integrate testing activities in design phases: Well accepted test-selection criteria and the derived test-case specifications making such test criteria operational can be easily adapted to generate test suites that are applicable on the level of executable models instead on the code level. For instance, coverage criteria like statement, decision or MC/DC coverage have been transferred to statechart models [55].

This way, development fully benefits from providing executable models early in the process: Relevant shortcomings in the requirements specification can be detected before detailed design and costly implementation efforts are started.

In the following, we shortly discuss three specific approaches to adapt testing to model-based development:

(1) Design models from a previous development stage build the specification from which test cases are constructed.
(2) Test models are built independently from the artifacts used in the development and serve as basis for test case generation.
(3) Models are built by (automated) abstraction from code.

Test Cases Generated from Specifications

In the first approach, the current model or implementation is tested with respect to its conformance to a specification from a previous phase. Generating test cases from a previous design model can be applied iteratively at each stage to uncover deviations of the behavior of the current model from that of preceding models or artifacts. Detected deviations may have several causes:

- A preceding design step is flawed, but the specification is correct with respect to the original requirements.
- The preceding model or requirements specification is ambiguous or incomplete. This may concern the function to be realized, the execution platform and its limited resources or the assumptions on the environment. At some point, such an aspect may come into focus because the latter, more detailed model requires to settle it.
- The current model integrates different views or parts of the system that have been developed separately so far[9]. In such a step, testing conformance to the preceding separated models may reveal inconsistencies and erroneous assumptions that have been introduced into one of the preceding models.

However, as the test cases are derived from the same source as the current design model, this approach may support verification, but no independent validation. In other words, this approach uncovers inconsistencies that are inherent in the requirements specification itself or introduced in functional refinement steps. Mismatches between the functional specification, the execution platform and the environment can be detected only if the integrated models address these issues. But if for any reason the issue is not faithfully reflected in the design models this approach will not reveal any hint to a problem [56].

---

[9] Examples are functional composition or resource allocation and deployment in a layered architecture.

Independent Test Models
The second approach is to construct dedicated test models independently from the line of design models. Independence of the test model from the design models opens up new views [57] for the obvious price of additional effort:

- The test model may represent the system under development (SUT), modeled from the testing perspective, but also the system operator or the environment. Between these positions, numerous combinations of SUT and environmental models are possible that may apply various abstraction principles. For instance, a SUT test model may be restricted to the most common usage scenarios or a functional kernel, an environmental model may consists of a stochastic profile on input values and loads and their admissible ranges.
- Modeling formalisms differ with respect to the handling of fundamental concepts like time, causality, determinism, etc. Additionally, they provide different views and follow different computation paradigms like functional, operational, probabilistic or data-flow-oriented models.

The key issue of an additional test model is to complement the knowledge on the SUT from an independent perspective. As the test model is not a step in the design, it may be optimised for validation and verification purposes with respect to the functional and extra-functional requirements addressed, but also in terms of the concepts, paradigms and notational elements used for modeling. Heimdahl [10] reported on experiences that complementing a specification by several alternative models is considered a major factor towards achieving completeness.

Often, test models directly support derivation of test cases; alternatively, more general test generation techniques via the definition of test selection criteria can be adopted. Thus, building independent test models is adequate for all verification purposes mentioned in the previous paragraph, and it adds a chance to uncover defects that are outside the focus of the design artifacts, and extends the scope towards validation.

Models as Code Abstractions
A third use of models for testing safety-critical systems is to deduce a formal behavioral model as an abstraction from code and - if needed - a machine model. An intermediate representation can be extracted from source code by standard parsing techniques. The intermediate representation is symbolically interpreted on an abstract machine model. Thereby, constraints on the variables are collected and simplified by various techniques from abstract and concrete interpretation. Solving the constraints enables the generation of input values for a test case that covers a particular run through the model. The method supports the efficient generation of test cases for structural coverage criteria and boundary value analysis, but also the precise construction of test cases for certain classes of runtime errors. This way, testing the software/hardware integration can be transferred partially to the model level when using a refined hardware model. An testing technology based on this kind of models has been proposed by Peleska [58].

All three approaches to integrate testing in model-based development provide new prospects of design verification at an earlier stage than code integration on the target platform. They do not eliminate the need for final tests by executing the implementation on the target system and showing that module, integration, system, and acceptance tests are passed. But they can shorten these activities and iterations in the design by uncovering errors earlier.

8.5.3 (Formal) Verification

In contrast to testing, the promise of formal verification is a hundred percent guarantee for compliance of an artifact with a certain claim. Though this may sound highly attractive, formal verification is still, even after forty years of thorough research, only used very rarely in the development of safety-critical systems. Some of the main obstacles can be summarized as follows.

(1) Incompatibility with the established design process.
(2) Limited scope and immaturity of the techniques and tools.
(3) Lack of skilled personnel.

We addressed the first and second point partly in Sec. 8.4. Here, we will elaborate more on the fundamental weakness of formal verification in practice, namely, that it firstly offers a proof in the mathematical sense and usually not in the juridical sense. This means that mathematical proofs are in most cases not easily usable in certification processes. This is due to several reasons.

(1) The proofs, if produced explicitly at all, are very large so that they cannot be checked manually.
(2) Tools which produce the proofs need to be verified or at least qualified themselves, what they usually are not.
(3) The statements proved are accessible only to specialists, and they are often difficult to interpret correctly.

We will exemplify these reasons by studying two proof techniques, model checking and theorem proving, and suggest remedies to these obstacles.

Model checking: Model checking is the name for mostly automatic proof routines which check whether the set of behaviors of a program (its runs) are a *model* for a specification formula, i.e., whether the runs satisfy the specification. There is a multitude of different model-checking algorithms and implementations. *Explicit* model checkers enumerate systematically execution states of the program, *symbolic* and *SAT-based* model checkers operate on logical representations of states. Common to most model checkers is the requirement that the examined program has only finitely many states (maybe after abstraction), and model checking consists in cleverly covering all relevant cases. The problem with this approach is that model checking is not intended to produce a proof – if the system is correct, its output may simply be "yes".

This is of course of not much help in convincing a certification authority of the correctness of a particular statement. Since model checkers are complex programs and efficiency is a major issue in their design, they are themselves hard to verify. A very appealing way to overcome this hurdle is the following.

(1) Extend the model checker to produce a proof. Such a proof might be rather large, but will likely employ only simple constructs – codes for finite sets, boolean representations and so on.
(2) Design a tool to check those proofs. It is easier to check proofs then to construct them, and checking has only to be performed on the proofs relevant to the safety case – namely on the final versions of design artifacts. Therefore, such a tool need not be as efficient and elaborate as the prover itself, so it can be much simpler and will be easier validated.
(3) Apply proof generation and proof checking for the program version to be certified. Previous versions which have been produced during the development need not be treated as thoroughly. Though formal verification may be applied to them, the verification itself need not be checked.

Examples of how to extend different model checkers can be found in [59, 60], though we are not aware of any realization used in practice. The reader may consider the experiences with model checking Rhapsody in Sec. 8.5.4 to see reasons for this state of affairs.

Theorem proving: Theorem proving offers a flexible way for machine assisted proof of complex verification problems, see for instance the impressive achievements of the VeriSoft project [61, 62]. In principle, theorem provers construct proofs, but there are two caveats: First, the proofs are constructed on the fly, that means, they are not intended to be stored. Second, usually a theorem prover for software verification employs automatic subroutines to increase efficiency. As neither the theorem prover implementation nor its automatic subroutines (nor the computer it runs on) are themselves verified, they face similar difficulties as model checkers when certification is concerned. And approaches to overcome the difficulties rely on similar means: Coq (http://coq.inria.fr/) has a small "certification kernel" to check proofs, as does the Boyer-Moore theorem prover.

Summarizing, while the confidence in a system's correctness may be greatly improved through formal verification, its practical value today is still limited: The effort involved is high, and the results, if achieved at all, are not readily usable for system certification. Considering the remarkable progress made in this field in the recent decade, we expect that these methods will gain importance in the future.

8.5.4 Tool Support

In this section we give two examples of tools offering V&V support for UML, the tool set ATG for automatic test generation [63] and the model checker RUVE (Rhapsody UML Verification Environment) [64]. They both are extensions to the "Rhapsody®[10] in Cpp" design environment. Rhapsody in Cpp has an elaborated `Code Generator` which produces Cpp code from models consisting of class diagrams (for structure) and state machines or activity diagrams (for behavior)

---

[10] Now IBM, formerly Telelogic, formerly i-Logix.

which carry conditions and statements formulated in Cpp. A `Simulator` permits to execute the resulting code with user input for external stimuli. There is no animation of models besides the one through Cpp, other for instance than in the case of Statemate®. Therefore, one may view the Cpp code as the semantics of a model.

We continue now with a presentation of the extensions and a discussion of their suitability for a safety-critical development.

Test Generation

The architecture of the test-generation extension (ATG for *Automatic Test Generation*)[11] to Rhapsody is depicted in Fig. 8.3. A simplified view of the Rhapsody environment is on the left of the figure, with `Model Constructor`, `Code Generator` and `Simulator` as its main components for our presentation. When an executable model has been constructed, the user can select a part of the model as a *System Under Test* (SUT) and provide test goals. These latter can be expressed on the level either of the Cpp code or of the model itself. The most important code-level testing goal is MC/DC, in model terms one may ask for covering all states and transitions of the state machines in the SUT. In our simplified description, we ignore details like code instrumentation (e.g. to observe coverage in terms of the model) and the issue of the environment of the SUT which is very important in practice. The goals are fed into the `Test Generator`. The generator outputs a set of test sequences driving the model according to the specified goals, together with expected reactions of the SUT. It also reports on the degree of coverage achieved by the set, which is not always hundred percent.

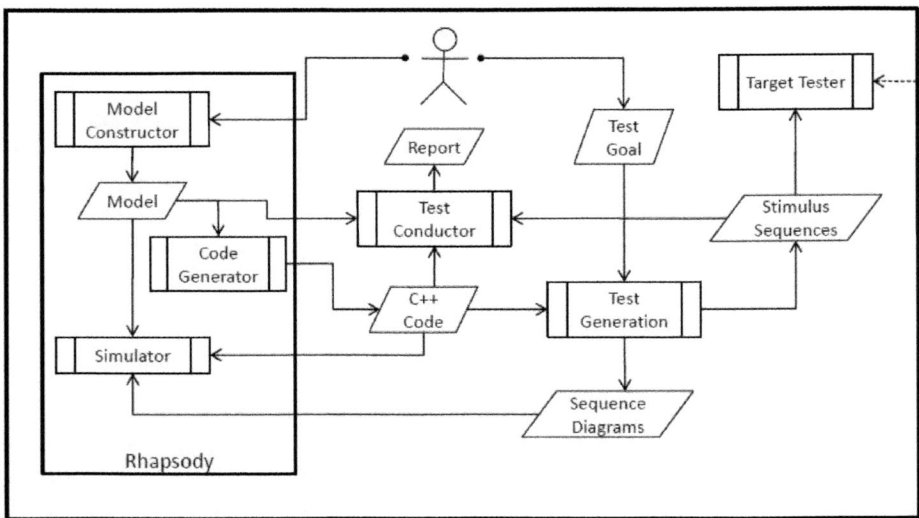

Fig. 8.3. Architecture of the Test-Generation Extension to Rhapsody

---

[11] Developed and marketed by BTC-ES.

The sequences may be exported in the form of sequence diagrams for visualization through the simulator. They can be applied to the model itself, or a different revision of the model, via the `Test Conductor` which, in the case of a regression test, will report any deviation from the expected behavior. Or they may be exported to other test-execution environments, e.g. for the purpose of testing compliance of target code or target system to the model. If the coverage is insufficient, the generated test suites may be completed by manual effort.

ATG can be applied in implementing the first two of the approaches to adapt testing to model-based development: Test cases generated from a model may be run on a model from a later stage, or even the target itself (first approach), or one may generate the test cases from specifically constructed model for covering certain aspects (second approach). With respect to standard requirements in safety-critical design, the generated test cases can be used for functional and black-box testing, or, with specific test models, also for non-functional aspects like timing.

Qualification is of course an issue. The test generator is the most advanced component of the ATG extension. It works by symbolically executing the Cpp code. Fortunately, the test generator itself does not have to be validated. Instead, one may independently – using a much simpler tool – validate the coverage achieved. The test-cases conductor is more critical. If one relies on automatic execution, the environment performing the tests has to be validated. Alternatively, one may add extensive reporting to provide evidence for correctness and completeness of the test execution.

Summarizing, the ATG extension to Rhapsody provides tool support to automate part of the testing activities which consume a substantial percentage of the development effort. Thus, it adds to the advantages of model-based design over traditional methods. Taking adequate provisions, ATG may even be applied when developing a safety-critical system.

Model Checking

The architecture of the model-checking extension is depicted in Fig. 8.4. Cpp code generated from Rhapsody models is translated into an automaton format and fed into a model-checking engine, whose other input may come from a number of different specification formats (graphical, pattern-based, temporal-logic formulae). The model-check engine is based on VIS, i.e. it is a symbolic model checker employing BDDs (Binary Decision Diagrams)[12]. Its output is either the message that the model satisfies the specified requirement (`"true"`) or an error path which may be animated on the model or visualized as a graphical specification.

Other than the ATG extension, RUVE is limited in the input it can process. Some of the limitations are inherent, others are founded in the experimental, not yet mature state of the tool set. Since semantically, the input to the model checker must be a finite automaton, dynamic object creation must be limited by static bounds. Also, the component and association structure cannot be changed

---

[12] A SAT-based engine may be used as an alternative.

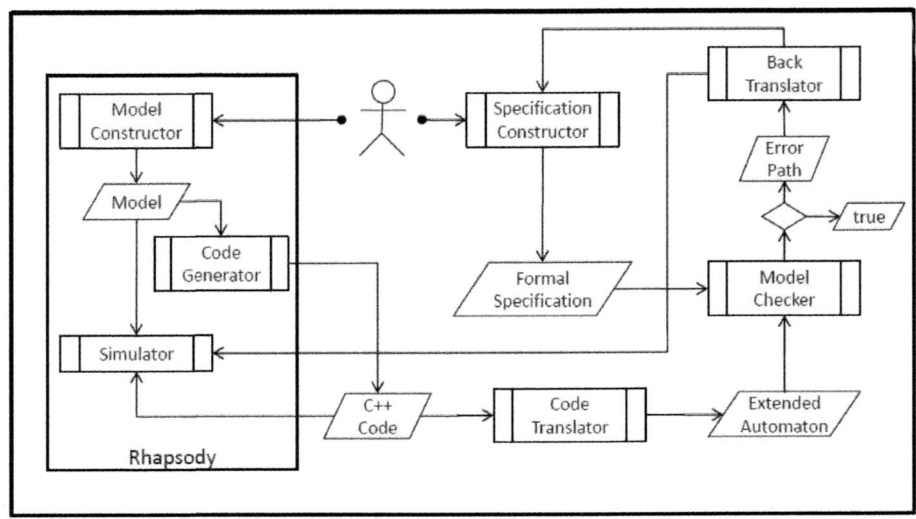

Fig. 8.4. Architecture of the Model-Checking Extension to Rhapsody

freely. Floating point numbers are not permitted for complexity reasons, and pointer operations are restricted. What remains is still a large and practically usable subset of the features of Rhapsody in Cpp.

The main usage of model checking is the verification of properties, which are, despite the different input formats available to the user, in their essence temporal-logic formulae. Model checking thus offers a way to verify properties which are predicative in nature. It is less suited to show refinements, e.g. that some model implements another, more abstract one. A second use case employs the error-path feature for design animation. For the specification "Always Not In(S)", the model checker will output a path leading to the state S if S is reachable. In that way, one may explore a model with automatic support. Further usages employ instrumented models.

If one wants to use the model checker for direct verification, the qualification question becomes important. There are two main issues.

First, the model checker does not work on the Cpp code (which, as already discussed, can be seen as the true semantics of the UML model), but on an automaton which has been generated by a nontrivial translation process. One way to ensure the correctness of this step is to employ methods from compiler validation. One could either validate the Code Translator, or independently verify that it worked correctly on those models appearing in the safety case.

Second, one will have to verify the operation of the Model Checker component. Since it seems impractical to certify this engine – the model checker works on advanced data structures and is geared towards efficiency – the best option is to verify that it worked correctly in the invocations relevant to the safety case. In Sec. 8.5.3 we explained that a promising way is to equip the model checker with a feature that produces on request a proof for the answer "true". Though this has not been done for a BDD-based model checker as far as we know, there is no fundamental difficulty involved. A proof will be a large object, but a rather simple procedure would be able to check its correctness.

These obstacles prevent the practical application of RUVE in a safety-critical development today. In addition, the resource consumption of RUVE is high already for rather small UML models. One may say that UML state machines – at least in this Rhapsody implementation – are a rather difficult object for formal verification. As a result, RUVE can be applied successfully only in specific cases, e.g. to aid the designer in analyzing a small model thoroughly. This may either be an abstract model at an early design stage, or a model of a component like one implementing a protocol for which model checking is particularly suited. Other than test generation, model checking does not seem mature enough for industrial application in UML-based design. Design for verification (i.e., language restrictions), dedicated procedures and additions for certification seem to be called for to enable profitable use of model checking on a larger scale.

8.6 Conclusion

UML and its extensions offer modeling elements for most aspects of interest in the context of safety-critical software and system development. A major advantage of UML is its wide dissemination for general purpose software development, while e.g. domain-specific languages face communication barriers due to their limited user basis. It is also clear that by introducing UML into the safety-critical systems domain not any UML expert will automatically become a software safety engineer.

Model-based software development is not yet considered in the safety-critical system standards. But statutory obligations and best practices recommend design qualities like structuredness, simplicity, and preciseness that are among the key promises of model-based development. Considering the plurality and complexity of UML, UML profiles and variants providing a well defined set of views and a restricted set of notational elements with precise (formally defined) semantics seem to be well suited for safety-critical systems. Upon the notational basis, a consistent set of techniques supported by qualified tools has to be defined that facilitates integration of UML models as key artifacts in a software safety design process. Limited maturity of techniques and tools as well as lack of elaborated methods supporting specific safety strategies are still severe hurdles for the proliferation of UML in industrial safety-critical software design.

Verification and validation benefit a lot from executable models in early design phases. Formal verification and testing contribute with a new quality of rigour and completeness to verification efforts. Recent research in this field has achieved substantial progress towards real-world size models and integration into industrial design processes. By customizing verification techniques to particular UML-based modeling frameworks, an equivalence between the verification model and the generated code executed on an abstract machine has been established for several tool environments. Such approaches are pathbreaking with regard to the rigour in establishing functional correctness. In addition, the adaptation of static analysis techniques to UML models enables assurance of most statically verifiable properties in principle. But in the whole, tool support is still fragmented, scalability is limited, and skilled personnel is rare.

Additionally, certification aspects have yet to be addressed adequately: A number of formally founded approaches still lack an explicit and conclusive argument on how mathematically proven facts relate to the properties of the real system, its software components, and the software design artifacts. Despite advances proposing solutions to particular aspects and sketches of integrating them with the software safety analysis sub-process, this still need to be improved for most UML-based techniques. This applies for design-centered methods as well as for V&V-centered ones.

Therefore, we may conclude that a lot of useful progress has been made. And while no mature, consistent methodology has been found yet, with prudent choice of techniques and tools, employing UML can improve the development of safety-critical systems in practice today.

References

[1] Leveson, N.: Safeware - System Safety and Computers. Addison-Wesley, Reading (1995)
[2] Lutz, R.: Software engineering for safety: A roadmap. In: FOSE 2000: Future of Software Engineering, Washington, DC, USA, pp. 137–152. IEEE Computer Society, Los Alamitos (2000)
[3] McDermid, J.A., Pumfrey, D.J.: Software safety: Why is there no consensus? In: 19th International System Safety Conference, System Safety Society (2001)
[4] European Committee for Electrotechnical Standardization: EN 50128: Railway applications - communications, signaling and processing systems - software for railway control and protection systems (2001)
[5] Radio Technical Commission for Aeronautics (RTCA): Software Considerations in Airborne Systems and Equipment Certification (December 1992)
[6] Intern. Electrotechnical Commission: IEC 61508: Functional safety of electrical / electronic / programmableelectronic safety-related systems (1998)
[7] Federal Aviation Administration: System Safety Handbook (2008)
[8] Bernardi, S., Merseguer, J., Petriu, D.C.: Adding dependability analysis capabilities to the MARTE profile. In: Czarnecki, K., Ober, I., Bruel, J.-M., Uhl, A., Völter, M. (eds.) MODELS 2008. LNCS, vol. 5301, pp. 736–750. Springer, Heidelberg (2008)
[9] Bernardi, S., Merseguer, J.: A UML profile for dependability analysis of real-time embedded systems. In: Proceedings of the 6th International Workshop on Software and Performance (WOSP), pp. 115–124 (2007)
[10] Heimdahl, M.P.E.: Safety and software intensive systems: Challenges old and new. In: FOSE 2007: Future of Software Engineering, Washington, DC, USA, pp. 137–152. IEEE Computer Society, Los Alamitos (2007)
[11] Intern. Electrotechnical Commission: 65A/524/CDV: IEC 61508-3: Functional safety of electrical/electronic/programmable electronic safety-related systems part 3: Software requirements, Committee Draft for Voting (2008)

[12] Esterel Technologies: Scade 6.0 (2008)
[13] McDermid, J.A., Nicholson, M., Pumfrey, D.J., Fenelon, P.: Experience with the application of HAZOP to computer-based systems. In: Haveraaen, M., Dahl, O.-J., Owe, O. (eds.) Abstract Data Types 1995 and COMPASS 1995. LNCS, vol. 1130, pp. 37–48. Springer, Heidelberg (1996)
[14] Gordon, M.J.C., Melham, T.F. (eds.): Introduction to HOL: A theorem proving environment for higher order logic. Cambridge University Press, Cambridge (1993)
[15] Hennessy, M.: Algebraic Theory of Processes. MIT Press, Cambridge (1988)
[16] Zoughbi, G., Briand, L., Labiche, Y.: A UML profile for developing airworthiness-compliant (RTCA-DO-178B) safety-critical systems. In: Engels, G., Opdyke, B., Schmidt, D.C., Weil, F. (eds.) MODELS 2007. LNCS, vol. 4735, pp. 574–588. Springer, Heidelberg (2007)
[17] Hooman, J., Kugler, H., Ober, I., Votintseva, A., Yushtein, Y.: Supporting UML-based development of embedded systems by formal techniques. Software and System Modeling 7(2), 131–155 (2008)
[18] Hungar, H., Robbe, O., Wirtz, B.: Safe-UML - Restricting UML for the development of safety-critical systems. In: Schnieder, E., Tarnai, G. (eds.) Proc. FORMS/FORMAT 2007, pp. 467–475 (2007)
[19] Object Management Group: UML Profile for Modeling and Analysis of Real-Time and Embedded systems (MARTE), Beta 2 (2008)
[20] Berkenkötter, K., Hannemann, U.: Modeling the railway control domain rigorously with a UML 2.0 profile. In: Górski, J. (ed.) SAFECOMP 2006. LNCS, vol. 4166, pp. 398–411. Springer, Heidelberg (2006)
[21] Object Management Group: SysML Specification Version 1.1 (2008-11-02) (November 2008), http://www.omg.org/spec/SysML/1.1/
[22] ATESST2: EAST-ADL2 Profile Specification (January 2008)
[23] Kelly, T.: Arguing Safety – A Systemic Approach to Managing Safety Cases. PhD thesis, University of York (September 1998)
[24] ISO TC22/SC3/WG16: Road Vehicles – Functional Safety. Committee Draft (September 2008)
[25] Telelogic: Rhapsody (2008)
[26] Eclipse Modeling Framework Project, EMF (2008), http://www.eclipse.org/modeling/emf/
[27] Graydon, P.J., Knight, J.C., Strunk, E.A.: Assurance based development of critical systems. In: The 37th Annual IEEE/IFIP International Conference on Dependable Systems and Networks (DSN), pp. 347–357. IEEE Computer Society, Los Alamitos (2007)
[28] Damm, W., Josko, B., Pnueli, A., Votintseva, A.: A discrete-time uml semantics for concurrency and communication in safety-critical applications. Sci. Comput. Program. 55(1-3), 81–115 (2005)
[29] Harel, D., Marelly, R.: Come, Let's Play - Scenario-Based Programming Using LSCs and the Play-Engine. Springer, Heidelberg (2003)
[30] Bozga, M., Graf, S., Mounier, L.: If-2.0: A validation environment for component-based real-time systems. In: Brinksma, E., Larsen, K.G. (eds.) CAV 2002. LNCS, vol. 2404, pp. 343–348. Springer, Heidelberg (2002)
[31] Owre, S., Rushby, J.M., Shankar, N.: PVS: A prototype verification system. In: Kapur, D. (ed.) CADE 1992. LNCS (LNAI), vol. 607, pp. 748–752. Springer, Heidelberg (1992)
[32] Object Management Group: UML2.0 superstructure specification (2005)
[33] Eisenbahn-Bundesamt: Technische Grundätze für die Zulassung von Sicherungsanlagen (1999)

[34] Guidelines for the use of the language C in critical systems (2004)
[35] Sanders, R.: Rhapsody 6.0 properties, Technical report, OSC-ES, Oldenburg, Germany (2006)
[36] Robbe, O.: Analysis of the Rhapsody C++-code and framework according to compliance with the EBA-guidelines 42720 and 42730. Technical report, OFFIS, Oldenburg, Germany (2005)
[37] Object Management Group: UML Profile for Schedulability, Performance, and Time (SPT), Version 1.1 (2005)
[38] Papyrus for UML (2009), http://www.papyrusuml.org
[39] Henia, R., Hamann, A., Jersak, M., Racu, R., Richter, K., Ernst, R.: System level performance analysis - the SymTA/S approach. IEEE Proceedings Computers and Digital Techniques 152(2), 148–166 (2005)
[40] Thiele, L., Chakraborty, S., Naedele, M.: Real-time calculus for scheduling hard real-time systems. In: International Symposium on Circuits and Systems (ISCAS), vol. 4, pp. 101–104 (2000)
[41] Hagner, M., Huhn, M., Zechner, A.: Timing analysis using the MARTE profile in the design of rail automation systems. In: 4th European Congress on Embedded Realtime Software, ERTS 2008 (2008)
[42] Object Management Group: UML Profile for Modeling Quality of Service and Fault Tolerance Characteristics and Mechanisms Specification, Version 1.1 (2008)
[43] Espinoza, H., Dubois, H., Gérard, S., Pasaje, J.L.M., Petriu, D.C., Woodside, C.M.: Annotating UML models with non-functional properties for quantitative analysis. In: Bruel, J.-M. (ed.) MoDELS 2005. LNCS, vol. 3844, pp. 79–90. Springer, Heidelberg (2006)
[44] Avizienis, A., Laprie, J.C., Randell, B.: Fundamental concepts of dependability. Technical Report LAAS Report no. 01-145, UCLA, LAAS-CNRS, Univ. of Newcastle upon Tyne (2001)
[45] Pataricza, A.: From the general ressource model to a general fault modeling paradigm? In: Jürjens, J., Cengarle, M.V., Fernandez, E.B., Rumpe, B., Sandner, R. (eds.) Critical Systems Development with UML – Proceedings of the UML 2002 workshop, TU München, Institut für Informatik, pp. 163–170 (2002)
[46] Thomas, F., Delatour, J., Terrier, F., Gérard, S.: Towards a framework for explicit platform-based transformations. In: 11th IEEE International Symposium on Object-Oriented Real-Time Distributed Computing (ISORC), pp. 211–218. IEEE Computer Society, Los Alamitos (2008)
[47] Graf, S., Ober, I., Ober, I.: A real-time profile for UML. International Journal on Software Tools for Technology Transfer (STTT) 8(2), 113–127 (2006)
[48] Berkenkötter, K.: OCL-based validation of a railway domain profile. In: Kühne, T. (ed.) MoDELS 2006. LNCS, vol. 4364, pp. 159–168. Springer, Heidelberg (2007)
[49] Haxthausen, A., Peleska, J., Große, D., Drechsler, R.: Automated verification of train control systems. In: Formal Methods for Automation and Safety in Railway and Automotive Systems (FORMS/FORMAT), pp. 252–265 (2004)
[50] Hungar, H., Bruhns, G., Plan, O., Lemke, O.: OPRAIL - Normenkonforme Entwicklung sicherheitsrelevanter Software unter Einsatz der UML. Signal + Draht 7 (2007)
[51] Cockburn, A.: Writing Effective Use Cases. Addison-Wesley, Reading (2000)
[52] Dwyer, M.B., Hatcliff, J., Robby, P.C.S., Visser, W.: Formal software analysis emerging trends in software model checking. In: Briand, L.C., Wolf, A.L. (eds.) Workshop on the Future of Software Engineering (FOSE), pp. 120–136 (2007)

[53] Lewis, J., Rierson, L.: Certification concerns with integrated modular avionics (IMA) projects. In: Digital Avionics Systems Conference (DASC). IEEE, Los Alamitos (2003)
[54] Kopetz, H., Grünsteidl, G.: TTP - a protocol for fault-tolerant real-time systems. IEEE Computer 27(1), 14–23 (1994)
[55] Mücke, T., Huhn, M.: Minimizing test execution time during test generation. In: IFIP Working Conference on Software Engineering Techniques (SET 2006). Springer, Heidelberg (2006)
[56] Pretschner, A., Philipps, J.: Methodological issues in model-based testing. In: Broy, M., Jonsson, B., Katoen, J.-P., Leucker, M., Pretschner, A. (eds.) Model-Based Testing of Reactive Systems. LNCS, vol. 3472, pp. 281–291. Springer, Heidelberg (2005)
[57] Utting, M., Pretschner, A., Legeard, B.: A taxonomy of model-based testing. Working Paper 04/2006, Department of Computer Science, The University of Waikato (2006)
[58] Peleska, J.: A unified approach to abstract interpretation, formal verification and testing of C/C++ modules. In: Fitzgerald, J.S., Haxthausen, A.E., Yenigün, H. (eds.) ICTAC 2008. LNCS, vol. 5160, pp. 3–22. Springer, Heidelberg (2008)
[59] Zhang, L., Malik, S.: Validating SAT solvers using an independent resolution-based checker: Practical implementations and other applications. In: DATE, pp. 10880–10885. IEEE, Los Alamitos (2003)
[60] Namjoshi, K.S.: Certifying model checkers. In: Berry, G., Comon, H., Finkel, A. (eds.) CAV 2001. LNCS, vol. 2102, pp. 2–13. Springer, Heidelberg (2001)
[61] Alkassar, E., Hillebrand, M.A., Leinenbach, D., Schirmer, N.W., Starostin, A.: The Verisoft approach to systems verification. In: Shankar, N., Woodcock, J. (eds.) VSTTE 2008. LNCS, vol. 5295, pp. 209–224. Springer, Heidelberg (2008)
[62] Beyer, S., Jacobi, C., Kroening, D., Leinenbach, D., Paul, W.: Putting it all together: Formal verification of the VAMP. International Journal on Software Tools for Technology Transfer 8(4-5), 411–430 (2006)
[63] Lettrari, M.: Using abstractions for heuristic state space exploration of reactive object-oriented systems. In: Araki, K., Gnesi, S., Mandrioli, D. (eds.) FME 2003. LNCS, vol. 2805, pp. 462–481. Springer, Heidelberg (2003)
[64] Schinz, I., Toben, T., Mrugalla, C., Westphal, B.: The Rhapsody UML Verification Environment. In: Proceedings of the 2nd International Conference on Software Engineering and Formal Methods (SEFM 2004), Bejing, China, pp. 174–183. IEEE, Los Alamitos (September 2004)

Part IV

Model Analysis

9 Model Evolution and Management

Tihamer Levendovszky[1], Bernhard Rumpe[2],
Bernhard Schätz[3], and Jonathan Sprinkle[4]

[1] Vanderbilt University, Nashville, TN, USA
tihamer@isis.vanderbilt.edu
[2] RWTH Aachen University, Germany
http://www.se-rwth.de
[3] fortiss GmbH, München, Germany
schaetz@fortiss.org
[4] University of Arizona, Tucson, AZ, USA
sprinkle@ECE.Arizona.Edu

Abstract. As complex software and systems development projects need models as an important planning, structuring and development technique, models now face issues resolved for software earlier: models need to be versioned, differences captured, syntactic and semantic correctness checked as early as possible, documented, presented in easily accessible forms, etc. Quality management needs to be established for models as well as their relationship to other models, to code and to requirement documents precisely clarified and tracked. Business and product requirements, product technologies as well as development tools evolve. This also means we need evolutionary technologies both for models within a language and if the language evolves also for an upgrade of the models.

This chapter discusses the state of the art in model management and evolution and sketches what is still necessary for models to become as usable and used as software.

9.1 Why Models Evolve and Need to Be Managed?

9.1.1 Introduction

Any complex set of artifacts needs to be managed, and models are certainly no exception—especially given that models are used to help manage complexity, and are therefore used in complex projects. Even though models do reduce the project's complexity, projects often have a complexity that even clever abstractions cannot transcend; thus, the models used in such a development project either become complex themselves or there are very many (perhaps, heterogeneous) models—or both.

This complexity also means that we cannot just create and forget models, but we must continuously evolve a model when adding new information, after quality reviews, redesigns against flaws and (in particular) according to changing requirements.

9.1.2 Model Management

Model management is the coordination between model-driven engineering (MDE) artifacts and resources, such as models, metamodels, transformations, correspondence, versioning, etc. [1]. Thinking in terms of these global (and entirely model-based) solutions has also been referred to as "megamodelling" [2].

Model management helps us to understand the status of models during the development and the maintenance process as well as how models relate to each other and to other artifacts of a development project. Proper model management is therefore a basic necessity to run a model-based development process (of a certain complexity). Most of the model management techniques are primarily for the developers themselves to simplify their life, increase efficiency when assessing or evolving models and ensure that less tedious work has to be done or even re-done when requirements or technical components evolve. Other techniques in particular high-level reports are for the project management— to understand and measure progress and project status.

Among model management techniques, we distinguish the following main categories:

- Checking for structural qualities of models, such as consistency and completeness on the one hand, but also guideline checking for additional quality attributes like readability or evolvability
- Transforming of models, including constructive and declarative descriptions of structural relations between models
- Versioning and version control of models, including the reporting of differences between model versions and merging of independently changed models with a common ancestor (so called "three way merge")

While we consider code generation, analysis and simulation tools the part of model management, we will not concentrate on these issues, but rather on the issues that treat the models as artifacts. Code generation, model-based analysis and simulation, etc., are undergoing heavy discussion and development right now, and there is a tremendous variety of approaches, from interactive one-shot generation of code-frames to repeatable fully automatic generation and customization/configuration of complete and deployable software components.

Model management for embedded systems includes additional twists on managing models for non-software artifacts. For example in product lines or other kinds of evolvable systems it is necessary to keep track of the connections between software and the mechanical, hydraulic or electric parts of an engine and their controlling software. This needs integrated models and thus integrated model management.

Model management for embedded systems includes additional twists on managing models for non-software artifacts. For example in product lines or other kinds of evolvable systems it is necessary to keep track of the connections between software and the mechanical, hydraulic or electric parts of an engine and their controlling software. This needs integrated models and thus integrated model management.

9.1.3 Model Evolution

According to our taxonomy, model evolution is one part of the more general model management issue. However, model evolution has many variations, and it is an important piece of model management. That is why we concentrate on model evolution, both from methodical as well as from a technical viewpoint in Section 9.3, because many of its concepts can be generalized to reflect versioning, interchange, tracking, consistency, etc.

In this paper, we use the term *model evolution* to refer to techniques to adapt existing models, as well as their related context, according to evolving needs. This context includes other models, code, tests, informal descriptions etc. that all might be affected when a model's content is changed. Evolution occurs when requirements or technology change as well as when improvements are made.

Model evolution may be necessary because of quality deficiencies according to two categories: if the model does not fit its context anymore; or if the representation of a model is bad and needs to be enhanced (e.g., to increase readability). As an important new problem, we also see the need of models to evolve together with the underlying language in which the model is expressed. As domain-specific languages (DSLs) [3] increase in popularity, and are often developed within or in parallel to the project, the evolution of a language quite often enforces the evolution of the models as well.

In this paper, we use *language evolution* to refer to techniques to evolve a modelling language according to domain or technology change, including the parallel evolution of that language's models and tools.

9.1.4 Chapter Outline

The rest of this chapter is organized as follows. Section 9.2 discusses the above mentioned techniques of model management. Section 9.3 deals with management of models, both from methodical as well as from a technical viewpoint, and in Section 9.4, we examine a particular instance of evolution driven by evolution of a domain-specific language. Furthermore, we examine in detail the case in which the evolution happens in small steps rather than abrupt changes.

9.2 Model Management

With a more widespread use of industrial-scale models throughout the development process, 'classical' problems found in code-oriented development start to impact a model-based development in a similar manner: The legacy of long-living models requires to address issues such as modeling standards and the quality of models, or model-versioning and -merging.

While for a code-based development many solutions have already been put in place to counteract these problems, in mode-based development these solutions are increasingly becoming available. For many issues, e.g., conformance or consistency analysis, the use of concept-rich, domain-specific models with a precise interpretation even allow to improve existing solutions for a code-based approach with its weaker-structured, more generic form of representation. For other issues, e.g., the merging or versioning, the additional complexity introduced by the rich structure of models leads to new challenges.

9.2.1 Model Quality and Modeling Standards

The application of quality constraints on the construction of software products – generally in the form of standards providing rules and guidelines for the construction of code – has repeatedly shown its merits in the development process concerning the quality of the product, both with respect to reliability as well as the maintainability. While the use of models in a development process provides an important constructive form of quality assurance, the rich and rigid structure and the domain-specific nature of the models used in the development allow to add new forms of quality constraints to the development process.

Therefore, quality constraints have increasingly gained attentions. Modeling tools provide mechanisms to ensure that the model under development respect modeling guidelines (e.g., the MAAB [4] guidelines for the Matlab/Simulink tool such [5]). Depending on the extent of these guidelines, they provide some constraints on the presentation (e.g., start state in top left area), the structure (e.g., number of interface elements), or even their interpretation (e.g., no lacking transitions). These constraints help to improve quality aspects like understandability, maintainability or even correctness.

Since obviously, not every syntactically correct model is a good model, there are many additional constraints that need to hold for a model to make it readable, changeable, or usable, etc. Depending on the nature of the constraints – and subsequently their implementation in a corresponding modeling framework – different kinds of conditions can be classified:

Syntactic constraints are immutable constraints enforced by construction on the structure of the model, ensuring that a model conforms to its metamodel and thus can be processed, stored and loaded, etc. Such a constraint e.g. ensures that a transition always is connected to a start and an end state.

Well-formedness constraints are constraints on the structure of the model enforced at specific steps in the development process, to ensure that the model is structurally sound. Such a constraint, e.g., ensures that used variables are actually defined and have the appropriate type. Such conditions are generally mechanically checked, e.g., upon processing or editing models.

Semantic constraints are constraints on the interpretation of a model, to ensure that a model is semantically correct. Such constraint, e.g., ensures that a state-transition model is not non-deterministic or incomplete. These conditions cannot always be effectively checked mechanically, facing the typical problems of model-checking approaches.

Note that shifting constraints between the first and the second class influences the rigidness of the development process.

Since model-based development increasingly deals with 'mega-modeling' issues [6] like large-scale, distributed models including linking models from heterogeneous domains – or meta-models – a second taxonomy builds around relations within and between models and domain-languages:

Intra-model conditions are defined over a single given model and thus can be formalized within the same domain-language and checked on a single model directly. A typical example is the type-correctness of a single dataflow model.

Inter-model conditions are conditions define for a set of models from the same domain, which still can be formalized within the same language but are checked on a set of models. A typical example is the interface consistency between models describing the environment of a system and its internal structure.

Inter-language conditions are conditions defined on a set of model of different modeling domains or languages. These conditions do not only require to check several models, but also can only be expressed in a super-language for these different languages. A typical examples is the consistency between different abstractions or viewpoints of a system, e.g., in case of the UML with its various sub-languages the consistency between a sequence diagram of a component, its state machine, as well as its interface view.

As both taxonomies are independent of each other, each combination has its relevance in the practical application.

For the practical application of conformance constraints in the development process, support for the definition of constraints on the models of the product under development as well the automated enforcement of these constraints proves to be an important asset, improving and front-loading this form of quality assurance in a model-based approach.

By adding a mechanism to automatically check for the validity of these constraints with respect to a product model and report violations on the level of the product model, conformance analysis can be supported by translating constraints of the guidelines to conformance conditions and using this mechanism for their validation.

Since analyzing the conformance of a model of a product to a certain modeling guideline is becoming increasingly relevant for the model-based development process, corresponding functionalities are added to tools supporting this kind of process. For widely used tools such as MatLab [5], additional mechanisms – integrated into the tool itself or provided as a stand-alone checker – ensure that the model under development respect conformance constraints. However, since often those mechanisms use the API of the tools (e.g., *M-Script* for *MINT* or *Model Advisor* [5]; *Java* for *MATE* [7]; *MDL* for ConQAT [8]), conformance constraints are rather defined on the level of the concepts of implementation language than at level of the concepts of the application domain.

Here the fact can be exploited that the conceptual domain model allows to define a *conceptual product model* and furthermore *provides a vocabulary capable*

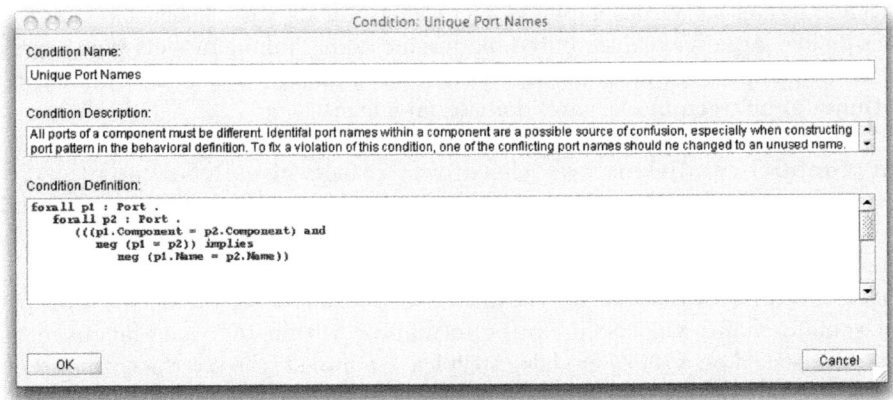

Fig. 9.1. Defining a Conformance Condition

of defining properties of such a product model. In order to enforce conformance constraints on products of a specific domain of application, this vocabulary must allow to *express these constraints as logical conditions* over the classes and associations of the concept model characterizing the domain of application. Obviously, this formalism should reflect the concept model as closely as possible, to abstract from the actual technical representation of the model of the product.

In [9], an approach is used for formalization of conformance constraints based on the OCL formalism [10]. [11] uses a simular formalism, based on predicate logic with the classes and associations of the concept model as first-class-citizens. Constraint checking can be performed by

- providing a checker separate from the tool for the construction of the product itself, generating a report as mentioned for the further improvement of the model; this technique is chosen in the former approach;
- integrating the checker into the construction tool, allowing direct navigation or direct application of improvement operations; this technique is chosen in the latter approach.

Independently of the degree of integration, this form of constraint analysis is performed in three steps:

(1) Definition of constraints, often combined with a grouping of similar constraints
(2) Checking of the validity of the constraints
(3) Inspection of the counterexamples for violated constraints.

In the following, these steps are demonstrated for the approach described in [11].

Although conformance analysis can be understood as checking the validity of conceptual conditions, for practical application, issues besides the evaluation have to be considered. Therefore, as shown in Figure 9.1, the definition of a consistency condition consists of a constraint name for selecting the the condition using the process interface, an informal constraint description, generally for the

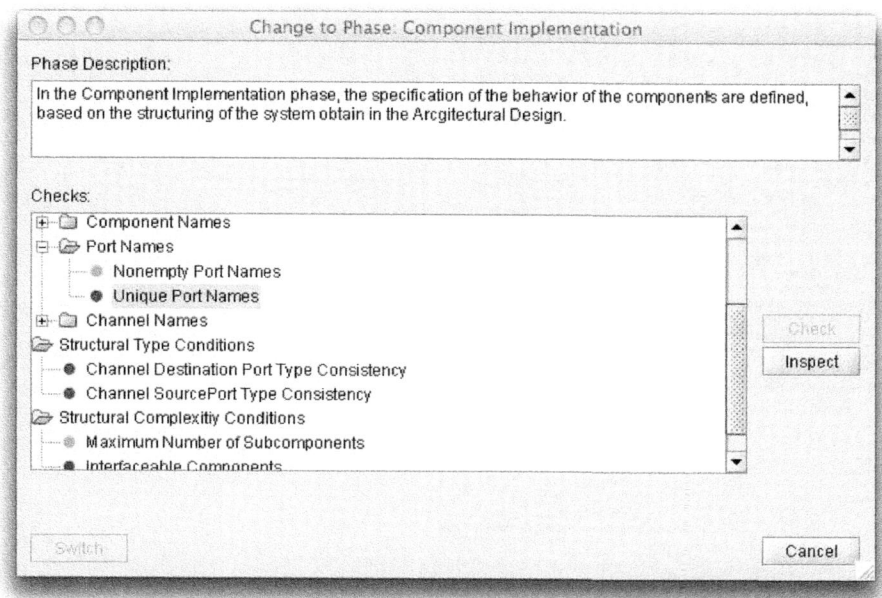

Fig. 9.2. Checked Conditions of a Phase

purpose of the condition and possible remedies in case the condition does not hold for the model, and an expression formalizing the condition definition.

Besides the validity of the consistency condition, the collection of all model elements not satisfying the condition are an important result in case the condition is not satisfied. Accordingly, the user is rather interested in in all model elements satisfying the complement of the consistency conditions. To check the validity of the formalized constraints in the next step, the constraints are evaluated. If all conditions are satisfied, the validity is stated. Otherwise – as shown in Figure 9.2 – violated entry conditions like 'Unique Port Names' are reported, and can be inspected as described below.

When evaluating a constraint, besides returning its validity, the checker returns the set of all unsatisfying assignments of the quantified variables. This collection of model elements not satisfying the consistency condition is the set of all counterexamples throughout the product model violating the consistency constraint imposed on the model

As shown in Figure 9.3, this result is returned as a list of assignments; additionally, the informal description of the consistency check is presented to support the user in correcting the inconsistency. To simplify correcting a violated consistency condition, a simple navigation mechanism from such an assignment to the graphical representation of the model is provided: by activating an assignment from the list of inconsistencies, the corresponding editors containing the violating element is presented.

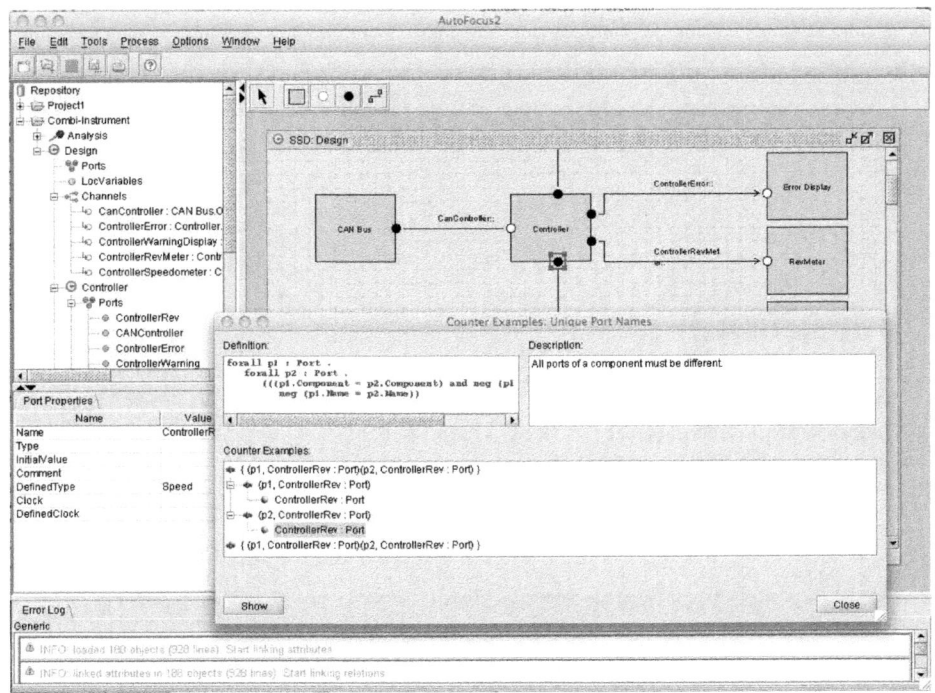

Fig. 9.3. Counterexamples of a Constraint Evaluation

Like the checking of modeling standards corresponding to the checking of coding standards, also other techniques found in a code-centric development are applicable for a model-based development process, e.g., the detection of model clones. In general, clones are *description fragments that are similar w.r.t. to some definition of similarity*. The employed notions of similarity are heavily influenced by the program representation on which clone detection is performed and the task for which it is used.

The central observation motivating clone detection research is that code clones normally implement a common concept. A change to this concept hence typically requires modification of all fragments that implement it, and therefore modifications of all clones, thus potentially increasing the maintenance effort. Additionally, clones increase description sizes and thus further increase maintenance efforts, since several maintenance-related activities are influenced by description size. Furthermore, bugs can be introduced, if not all impacted clones are changed consistently.

Here, detection of model clones like in [12] can improve the maintainability of evolving models, helping to identify redundant model fragments. Figure 9.4 shows the application of clone detection to dataflow languages as used in Simulink.

Model Evolution and Management 249

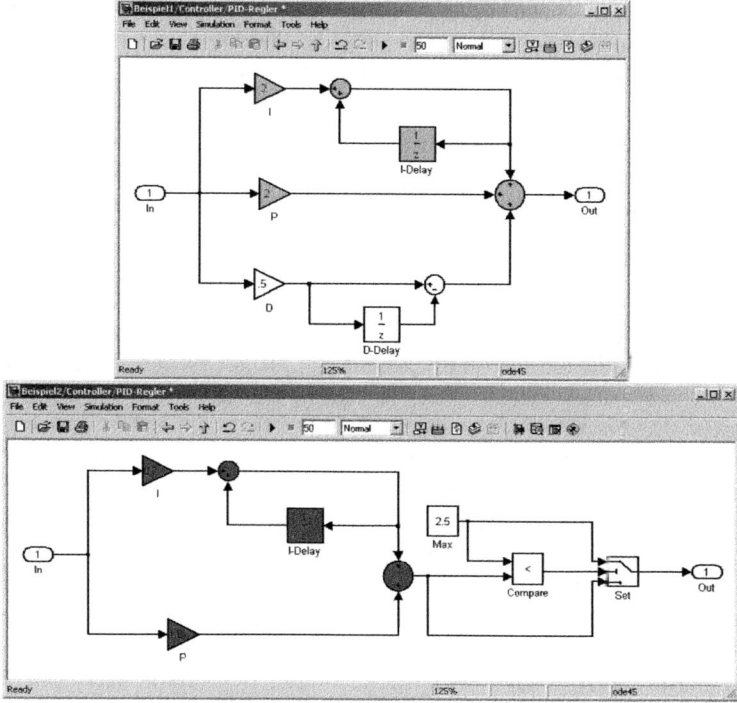

Fig. 9.4. Example for Clones in Dataflow Models

9.2.2 Model Transformation

Especially in a model-based approach with structure-rich system descriptions, automated development steps, – using mechanized transformations – have the potential to provide an important technique to improve the efficiency of the development process. Besides increasing efficiency, these structural transformations can offer consistency ensuring modification of models. There is a range of different applications for model transformation:

- Refactoring of models, e.g, to improve the architecture of a system, operating on a single model
- Merging of models, e.g., to consistently weave in standard behavior, operating on a set of model of the same language
- Translation of models, e.g., to generate a platform-specific model from a platform-independent model, operating on a set of models of different languages

However, for their effective application, frameworks providing these transformations should use formalisms to enable sufficiently abstract yet executable descriptions, support their modular definition by simple composition, and supply mechanisms for parameterization. Generally, these transformations are executed on the internal representation, called *abstract syntax*, of the models, but often defined on the representation of the model, called the *concrete syntax*.

Similarly to the case of the structural analysis of conceptual product models, the principle of transformation on the internal model representation makes use of the fact that, in a model-based approach, a product model comes with an *explicit representation of the abstract syntax* composed of domain-specific concepts and their associations; therefore, *transforming this structure* corresponds to transforming the product model.

By providing a *language capable of relating properties of those structures* of concepts and associations, a transformation can be understood as a *relation between a product before and after the transformation*. Then, by applying one argument of this relation to the model of the product under development and by providing a mechanism to *constructively compute* the other argument, the relation creates the transformed product. Thus, by formalizing standard operations of a development process as transformation relations, the process can be supported by mechanized operations. Examples for these operations are architectural refactorings of systems consisting of hierarchical components and connected via ports linked by channels; e.g., the pushing down of a component into a container component, making it a subcomponent of that container and requiring to split or merge channels crossing the boundary of the container component via the introduction or elimination of intermediate ports. Transformations like this structural refactoring or the semantically equivalent refactorings of state machines found in [13] are especially important since they leave the behavior of the modeled system unchanged.

Due to their generality, model transformations form the basis for many model-driven approaches ([14, 6, 15, 16]). In contrast to other development environment such as the Eclipse Refactoring plug-in, providing a fixed set of refactoring rules, a generic transformation mechanism allows the tool-user or tool-adaptor to enhance the functionality of the tool by defining domain-specific operations such as safe refactoring rules. Since models can be interpreted as graphs, transformation frameworks generally define operations as graph transformations, providing constructs to manipulate nodes (elements) and edges (relations) of a product model.

This generic approach is used in graph-based frameworks such as MOFLON/TGG [17], VIATZRA [18], FuJaBa [19], DIAGEN [20], AToM^3 [21], or GME [22]. These approaches are based on graph-grammars or graphical, rule-based descriptions [23]. Basically, for the declaration of basic transformations the transformations are described in a pre-model/post-model style using graph-patterns. In triple-graph-grammar approaches [24] additionally a correspondence graph [25] is added to explicitly model mappings between (parts of) the pre- and post-model during transformation. Furthermore, those approaches often use extensions to enhance the patterns as well as to describe their compositions, such as OCL expressions, and state machines. Figure 9.5 shows the formalization of a rule for the merging of a channel in the push-down refactoring in the FuJaBa approach, using an extended object diagram notation with annotations to specify the creation or destruction of objects and links in a product model.

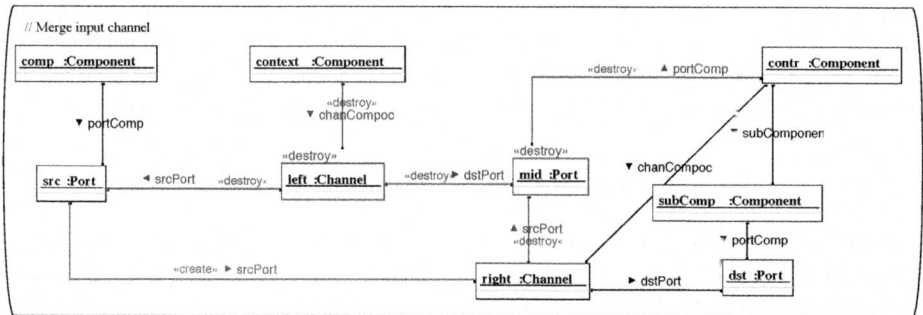

Fig. 9.5. Pattern-Based Specification of the Merge Rule of the Push-Down Refactoring

Since transformations basically correspond to relations between graphs, besides using these object pattern diagrams operations can be described directly in relational formalisms similar to the QVT approach [26] and its respective implementations such as ATLAS [27], F-Logics based transformation [28], or TEFKAT [29]. Furthermore, these rule-based relational approaches allow to use more declarative as well as more imperative forms of specification, e.g., providing a description of a specification in a purely declarative fashion, and an alternative, more imperative and efficient form. Strictly declarative, relational, and rule-based approaches as in [30] even allow to use a single homogenous formalism to describe the basic transformation rules and their composition. Complex analysis or transformation steps can be easily modularized since there are no side-effects or incremental changes during the transformation. Additionally, declarative approaches allow to use loose characterizations of the resulting model, supporting the exploration of a set of possible solutions to automatically search for an optimized solution, e.g., balanced component hierarchies, using guiding metrics; the set of possibile solutions can also be incrementally generated to allow the user to interactively identify and select the appropriate solution.

Technically, often a distinction between *exogenous* and *endogenous* transformations is used, depending on the characteristics of the metamodels, the source and the target of the transformation conforms to ([31], [32]). While in endogenous transformations, the source and target models are instances of the same metamodel, in the exogenous case they are instances of different metamodels. Besides the endogenous model refactorings and the exogenous model translations, model transformations are particularly helpful in the in between case of metamodel evolution. Large overlaps between source and the target domain lead to similar but differing metamodels. Here, model transformations can support the migration of models during the evolution of the metamodel, as discussed in Section 9.4.

Besides these fundamental issues of model transformation – see [31] for an overview – for the practical application also further aspects are of importance. Specifically, aspects like debugging support to trace the application of rules, analysis support to ensure syntactic and semantic correctness of transformation rules, the understandability of rules and their changeability with respect to size, complexity and degree of modularization,or the efficiency of transformations both with respect to the framework to import and export models as well as the execution of the transformation rules are gaining increasing attention.

9.2.3 Model Versioning and Model Merging

Model-based software engineering improves the development process by lifting the level of description from the solution domain – i.e., the domain of the implementation – to the problem domain – i.e., the domain of application – raising the level of abstraction to reduce the accidental complexity of the engineering task to focus on the essential complexity. Model analysis, e.g., conformance checking, and model synthesis, e.g., model transformation increase the degree of automation. However, raising the level of abstraction also introduces new problems.

A core aspect in the development of complex and long-lasting systems, as, e.g., generally found in embedded software systems, is the construction of those systems in incremental and often parallel steps. In traditional, code-based approaches, techniques like versioning and merging support the step-wise and distributed implementation. In a model-based development process, corresponding techniques must be supplied on model-level. While the linear structure of program code simplifies the task of comparing different fragments or merging them, the same problem of comparing or merging models is more complex due to their more general, graph-like structure.

To compute the difference between two models or to obtain a merge version of two models, the commonalities between those two models are identified via matching. To construct this matching, two different approaches are possible. If a model is described via its edit-history, consisting of the basic operations – like introducing or deleting elements or relations, changing their attributes, etc – to obtain this model, the matching essentially corresponds to identify the common operations.

However, in many cases models do not include those edit histories. Here, the matching has to be constructed by directly comparing these models; since elements of these models generally also do not maintain unique identifiers under modification – especially during deletion and re-insertion – matching has to be based on some notion of correspondence over model elements, generally based on similarities of attribute values. The most general approach to construct a matching for that case is based on a fixed-point iteration, starting with a pair-wise correspondance between the elements of two models, and extending this correspondance through the relations of each model. Since this general approach is rather complex, generally heuristics are used to improve the efficiency of the matching.

The latter matching approach is, e.g., implemented in the *SiDiff* algorithm [33]. For the construction of differences on the model level, *SiDiff* has, e.g., been integrated in the MatLab environment *MATE* [7], or the UML-like FuJaBa framework [19].

9.3 Evolution

Model management generally handles the operations necessary to deal with models on a large scale in large projects. Equal in importance, though perhaps not equal in scale, is the need to manage models as incremental evolution is required. We discuss these issues in this section.

We use the term *evolution* here in the same sense that it is commonly used in discussions of science: incremental changes brought about as external factors change. As we discuss in this section, these external factors can include changes to the system requirements, the language used to describe requirements and models, as well as changes of style. In each of these cases, the technology used to evolve the models is largely the same. However, the techniques to evolve the models may vary.

The problem of evolution is not new to software engineering. Various approaches have been suggested to address the evolution problem in various software domains, the most prominent being schema evolution in object oriented databases. While there have been some attempts to extend these techniques to other areas such as model based software [34] [35], the nature of DSMLs and their evolution suggest that there is a need for a dedicated solution.

9.3.1 Evolutionary Model Development

When categorizing development processes that we use today, we find two basically different approaches with respects to models: The document / waterfall like approaches use chains of models from early requirements down to running code. When changes appear they are usually applied on the current level only (i.e., the code level) and models of earlier phases are not touched anymore. These approaches need tracking of decisions between their artifacts that then allow co-evolution of models and code with respects to new requirements etc.

The other type of software development approaches, namely the agile ones, do not use models at all. They rely on code from the beginning, and this has several advantages: Code is executable and thus provides a form of immediate feedback that non-executable models couldn't. Furthermore, tests can be written in code to and automatically rerun every single adaptation of the code. Regression tests such as these give confidence that changes to the code do not violate the requirements that are encoded in such tests.

Many users of modelling technologies desire to raise the level of abstraction at which they are programming. This requires code generators to produce software based on the semantics of the models. Assuming that such code generators provide us with the ability to "program" on the higher level using models, we can use these models to create tests as well as our system specifications. Such models can be considered "executable" (i.e., not used simply for analysis or documentation), and are therefore the principal artifacts of our construction phase.

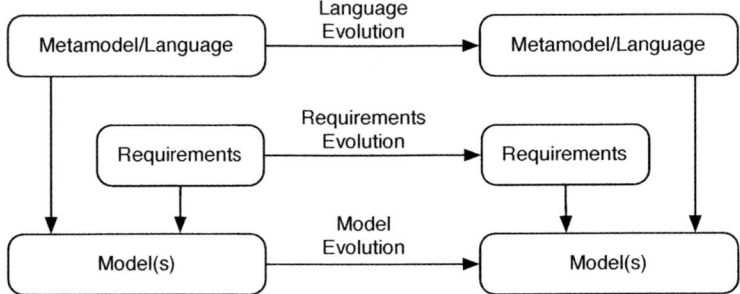

Fig. 9.6. There are two major preconditions for the need to evolve models. The evolution of the modelling language may invalidate structures, types, or constraints that are used and assumed by the original models. The evolution of requirements may invalidate design choices made by the original modeler. Either, or both, of these evolution types may trigger the need for model evolution.

Elements of Model Evolution

When discussing model evolution there are a few terms that we use with a specific meaning. We discuss mainly the evolution of models that conform to a *metamodel*. This metamodel describes the language used to specify the models.

Models are frequently transformed in order to be used as an artifact later in the design chain. We use the term *model transformation* to describe rules for rewriting models. Model transformation is one technique for *model interpretation* which is a generic term that describes how meaning is given to some model.

In model transformation, we talk about transforming some *source model* (SM) into a *destination model* (DM). Rules that describe how these transformations are specified are discussed later. Destination models may conform to a different metamodel than the source models.

Now, any of these elements may change, and thus require evolutionary changes to the models. If the destination metamodel changes, then existing rules to transform source models may need to be updated. If changes are made to the source metamodel, these transformation rules may also be invalid. Similar changes may need to be made if new constraints on the structure of either the source or destination languages is anticipated.

Actors in Model Evolution

There are several actors specific to model evolution:

- Model Designer (a.k.a., Original Modeler): This actor utilizes a modelling language to develop a model. This model is the *original* or *source* model in the discussions of this section, and represents an authoritative version of the intent of the model designer.
- Model Evolver: This actor takes the *original* model and evolves it as necessary. Reasons for this evolution may include bug fixes, changes in requirements, changes in the language, etc.
- Language Designer: This actor created the modelling language used by the original modeler (the *original modelling language*).

- Language Evolver: This actor modified the original modelling language.
- Requirements Specifier: This actor created the *original* requirements against which the original model was designed or tested.
- Requirements Evolver: This actor modified the original requirements.

These roles may be played by the same person for small projects, but for many large projects there will be several persons playing the various roles, or in fact several persons playing the same role. As such, the issue of maintaining intent throughout the design and construction phases, as well as subject to various evolutionary paths, is of great importance. However, not all evolutionary transformations need human intervention, as we discuss next.

9.3.2 Automating Evolutionary Transformations

Although there are many motivators for transformations, we examine here a few of the most common reasons for transforming models, namely changes in requirements, and changes in style that fall under the category of "refactoring." As discussed earlier, the changes in requirements may result in changes of semantics, while refactoring changes are (by definition) limited to behavior-preserving transformations. We also will discuss how transformations of a domain-specific modeling environment can be further automated, based on the strong typing in their metamodel. Another major reason for model transformations is changes in the domain itself, but the complexity of this issue deserves additional explanation, and it is covered in Section 9.4.

How can we automate the transformation of models, or automate the creation of transformation specifications, to aid model evolution? In fact, the former is achievable if the burden of creating the patterns and other assorted rules is placed on a model transformation expert. The automatic creation of these specifications (rules) is computationally feasible, though it brings into question whether semantics are maintained, constraints are satisfied, etc.

Evolution through Refactoring

In an agile development process (where only executable artifacts are created) we must accept that the executable artifacts—in our case, the models—need to be capable of modifications similar to those of software refactoring [36].

Model refactoring is a form of model evolution, where the semantics of the model remain the same, but the structure changes. We note that this definition is not always used for refactoring (specifically the clause that semantics are preserved), but we use this definition to avoid confusion with other kinds of model evolution where semantic preservation is not the goal. It may also be useful for the reader to consider the notion of software refactoring, where automatic changes are made to a software project based on renaming a variable, or changing a class name (i.e., all dependencies and uses are appropriately renamed throughout the software project). A comprehensive listing of refactoring types for *software* is presented by Fowler et al. in [36].

Similar changes in the domain of model refactoring are possible. For example, in [37] a tool is discussed that shows how to specify model transformations that will extract a superclass from a set of selected classes on the metamodel level. This application towards domain-specific modelling languages provides an avenue to maintain the types of a modelling language, while streamlining the metamodel definition.

There are also many applications to model refactoring for domain-independent models created using UML modelling tools. In [38], the authors show how the development of new metamodels (representing the "source" patterns to be matched, and "destination" patterns to be the resultant models) can work with existing UML models.

Evolution for Requirements Changes

The application of agile techniques in a model-driven sense means that models are used in the beginning of development, and continued to be used throughout the project lifecycle. Since the definition of agile development is to discover requirements along the way, or refine them as they are clarified by some customer, a model-driven approach *must* be robust to changes in the requirements during the development phase [39].

This means that some models, though correct when they were built, may be subject to new requirements now, or in the future. Automating these transformations based on updated specifications changes is not feasible, and many specifications languages such as $\mathbb{Z}$ [40] are somewhat infamous for an inability to synthesize the system for which they describe requirements. This is not as much a limitation of those requirements languages, but rather a reminder that updates from changes in formal requirements should be made by a knowledgeable actor—namely the model evolver, consulting with the requirements evolver.

Consider a model-based design of a controller for an unmanned aerial vehicle [41]. The controller is designed to satisfy a certain requirement for rise-time and overshoot of the vehicle state. However, if that requirement changes, the controller design must also change. This may be simply a change in values (changing the rise time, for example), but if the change in requirements is significant, or disruptive, the design may also need to change.

In any case, one major benefit of model-based engineering is that the controller is synthesized from the model (either in software, configuration for a runtime tool, etc.). However, there is the question of *are any existing requirements unmet, after the evolution performed for requirements changes?* In order to answer these questions, the model must be subjected to regression tests that verify requirements satisfaction for the model. Previous work in regression tests for models [42] concentrates on the differences between two models, thus reducing the number of regression tests that need to be run. If models are appropriately tracked, then work such as this can dramatically reduce the time to confirm that the models still conform to the requirements.

It is important to point out, though, that without existing tests for meeting requirements, that there can be no certainty that the models *as built* even conform to the specified requirements.

Domain-Specific Model Evolutions

A domain-specific modeling language provides special advantages to evolving models within the same language. This is because the specific goal of domain-specific modeling is to provide a programming environment (language) that represents the domain concepts as programming primitives. This is true both for domain-specific models that are in a metaconfigurable environment such as GME [43] or AToM[3] [21], but also for sophisticated user environments such as LabView and MATLAB/Simulink, who present domain-specific toolboxes for creating models.

9.3.3 Semantics of Evolution

A conflict of intentions comes to the forefront when evolving models. Specifically, the following question must be answered: *as these models evolve, should the original intent of the modeler be preserved, or does this evolution overrule any original intent?* Answering this question can be difficult, and in many cases requires judgement to be made by the model evolver.

There are many changes to the language, and some in the requirements, that can be automated such that no model evolver need be "in the loop" to confirm semantic interaction. However, there are many other evolutionary transformations that can be automated, but *certainly* need intervention by a modeler to confirm intent of evolution.

Consider the example of port specialization, where an object of kind `Port`, specialized into two types, `InPort` and `OutPort`, should be rewritten as either one of these types. The metamodel is shown in Figure 9.7a, and explicitly shows that a `Component` can contain zero or more *Port* objects. These `Port` objects can be associated through a `BufferedConnection`. As the metamodel shows, this connection can either be made to any to `Port` objects (who would then play the role of `src` or `dst` in that association), or explicitly between an `OutPort` (playing the role of `src`) and an `InPort` (playing the role of `dst`).

In order to rewrite models to (essentially) make `Port` an abstract type, some context is needed to determine whether existing `Port` objects are likely to be playing the role of `src` or `dst`. Certainly a patten could be written such that all models playing the role of `src` become `OutPort` models. However, this gives rise to two obvious problems: (1) what if there exists a `Port` that is not playing *any* role in an association, and (2) what if there exists a `Port` object that is playing *both* the `src` and `dst` role in two separate associations? These possibilities are shown in Figure 9.7b.

Port p12 of C1 exemplifies the issue of determining the type of a `Port` that plays *no* role in any association. A casual human observer may infer that the placement near another `InPort`, or its proximity to the left of C1 would imply an `InPort`, but at this point, some human must enter the decision loop to help determine this, or a policy of "all unmatched ports become `InPort` models" must be adopted.

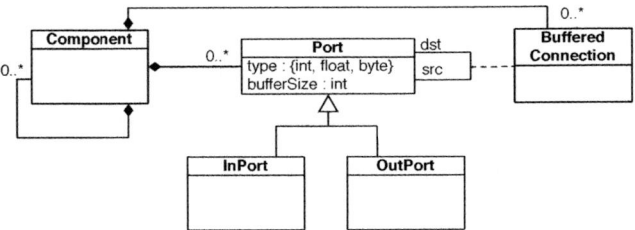

(a) The metamodel allows objects of kind Port, which is specialized as InPort and OutPort.

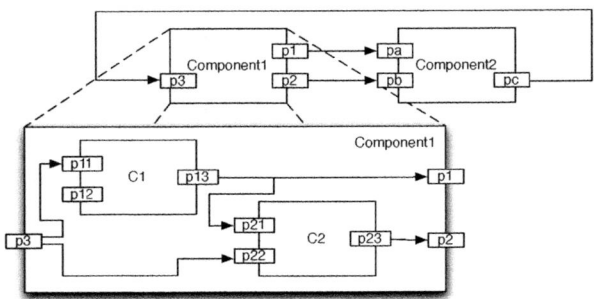

(b) A model built using the metamodel in (a). The contents of Component1 are shown to display the additional associations in which its Port objects play a role. The "arrow" end of the connections represent the dst role.

Fig. 9.7. (a) A metamodel allowing port interconnection between components. (b) A model built using the metamodel in (a). The "arrow" end of the connections represent the dst role. If all ports in the various components are of type Port, then how to automatically convert them into InPort or OutPort while maintaining the original modeler's intent?

Port p3 of Component1 plays the role of dst in its connection with port pc of Component2[1], and plays the role of src in its connection with ports p11 of C1, and p22 of C2. It is clear to a casual human observer that p3 is an InPort, but making this determination based on context requires careful specification of many partially ordered matches.

In [44] the issue of additional context is solved, where the containment relationships of various Port objects, the Component to which they belong, and the Component to which the BufferedConnection belongs all play a role in matching the pattern. However, as that work also states, the first problem (Port objects that play no role at all) still requires some actor, namely the model evolver, to make a decision on which type the Port should assume upon transformation.

Although there are a significant number of corner cases such as these, where the difficulty of creating an evolution strategy that preserves the original intent

---

[1] As a shorthand, we use directed connections to show src and dst roles, where the arrow resides on the dst role.

is called into question, there are a significant number of evolutions that can be created computationally. We will discuss the techniques useful for automating such transformations in Section 9.3.2, but now we turn to the mechanics most commonly used in performing the transformations.

9.4 Modelling Language Evolution

Development environments evolve as tool vendors constantly improve their tools. Although programming languages have become quite stable, we cannot claim the same about modelling languages: OMG languages are still the subjects of major upgrades and the UML models created so far need to be upgraded as well. If models are used to program against libraries or components, we have another source of model upgrades, although they do not change the structure of the models, but their vocabulary. With domain-specific languages, [3], we will need even more profound evolution techniques, as DSLs are usually made for single domains/companies or even projects and thus have the tendency to strongly evolve over time.

Model evolution is the transformation of domain models that were created under a language, L, to be well-formed and conform in the successor language, L'. Of particular importance is the question of the semantics of the models under each language. Existing work in the area of domain model evolution focuses on the techniques and methods for synthesizing transformations based on changes in the metamodel. Sprinkle's thesis [44] provides an academic perspective (for the mechanics of synthesizing such transformations, see [45]). Techniques for the graphical specification of the semantics of a modeling language (i.e., the code generator associated with a metamodel) can be found in [46]. A proposal by Bell [47] advocates the creation of a catalog of grammar transformations that are capable of automating the evolution of DSL programs.

We divide these kinds of model evolution tasks into two categories: *syntactic model evolution* and *semantic model evolution*.

9.4.1 Syntactic Model Evolution

Syntactic model evolution is a transformation or set of transformations that rewrites a model to conform to its new metamodel. It is useful for this set of transformations to be partially ordered, to permit deterministic results of the application. We do not require syntactic model evolution to be an atomic translation, but we instead depend upon the definition of a deterministic syntactic transform set to produce a logically atomic translation (though perhaps in several phases which produce intermediate or temporary artifacts).

Syntactic model evolution *only* guarantees that the model as transformed will be syntactically valid (i.e., conform to the new metamodel). As such, a trivial solution is to delete all models in the repository, but such a solution is clearly not acceptable. However, it does present the difficulty of using syntactically driven transformations to a model evolver after the language has evolved. Consider the frustration of loading a model into a modeling environment, only to realize that one single model is causing an exception. If deleting that model allows the model evolver to load the models, they may decide that they have completed evolution, but may have violated a large set of requirements in deleting that model.

There is a concrete example, which we can draw from our previous discussion of Figure 9.7. If we interpret this issue as removing the type `Port` from the modeling language, and replacing it with two types, `InPort` and `OutPort`, we are now dealing with a model that does not conform to its evolved metamodel. Namely, the existence of objects of type `Port` violates the abstract syntax requirements.

There are two trivial solutions which satisfy the requirements for syntactic model evolution; (1) transform all models of type `Port` to `InPort`, and delete all `BufferedConnections`; (2) transform all models of type `Port` to `OutPort`, and delete all `BufferedConnections`. Of course, nontrivial solutions will yield a "correct" result, which we discuss in the next subsection.

Nonetheless, syntactic model evolution is a powerful tool for an advanced model evolver. With expert knowledge of the metamodel, and of the state of the model, syntactic model evolution can provide a rapid way to reload existing models, regardless of their semantics. One reason for this might be that changes to the language were to remove types that were no longer relevant: so, deleting those types is appropriate. Another reason might be that the models were developed in the very early stages of the project, and they will all have to be examined anyway, so any models preserved will be used, but deleting models that violate new language conditions is not disastrous, as they will be recreated with new types.

Of course, much of this depends on the size of the database of models as well. For model databases of size 10-20, a careful, complete, model evolution may take weeks to create, but the models can be rebuilt in a few hours. All of these considerations are relevant to the decision of the model evolver.

9.4.2 Semantic Model Evolution

Semantic model evolution is a transformation or set of transformations that rewrites a model to have the same semantics in its new language that it had in its original language. The observant reader will note that syntactic model evolution is guaranteed in a semantic model evolution process, because for preserved semantics, the evolved model must conform to the evolved language.

It is undoubtedly true that syntactic model evolution can result in a semantic model evolution, if properly applied by the model evolver. This allows standard model transformation techniques to be applied to evolving models, if the transformation patterns are appropriately designed. Such is the work by Sprinkle in [45], and by Karsai et al. in [46].

The semantic model evolution problem is also similar to the tool integration problem. In [48], Tratt motivates the benefits that model transformation offers for tool integration. The two issues of syntactic and semantic interoperability of tools is discussed in [49], which also advocates model transformation as the conversion mechanism between tool models.

Questions of Semantics

What happens when more than one semantic domain exists for a particular language? If multiple semantic interpretations exist, then each member of the set of semantics *must* be satisfied in order to claim that a semantic model evolution has taken place. This issue can be extremely difficult, as changes in the language may negate the ability to attach a models semantics to a particular semantic domain.

Consider the domain of hybrid systems, where transitions between states represent discrete switches in the continuous dynamics of a system. If two particular semantic domains, simulation using one tool, and analysis using another, depend on portions of the modeling space not used by the other tool, than any removal of those portions of the modeling space may affect one tool, but not the other. A concrete example is to remove the invariant set from the modeling language: analysis tools require this set in order to verify whether the system state travels outside this set, while simulation tools can still operate without that set. It is possible, therefore, to still utilize one semantic domain, but not the other, with existing models by just deleting the invariant set from all objects.

The lesson here is that the more semantic domains to which a modeling language attaches, the more difficult it is to evolve that modeling language. For domain-specific languages, the issue is both more complex, and simpler, in that by constraining the language to a particular domain, the risk of that domain changing is reduced: however, if changes do propagate to that domain, the language *must* be evolved in order to maintain its intuitive relationship with the domain types.

There are additional difficulties introduced when changes in the constraints of a language (only) may in fact cause certain models to no longer satisfy those constraints. However, this problem can easily be checked by loading the models and running the constraint checker to determine any violations. It is still an open problem to understand what changes of constraints can be directly used to transform models where violations occur, or to predict that no violations will occur.

9.4.3 Techniques for Automated Model Evolution

Automating model evolution in the face of language evolution is tricky, if the goal is semantic model evolution. Nonetheless, there exist techniques for helping to determine how models should evolve in order to maintain semantics across evolution.

Differences between the original and evolved metamodel can help identify elements that have changed. This does require, however, some fairly advanced algorithms for detecting changes [50], unless such changes are recorded as they are made. In this sense, the correspondence models of a triple-graph grammar may provide sufficient indication of change, but may not provide a sufficient indication of what transformations are required for a semantic model evolution.

Examining the semantic domains to which the modeling language attaches, and how that semantic domain has changed between the original and evolved

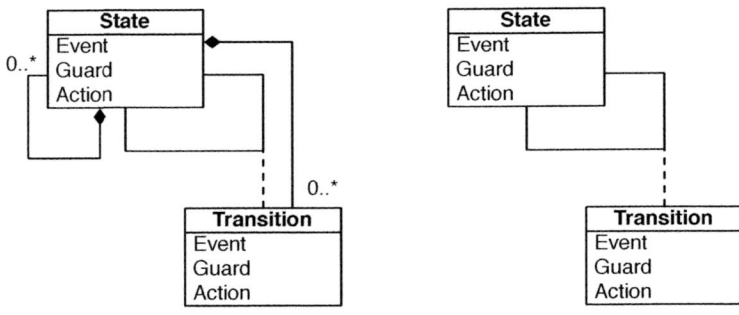

(a) A metamodel for a hierarchical finite state machine.

(b) A metamodel for a "flattened" finite state machine.

Fig. 9.8. (a) This metamodel allows containment of states by other states.(b) This evolved metamodel removes the ability to contain states hierarchically. Looking at the code generator, which removes checking for State containment, may imply that such semantics are no longer important, rather than implying that such semantics should be reapplied without hierarchy.

metamodel, is another aid. For example, if the algorithm to generate code or models has simply renamed Type1 to Type2, then this may be sufficient to evolve the models (change all models of type Type1 to be of type Type2).

However, there may be subtle issues even with this approach, as we show with the metamodel and evolved metamodel shown in Figure 9.8. If the algorithm to generate code removes the check for State objects contained within a State, and the metamodel indicates that containment of States within each other is no longer allowed, then a naive approach could simply remove all State objects contained within another State. Unfortunately, this can also be interpreted as a requirement that an existing hierarchical state machine must be flattened. Algorithms exist that can refactor state machines [13] to be semantically equivalent, but the model evolver must realize that this is the requirement.

9.4.4 Step-By-Step Model Evolution

In the previous sections, we discussed modeling language evolution methods that are able to handle arbitrarily large gap between the original and the evolved language. However, in most of the practical cases, modeling language evolution does not happen as an abrupt change in a modeling language, but in small steps instead. This also holds for UML: apart from adding completely new languages to the standard, the language has been changing in rather small steps since its first release.

This assumption facilitates further automation of the model evolution by tools for metamodeled visual languages [51] [52]. The main concepts of a step-by-step evolution method is depicted in Figure 9.9.

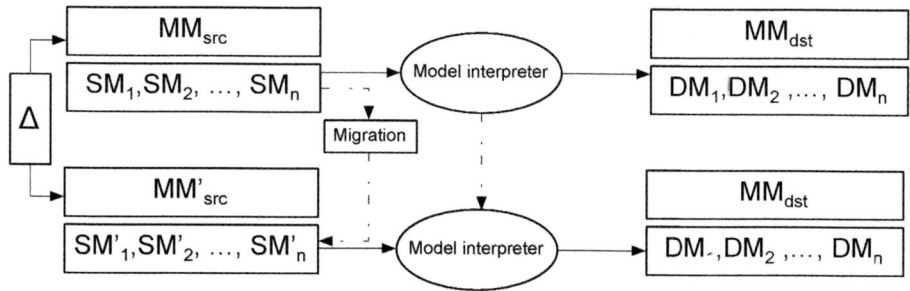

Fig. 9.9. Step-By-Step Evolution Concepts

The backbone of the diagram is a well-known DSL scenario depicted in the upper half of the figure. When a domain-specific environment is created, it consists of a metamodel (MM_{src}), which may have arbitrary number of instance models (SM_1, SM_2, ...,SM_n. The models need to be processed or transformed ("interpreted"), therefore, an interpreter is built. The interpreter expects that its input models are compliant with MM_{src}. In parallel, the output models of the interpreter must be compliant with the target metamodel MM_{dst}. The inputs of the interpreter are MM_{src}, MM_{dst}, and an input model SM_i, and produces an output model DM_i.

The objective is to migrate the the existing models and interpreters to the evolved language. For the sake of simplicity, we assume that only the input modeling language evolves, de output model remains the same. The evolved counterparts are denoted by adding a prime to the original notation. In the evolution process, we create the new (evolved) metamodel (MM'_{src}). We assume that the changes are minor enough both in size and nature, such that they are worth being modeled and processed by a tool, rather than writing a transformation from scratch to convert the models in the old language to models in the evolved language. This is a key point in the approach.

Having created the new language by the evolved metamodel, we describe the changes in a separate migration DSL (Model Change Language, MCL). This is denoted by Δ, and it represents the differences between MM_{src} and MM'_{src}. Besides the changes, this model contains the actual mappings from the old models to the evolved ones, providing more information how to evolve the models of the old language to models of the new language. Given (MM_{src}), (MM'_{src}), and MCL, a tool can automatically migrate the models of the old language to models of the evolved language.

Furthermore, also based on the (MM_{src}), (MM'_{src}), and MCL, it is possible to migrate the model interpreter. Since it cannot be expected that the way of processing the new concepts added by the evolution can be invented without human interaction, the tool can produce an initial version of the evolved interpreter only. A usual implicit assumption here is that the language elements appearing in both the old an the evolved model should be processed in the same way. Moreover, using this assumption and the MCL model, the interpreter for the parts of the old language that have been unambiguously changed by the evolution can also be generated.

In the following two sections, we present a possible realization for both the change description and the interpreter evolution.

Describing the Changes

Recall that our approach uses a DSL to describe the changes between the original and the evolved metamodels. This raises a a natural question: what sort of changes should be described and how? The second part of the question is partly answered: one can use a DSL to describe these changes. However, there is another criterion: the change description must hold enough information to facilitate the automated evolution of the already existing models (SM_n).

The first part of the question can only be answered by the practice. Below we show the construct we distilled by the experience gained in one of our research project and described in detail in [51].

Figure 9.10 outlines the structure of an MCL rule. For the sake of simplicity, we use the convention that elements on the left-hand-side of the *MapsTo* relationship belong to the original metamodel (MM_{src}), and the right-hand-side elements are taken from the evolved metamodel (MM'_{src}). The most important concept in MCL is the *MapsTo* connection. This connection originates from a class in the original metamodel, and points to a class in the evolved metamodel. One can assign conditions and commands written in imperative code to a mapping.

The basic operations provided by MCL are as follows:

(i) Adding elements, such as class, associations, and attributes. In Figure 9.11, we add a new element called *Thread* within a Component, along with a constraint that every *Component* must contain at least one *Thread*. The old models can then be migrated by creating a new *Thread* within each *Component*.

(ii) Deleting elements: classes, attributes, or associations. It is important that deleting elements means that we do not need that information anymore, we can lose it. If the information contained by an element is used in the evolved model, the element should not be deleted. The operation needed in this case falls into the next category. As Figure 9.12 shows, deletion is modeled with mapping a class to the null class.

(iii) Modifying elements, such as attributes and class names. The conditional mapping to new or other model elements also falls into this category. Figure 9.13 depicts a migration rule for a prevalent model refactoring case: a class becomes abstract base, and the existing instances are migrated as the instances of the new subclasses, based on certain conditions, typically, the attribute values of the instances. The conditions assigned to the *MapsTo* connections specify which mapping must be performed. The attribute calculations and other projections from the old class to the new ones are described by the commands assigned to the connection *MapsTo*.

(iv) Local structural modifications. If the operations detailed above need to be performed in a certain context, it can be defined by the *WasMappedTo*

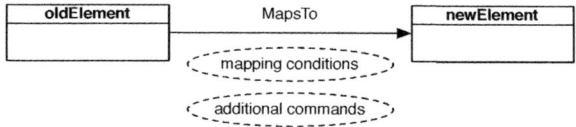

Fig. 9.10. Schematic description of an MCL rule

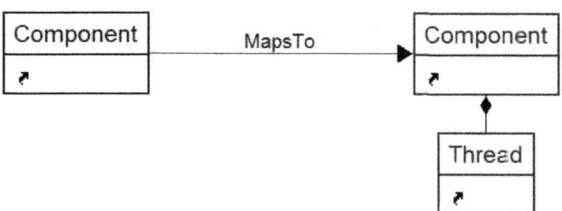

Fig. 9.11. Addition rule

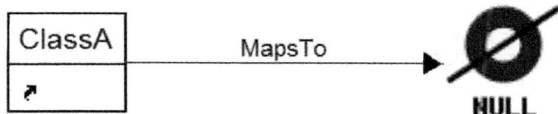

Fig. 9.12. Deletion rule

connection. Figure 9.14 shows an example, where we short-circuit the containment hierarchy. The intent of the migration is to move all instances of *Class* up in the containment hierarchy: the instances should be contained by *ParentParent* instead of *Parent*. *WasMappedTo* does not specify an operation, it ensures that that *ParentParent* originally containing the *Parent* should be the parent of the *Parent*'s children. Recall that the left-hand-side of the figure references classes from the source metamodel, whereas the other side references classes in the evolved metamodel, thus, the name conflict does not matter in this case.

Given a the old metamodel, the evolved metamodel and the MCL description, a code generator is able to create executable code that migrated the (SM_n) models to the new DSL defined by the evolved metamodel.

Evolving the Model Transformations
As it is shown in Figure 9.10, not only the models, but the model interpreter must also evolve. The Universal Model Migrator Interpreter Evolver (UM-MIE) package is a tool to semi-automate the interpreter migration process. The tool takes the old metamodel (MM_{src}), the evolved metamodel (MM'_{src}), the Δ model in MCL, the destination metamodel (MM_{dst}) and the old model

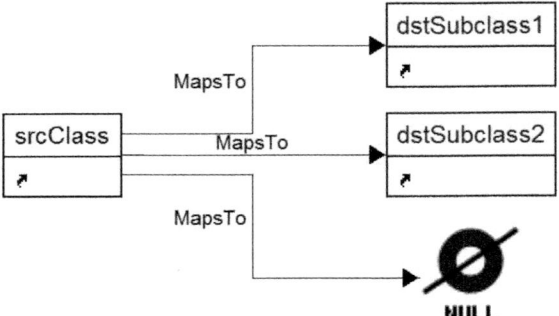

Fig. 9.13. Modification rule

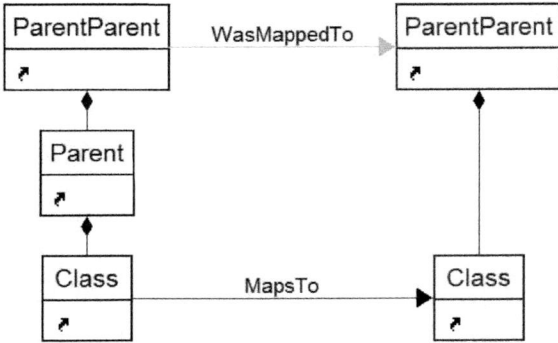

Fig. 9.14. Rule with context

transformation rules under the assumption that the destination model does not evolve (MM_{dst} is identical to MM'_{dst}).

We assume that the rule nodes in the transformation reference the input and output metamodel classes. The tool traverses the transformation rules, and takes each rule node to process the referenced metamodel classes. If these classes are in the destination model (MM_{dst}), or they were not changed by the MCL model, they remain intact. If a class has been deleted, the reference in the rule is set to null reference. Moreover, a warning is emitted that the null reference in the rule must be resolved manually. If a class has changed unambiguously by modification, such as renaming attributes, the tool automatically updates the rules. If there are multiple mappings such as in Figure 9.13, the tool emits a warning that the mapping should be done manually. Since the tool traverses the old transformation rules, the additions are not handled by the tool, their evolution must be performed by hand.

The UMMIE tool performs all the changes that must always be made. There are cases, in which there are several options, it depends on the intentions of the transformation developer. The main future direction of the tool is to provide "design patterns" for these cases, exposing the options to the developer, and after the selection, the evolution step is completed automatically.

Acknowledgements

The work presented is partially sponsored by DARPA, under its Disruptive Manufacturing Program. The views and conclusions presented are those of the authors and should not be interpreted as representing official policies or endorsements of DARPA or the US government.

References

[1] Bézivin, J., Favre, J.M., Rumpe, B.: Introduction to gamma 2006 first international workshop on global integrated model management. In: GaMMa 2006: Proceedings of the 2006 International Workshop on Global Integrated Model Management, pp. 1–3. ACM, New York (2006)
[2] Brunet, G., Chechik, M., Easterbrook, S., Nejati, S., Niu, N., Sabetzadeh, M.: A manifesto for model merging. In: GaMMa 2006: Proceedings of the 2006 International Workshop on Global Integrated Model Management, pp. 5–12. ACM, New York (2006)
[3] Tolvanen, J.P., Gray, J., Sprinkle, J. (eds.): 6th OOPSLA Workshop on Domain-Specific Modeling (DSM 2006). University of Jyväskylä, Jyväskylä, Finland, OOPSLA (October 2006), ISBN: 951-39-2631-1
[4] MathWorks Automotive Advisory Board: Control Algorithm Modeling Guidelines Using MATLAB Simulink, Simulink, and Stateflow (2007)
[5] The MathWorks: Using MATLAB (2002)
[6] Bézivin, J.: On the unification power of models. Software and Systems Modeling 4(2), 171–188 (2005)
[7] Stürmer, I., Dziobek, C., Pohlheim, H.: Modeling Guidelines and Model Analysis Tools in Embedded Automotive Software Development. In: Rumpe, B., Giese, H., Klein, T., Schätz, B. (eds.) Modellbasierte Entwicklung eingebetteter Systeme (2008)
[8] Deissenboeck, F., Juergens, E., Hummel, B., Wagner, S., Parareda, B.M., Pizka, M.: Tool Support for Continuous Quality Control. IEEE Software 25(5) (2008)
[9] Farkas, T., Röbig, H.: Automatisierte, werkzeugübergreifende Richtlinienprüfung zur Unterstützung des Automotive-Entwicklungsprozesses. In: Rumpe, B., Giese, H., Klein, T., Schätz, B. (eds.) Modellbasierte Entwicklung eingebetteter Systeme (2006)
[10] OMG: Object Constraint Language Specification. Technical Report 1.1, ad/97-08-08, Object Management Group, OMG (1997), http://www.omg.org
[11] Schätz, B.: Mastering the Complexity of Embedded Systems - The AutoFocus Approach. In: Kordon, F., Lemoine, M. (eds.) Formal Techniques for Embedded Distributed Systems: From Requirements to Detailed Design. Kluwer, Dordrecht (2004)

[12] Deissenboeck, F., Hummel, B., Jürgens, E., Schätz, B., Wagner, S., Girard, J.F., Teuchert, S.: Clone detection in automotive model-based development. In: Schäfer, W., Dwyer, M.B., Gruhn, V. (eds.) 30th International Conference on Software Engineering (ICSE 2008), Leipzig, Germany, May 10-18, pp. 603–612. ACM, New York (2008)

[13] Pretschner, A., Prenninger, W.: Computing refactorings of state machines. Software and Systems Modeling 6(4), 381–399 (2007)

[14] Fabro, M.D., Valduriez, P.: Semi-automatic model integration using matching transformations and weaving models. In: MT 2007 - Model Transformation Track, The 22nd Annual ACM SAC, pp. 963–970 (2007)

[15] Porres, I.: Rule-based update transformations and their application to model refactorings. Software and Systems Modeling 4(5), 368–385 (2005)

[16] Mens, T., Taentzer, G., Runge, O.: Analysing refactoring dependencies using graph transformation. Software and Systems Modeling 6(3), 269–285 (2007)

[17] Klar, F., Königs, A., Schürr, A.: Model transformation in the large. In: ESEC/FSE 2007. ACM Press, New York (2007)

[18] Varro, D., Pataricza, A.: Generic and meta-transformations for model transformation engineering. In: Baar, T., Strohmeier, A., Moreira, A., Mellor, S.J. (eds.) UML 2004. LNCS, vol. 3273, pp. 290–304. Springer, Heidelberg (2004)

[19] Grunske, L., Geiger, L., Lawley, M.: A graphical specification of model transformations with triple graph grammars. In: Hartman, A., Kreische, D. (eds.) ECMDA-FA 2005. LNCS, vol. 3748, pp. 284–298. Springer, Heidelberg (2005)

[20] Minas, M.: Spezifikation und Generierung graphischer Diagrammeditoren. Habilitation, Universität Erlangen-Nürnberg (2001)

[21] de Lara, J., Vangheluwe, H., Alfonseca, M.: Meta-modelling and graph grammars for multi-paradigm modelling in AToM^3. Software and Systems Modeling 3(3), 194–209 (2004)

[22] Sprinkle, J., Agrawal, A., Levendovszky, T.: Domain Model Translation Using Graph Transformations. In: ECBS 2003 - Engineering of Computer-Based Systems (2003)

[23] Rozenberg, G. (ed.): Handbook on Graph Grammars and Computing by Graph Transformation: Foundations. World Scientific, Singapore (1997)

[24] Klar, F., Königs, A., Schürr, A.: Model Transformation in the Large. In: Proceedings of the 6th joint meeting of the European Software Engineering Conference and the ACM SIGSOFT Symposium on the Foundations of Software Engineering. ACM Digital Library Proceedings, pp. 285–294. ACM Press, New York (2007)

[25] Schürr, A.: Specification of graph translators with triple graph grammars. In: Mayr, E.W., Schmidt, G., Tinhofer, G. (eds.) WG 1994. LNCS, vol. 903. Springer, Heidelberg (1995)

[26] OMG: Initial submisison to the MOF 2.0 Q/V/T RFP. Technical Report ad/03-03-27, Object Management Group (OMG) (2003), http://www.omg.org

[27] Jouault, F., Allilaire, F., Bézivin, J., Kurtev, I., Valduriez, P.: ATL: a QVT-like transformation language. In: OOPSLA 2006, pp. 719–720. ACM Press, New York (2006)

[28] Gerber, A., Lawley, M., Raymond, K., Steel, J., Wood, A.: Transformation: The Missing Link of MDA. In: Corradini, A., Ehrig, H., Kreowski, H.-J., Rozenberg, G. (eds.) ICGT 2002. LNCS, vol. 2505, pp. 90–105. Springer, Heidelberg (2002)

[29] Lawley, M., Steel, J.: Practical declarative model transformation with tefkat. In: Bruel, J.-M. (ed.) MoDELS 2005. LNCS, vol. 3844, pp. 139–150. Springer, Heidelberg (2006)

[30] Schätz, B.: Formalization and Rule-Based Transformation of EMF Ecore-Based Models. In: Gašević, D., Lämmel, R., Van Wyk, E. (eds.) SLE 2008. LNCS, vol. 5452, pp. 227–244. Springer, Heidelberg (2009)
[31] Czarnecki, K., Helsen, S.: Feature-based survey of model transformation approaches. IBM Systems Journal 45(3), 621–646 (2006)
[32] Mens, T., Czarnecki, K., Gorp, P.V.: A taxonomy of model transformations. In: Dasgstuhl Proceedings of the Seminar on Language Engineering for Model-Driven Software Development, vol. 04101 (March 2004)
[33] Schmidt, M.: Generische, auf Ähnlichkeiten basierende Berechnung von Modelldifferenzen. SiDiff 27(2) (2007)
[34] Bernstein, P.A., Melnik, S.: Model Management 2.0: Manipulating Richer Mappings. In: SIGMOD 2007 (2007)
[35] Banerjee, J., Kim, W., Kim, H.J., Korth, H.F.: Semantics and Implementation of Schema Evolution in Object-Oriented Databases. In: Proceedings of the Association for Computing Machinery Special Interest Group on Management of Data, pp. 311–322 (1987)
[36] Fowler, M., Beck, K., Brant, J., Opdyke, W., Roberts, D.: Refactoring: Improving the Design of Existing Code. Addison-Wesley Object Technology. Addison-Wesley Professional, Reading (1999), ISBN: 978-0201485677
[37] Zhang, J., Lin, Y., Gray, J.: Generic and domain-specific model refactoring using a model transformation engine. Research and Practice in Software Engineering, vol. II, pp. 199–218. Springer, Heidelberg (2005)
[38] France, R., Ghosh, S., Song, E., Kim, D.K.: A metamodeling approach to pattern-based model refactoring. IEEE Softw. 20(5), 52–58 (2003)
[39] Ambler, S.W.: Agile Modeling: Effective Practices for Extreme Programming and the Unified Process. Wiley, Chichester (2002), ISBN: 978-0471202820
[40] Woodcock, J., Davies, J.: Using Z: Specification, Refinment and Proof. Prentice-Hall, Englewood Cliffs (1996), ISBN: 0-13-948472-8
[41] Hoffmann, G.M., Huang, H., Wasl, S.L., Tomlin, C.J.: Quadrotor helicopter flight dynamics and control: Theory and experiment. In: Proc. AIAA Guidance, Navigation, and Control Conf. (2007)
[42] Korel, B., Tahat, L., Vaysburg, B.: Model based regression test reduction using dependence analysis. In: Proceedings of International Conference on Software Maintenance, pp. 214–223 (2002)
[43] Ledeczi, A., Bakay, A., Maroti, M., Volgyesi, P., Nordstrom, G., Sprinkle, J., Karsai, G.: Composing domain-specific design environments. Computer 34(11), 44–51 (2001)
[44] Sprinkle, J.: Metamodel Driven Model Migration. PhD thesis, Vanderbilt University, Nashville, TN 37203 (August 2003)
[45] Sprinkle, J., Karsai, G.: A domain-specific visual language for domain model evolution. Journal of Visual Languages and Computing 15(3-4), 291–307 (2004); Special Issue: Domain-Specific Modeling with Visual Languages
[46] Karsai, G., Agrawal, A., Shi, F., Sprinkle, J.: On the use of graph transformation in the formal specification of model interpreters. Journal of Universal Computer Science 9(11), 1296–1321 (2003)
[47] Bell, P.: Automated transformation of statements within in evolving domain-specific languages. In: Sprinkle, J., Gray, J., Rossi, M., Tolvanen, J.P. (eds.) 7th OOPSLA Workshop on Domain-Specific Modeling (DSM 2007), Montreal, pp. 172–177 (October 2007)
[48] Tratt, L.: Model transformations and tool integration. Software and Systems Modeling 4(2), 112–122 (2005)

[49] Bézivin, J., Brunelière, H., Jouault, F., Kurtev, I.: Model engineering support for tool interoperability. In: MODELS Workshop in Software Model Engineering (WiSME), Montego Bay, Jamaica (September 2005)
[50] Mens, T.: A state-of-the-art survey on software merging. IEEE Transactions on Software Engineering 28(5), 449–462 (2002)
[51] Balasubramanian, D., van Buskirk, C., Karsai, G., Narayanan, A., Ness, S.N.B., Shi, F.: Evolving paradigms and models in multi-paradigm modeling. Technical Report ISIS-08-912-2008, Institute for Software Integrated Systems (December 2008)
[52] Narayanan, A., Levendovszky, T., Balasubramanian, D., Karsai, G.: Automatic domain model migration to manage meta-model evolution. In: Schürr, A., Selic, B. (eds.) MODELS 2009. LNCS, vol. 5795, pp. 706–711. Springer, Heidelberg (2009)

10 Model-Based Analysis and Development of Dependable Systems

Christian Buckl[1], Alois Knoll[2], Ina Schieferdecker[3], and Justyna Zander[3]

[1] fortiss GmbH, Germany
buckl@fortiss.org
[2] Technische Universität München, Germany
knoll@in.tum.de
[3] Technical University Berlin, Germany, Fraunhofer FOKUS, Germany
{ina.schieferdecker,justyna.zander}@fokus.fraunhofer.de

Abstract. The term dependability was defined in the 1980s to encompass aspects like fault tolerance and system reliability. According to IFIP, it is defined as the trustworthiness of a computing system which allows reliance to be justifiably placed on the service it delivers. Hence, dependability is the capability of a system to successfully and safely complete its mission. This chapter concentrates on safety and reliability aspects. It starts with a review of the basic terminology including, for example, fault, failure, availability, and integrity. In the following, a mathematical model of fault-tolerant systems is defined. It is used in the further sections for comparison with different techniques for safety and reliability analysis. Also selected currently available model-based development tools are reviewed. A summary and identification of future research challenges conclude the chapter.

10.1 Introduction

In the last years, the trend to replace mechanical/electrical solutions by software centric solutions has been intensified. Even systems with strong requirements on safety and reliability are automated by the use of computer systems. Although the term dependability comprises several other aspects as well, the focus of this chapter is set on safe and/or reliable systems. These systems have to be designed fault-tolerant to fulfill the targeted requirements.

Typically, fault-tolerance is achieved by running the applications on replicated hardware and/or software components. The resulting complexity of the considered systems raises major concerns, in particular with respect to the validation and analysis of performance, timing, and dependability-related requirements, but also with respect to development times. Model-driven engineering addresses the problem of complexity by increasing the level of abstraction and by partial or total automation of selected phases within the development process.

Several tools are available for modeling dependable systems, many of them based on the Unified Modeling Language [1]. This chapter, however, focuses on model-driven methods that automate phases in the development process either by automating the dependability analysis for certain system architectures, or by automating the code generation process.

The chapter starts with a definition of terms relevant for dependable systems in Sec. 10.2. Section 10.3 presents a generic model of fault-tolerant systems based on the work of Arora and Kulkarni [2, 3]. Section 10.4 discusses existing approaches for reliability and safety analysis. Subsequently, three examples of model-driven tools in the area of safety-critical systems are analyzed in Sec. 10.5. The chapter is concluded by a summary and the identification of some research challenges in Sec. 10.6.

10.2 An Overview on Dependability

The term dependability was defined in the 1980s to unite relevant aspects [4]. Laprie defined computer system dependability as the quality of the delivered service such that reliance can justifiable be placed on this service. Avizienis et al. [5] defined six attributes of dependable systems as depicted in Figure 10.1: availability, reliability, safety, confidentiality, integrity, and maintainability. As mentioned before, this chapter focuses mainly on safety and reliability aspects.

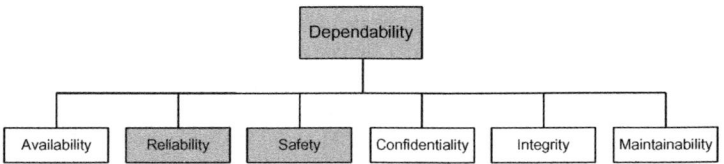

Fig. 10.1. Dependability Aspects

Definition 10.1. *Safety is defined as the freedom from those conditions that can cause death, injury, occupational illness, or damage to or loss of equipment or property [6]. It is also the expectation that a system does not, under defined conditions, lead to a state in which human life is endangered [7].*

Definition 10.2. *Functional safety is part of the overall safety that depends on a system or equipment operating correctly in response to its inputs [8].*

In case when the correct behavior cannot be guaranteed a safety-critical system should be brought into a safe mode (e.g., an emergency stop) instead of continuing to deliver the specified function. This is the main difference in comparison to reliability:

Definition 10.3. *Reliability is the ability of a device, system, or process to perform a required function under given environmental and operational conditions, and for a stated period of time [9].*

Both, safety and reliability of a system can be impacted by **faults**, **errors**, and **failures**. The system must handle them appropriately to achieve safety and reliability (i.e., it must be designed as fault-tolerant). The terms fault, error, and failure can be explained best by using a three-universe model of Pradhan [10]. This model, an adaptation of the four-universe model introduced by Avizienis [11], describes the different phases of the evolution from a fault to a failure.

The first universe is the **physical universe**, where faults occur.

Definition 10.4. *A **fault** is a physical defect, imperfection, or flaw that occurs within some hardware or software component [10].*

Faults can be **dormant** for a long time and not influence the execution of the component. When a fault is activated, the effects can be observed in the **informational universe**, classified as the second universe in [10].

Definition 10.5. *An **error** is the manifestation of a fault [10].*

Errors can be detected by the component itself, if some rules are defined to evaluate the state of the component. However, these tests may not be able to identify the cause of the error (i.e., the fault). Initially, errors are only reflected in parts of the component's state. If the error is not detected early enough by the component, the error may cause a subsequent failure.

It is important to notice that different definitions for fault and error are used in the literature as they are closely related concepts. Throughout this chapter, we will try to use the terms as defined in the previous definitions. However, if in the described projects or approaches the terms are used differently, we will use the terminology of the projects.

Definition 10.6. *A **failure** of a component occurs when the component deviates from the specified behavior [12].*

Hence, the third universe is the **external universe**, where the deviation from the expected behavior of a component can be observed. Consequently, a failure is the event that can be detected by interacting components. Thereby, a failure of a component can be a fault to its environment.

There are various reasons for faults. For instance, a fault can be a design fault, a physical fault, or an operational fault. While design faults are always active, physical faults are activated spontaneously with a certain probability. Faults can be classified according to their effect, as well. The effect can either be in the value domain or in the time domain [13]. Faults in the time domain are, for example, lost or delayed messages in a communication channel, but also replicated messages. Faults in the value domain are, for instance, erroneous results or bit flips in a message.

Fault-tolerance is the technique to guarantee that despite the presence of faults a system provides the specified behavior to its outer boundaries [4]. Fault-tolerance is always based on the effective deployment of redundancy, additional means that are not required to provide the specified behavior in the absence of faults. It is important to note, that redundancy is not only restricted to replicating hardware: the type of redundancy ranges from software or data redundancy to time and hardware redundancy.

A concrete selection and implementation of fault-tolerance mechanism depends on the number and types of the expected faults. These assumptions are summarized in the fault hypothesis.

Definition 10.7. *The **fault hypothesis** contains the assumptions about possible faults, their probability, and their effects to the components of a system.*

Based on the concrete fault hypothesis, the developer has to select appropriate mechanisms to tolerate these faults. Most of the different mechanisms are known since the 1950's due to the unreliability of the components at that time [5]. In general, one can divide the applied fault-tolerance mechanisms into four groups: error detection, error recovery, error handling / masking, and integration.

Definition 10.8. ***Error detection*** *allows the detection and localization of errors.*

Detecting an error is the first step to achieve the fault-tolerance. After an error is detected, the component has to analyze the affected subcomponents and the error type. This is essential to perform error recovery.

Definition 10.9. ***Error recovery*** *transforms a system state that contains one or more errors into a state without detected errors [5].*

There are different mechanisms to perform error recovery. The two most prominent types are **rollback** and **rollforward** recovery. Rollback is realized by restoring a previous state of the component [10]. This state is saved in a **checkpoint** before the component detects the error. The difficulty of a rollback recovery arises from designing and generating the checkpoints. Especially, if several components must be set back the realization may demand more efforts. The rollforward recovery uses application knowledge to compute a new, correct state out of the erroneous state. Usually, this transformation implicates a reduced quality of service.

Regardless of the concrete error recovery mechanism it is essential to ensure that the same fault is not activated again.

Definition 10.10. ***Error handling*** *prevents system's state corruption after the detection of a fault.*

To correctly perform the error handling the first step is the localization of the error and the identification of its cause. Within the second step, the fault is isolated by excluding the affected component from further interactions with other components. The affected component might be replaced by spare components. Further possibilities are to use other components to deliver the functionality in addition to the already delivered functionality or to degrade the system (i.e., **graceful degradation**). The isolated component can then be repaired, typically by an external agent.

If a sufficient level of redundancy is employed in the system, explicit error detection is not required. Instead one can use error masking.

Definition 10.11. *Error masking* guarantees that programs continually satisfy their intended specification, even in the presence of faults [14].

Typical examples for error masking are hot-redundant systems, where several redundant units are executed in parallel. Errors can be detected by comparing the results. If the master unit is affected by an error, another correct unit immediately takes over the master's task. The erroneous unit is excluded and can be repaired in the following. After a successful repair, it is necessary to reintegrate the repaired unit into the system to preserve the intended dependability:

Definition 10.12. *Integration* allows a repaired component to resume with its intended behavior and interaction.

For a successful integration, the **state synchronization** is essential. All participating units must agree on a new system state. A correct implementation of the state synchronization is an important, though complicated step.

The dependency goals and fault assumption determine the type of the fault-tolerance mechanism applied in a system.

10.3 A Generic Model of Fault-Tolerant Systems

The development of dependable systems can be supported by modeling. On the one hand, models are used to analyze the dependability of the system, (e.g., by using fault trees as discussed in Sec. 10.4. On the other hand, model-driven approaches can be used for generation of code related to fault-tolerance mechanisms.

In this section, a generic mathematical model of dependable systems is given. It is based on the work of Arora et al. [2, 3]. Based on this model, we can identify the basic aspects required to describe dependable system and compare the different models used by the tools discussed in this paper.

In the following, we start by specifying the execution of a system. Subsequently, we introduce the effects of faults and of fault-tolerance mechanisms.

10.3.1 System Operation without Faults

Definition 10.13. *A system* $S = (V, \Pi)$ *can be described by a finite set of variables* $V = \{v_1, ..., v_n\}$ *and a finite set of processes* $\Pi = \{\pi_1, ..., \pi_m\}$. *The domain* D_i *is finite for each variable* v_i. *A state* s *of system* S *is the valuation* $(d_1, ...d_n)$ *with* $d_i \in D_i$ *of the program variables in* V. *A transition is a function* $tr : V_{in} \to V_{out}$ *that transforms a state* s *into the resulting state* s' *by changing the values of the variables in the set* $V_{out} \subseteq V$ *based on the values of the variables in the set* $V_{in} \subseteq V$.

Definition 10.14. *The system is build up from a set of components* C. *A set of variables* $V_c \subseteq V$ *is associated with each component* $c \in C$. $V_c = V_{c,\text{internal}} \cup V_{c,\text{interface}} \cup V_{c,\text{environment}}$ *is composed by three disjoint variable sets: the set of internal variables* $V_{c,\text{internal}}$, *the set of interface variables* $V_{c,\text{interface}}$, *and the* set of environment variables $V_{c,environment}$. *Internal variables can only be accessed and altered by the set of processes associated with C:* $\Pi_c \subseteq \Pi$. *Interface variables are used for component interaction and can be accessed by all interacting processes. Environment variables are variables that are shared between the component and the environment of the system. Note that environment variables can only accessed by exactly one component. This set can be again divided into the input variables* $V_{c,input}$ *that are read from the environment and the output variables that are written to the environment* $V_{c,output}$.

Components can also be structured in a hierarchical way. A component $c \in C$ may consist of several subcomponents $c_1, ..., c_n \subset C$. The set of interface variables $V_{c,interface} \subseteq \bigcup_{1 \leq i \leq n} V_{c_i,interface}$ of c is a subset of the interface variables of its subcomponents $c_1...c_n$. The set of environment variables $V_{c,environment} = \bigcup_{1 \leq i \leq n} V_{c_i,interface}$ is the union set of all environment variables of the subcomponents.

Definition 10.15. *The functional behavior of a component* $c \in C$ *is reflected by the corresponding* **processes** Π_c. *Let* $V_{interface} = \{v | v \in V_{c',interface} \wedge c' \in C\}$ *be the set of all interface variables.* Π_c *is specified as a finite set of operations of the form* **guard** $\rightarrow$ **transition**, *where* $\text{guard} : V_{guard} \rightarrow \mathbb{B}$ *is a Boolean expression over a subset* $V_{guard} \subseteq V_c \cup V_{interface} \cup V_{c,input}$ *and* $\text{transition} : V_{in} \rightarrow V_{out}$ *is the appendant transition with* $V_{in} \subseteq V_c \cup V_{interface} \cup V_{c,input}$ *and* $V_{out} \subseteq V_c \cup V_{interface} \cup V_{c,output}$. *We refer to the old value of a variable by* v *and to the new value by the primed variable* v′.

The processes are expected to be deterministic, meaning that for each state s at most one **guard** can evaluate to **true**. This condition can be easily implemented by using one variable as a program counter and including this variable into the **guard** expression.

However, by allowing different processes to coexist simultaneously, non-determinism is introduced. There is no semantics which process will perform its operation, if several processes have an enabled operation. While non-determinism is not desirable for modeling the normal execution of the system, it is required to model faults due to the non-deterministic behavior of faults. To achieve determinism of the system execution, the interplay between different processes can be implemented in a deterministic way by specifying adequate guards in such a manner that for each possible state at most one process is enabled. To reach this goal, one might need to introduce auxiliary interface variables or use the value of time for time-triggered systems.

Definition 10.16. **Time** *is modeled similar to a component and represented by one process* Π_{Time} *realizing the time progress and a variable* v_{time} *containing the current time.* Π_{Time} *reflects the logical time and cannot be affected by any faults. In contrast, the local time on the individual computational nodes of the distributed system is derived from the components describing the behavior of the clocks used in the system, the related process, and its variables. The transitions can describe their temporal behavior by adapting the local time variable.*

Until now, the system has been considered in the absence of faults and without any fault-tolerance mechanisms. The first step to reach fault-tolerance is to translate the safety specification into a set of properties that must be valid for the application. While Arora et al. use computations and sequences of subsequent states to express safety properties we use state predicates P to express properties.

Definition 10.17. *A **state predicate** P is a Boolean function over a set of variables* $V_P \subset V$. *The set of state predicates represents the specification of the system and is therefore defined implementation-independent. Hence, the set of variables* $V_P \subseteq \bigcup_{c \in C} V_{c,\text{environment}}$ *is a subset of all variables that can be observed by the environment of the system.*

It may be necessary to define auxiliary variables that record the progress of the environment variables over time to express temporal properties. Establishing these variables explicitly, a potentially unnecessary tracking of all variables can be avoided. In general, only very few variables are needed for the history state [14]. Liveness specifications can be expressed by state predicates using the time process Π_{time}.

The transitive closure of the transitions of all processes defines the fault-free system as depicted in Fig. 10.2. It is defined as all states that can be reached beginning from some start states s_{start}. The state predicates P describing the intended operation must be true for all states within this transitive closure.

10.3.2 Faults

The introduction of faults into our model of fault-tolerant system is straightforward and can be designed as a component FH.

Definition 10.18. *The **fault component** is described as a set of variables* $V_{c,\text{FH}}$ *and processes* $\Pi_{c,\text{FH}}$ *that perform actions in accordance to the fault hypothesis.*

Due to the non-deterministic behavior of processes, the non-deterministic behavior of certain fault types appears. The propagation of an error depends, in turn, on the interaction between different components and their implementation. Therefore, it is necessary to define the behavior of a component in the presence of faults. This can be done by changing the actions of Π_c for a specific component. These could be the introduction of new actions or the addition of conditions to a guard. Both, additional elements and new actions can be based on the variables V_c and $V_{c,\text{FH}}$.

A good example is a fail-stop [14], where an auxiliary variable up_c denoting the fault status of a component c can be introduced. For all actions of P_c, the guard is expanded with a condition $!up_c$ to restrict the execution to such states where the component is not affected by a fail-stop fault.

10.3.3 Fault-Tolerance Mechanism

Kulkarni and Arora [15, 3, 14] pointed out that it is sufficient to use **detectors** and **correctors** to reach fault-tolerance.

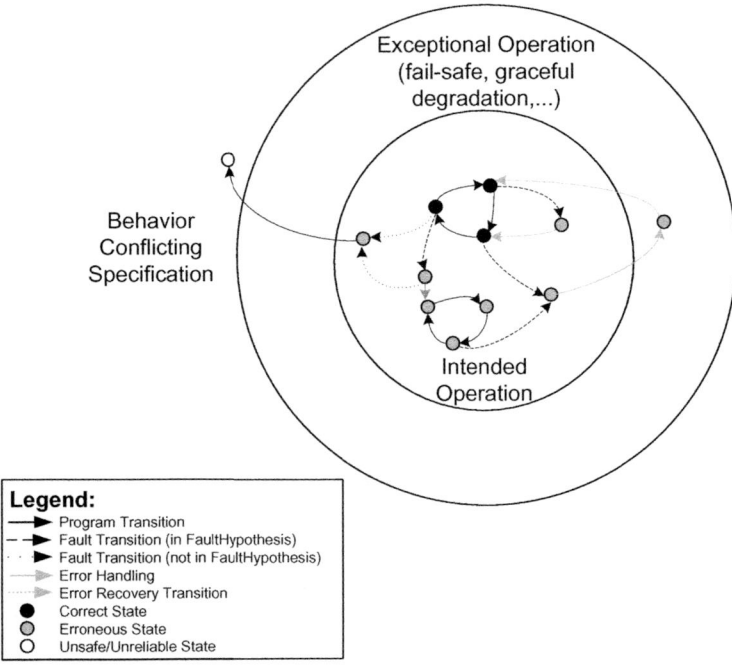

Fig. 10.2. Fault-Tolerance Concepts

Definition 10.19. Detectors $d_e : V' \subseteq V \to \mathbb{B}$ *are Boolean functions that monitor the variables of a system and can detect errors. Using the definition of predicates, a predicate* d *(Detector) detects a predicate* e *(erroneous state), if the following conditions are satisfied for each possible sequences* $s_0, s_1, ...$:

- **Safeness:** $\forall i \geq 0 : d_e(s_i) \Rightarrow e(s_i)$. *This condition requires that, if the detector detects an erroneous state, the decision has to be correct. False positives are not accepted.*
- **Eventual Detection:** $\forall i \geq 0 : e(s_i) \Rightarrow \exists j \geq i : d_e(s_j) \vee !e(s_j)$. *This condition requires that a detector will eventually detect a permanent erroneous state.*
- **Stability:** $\forall i \geq 0 : d_e(s_i) \Rightarrow d_e(s_{i+1}) \vee !e(s_{i+1})$. *The detector is also required to be stable, that is, it should not signal the disappearance of an error if the error is still present.*

Definition 10.20. Correctors *are actions of the form* guard $\to$ transition *that transform an erroneous system into a correct system. The actions are triggered by a detected error.*

This notion is, however, very abstract and it is useful to distinguish between different types of correctors. We will differentiate between operations for error treatment, error recovery, proactive operations, and integration. Operations for error treatment describe the reactions of the system when new errors are detected. A classic error treatment is the switch to a correct backup unit or a rollback recovery operation [16]. The operation may be based on previously executed **proactive operations** that are executed to generate information redundancy (e.g., in the form of checkpoints). The introduction of proactive operations allow a separation of fault-tolerance concepts and application logic. Erroneous components are usually excluded from the system operation and can perform **error recovery** operations offline. After a successful completion of the recovery operations, the erroneous components can be integrated to guarantee the achievement of the reliability goals. The **integration** operations perform the state synchronization.

10.3.4 Summary: Modeling of Dependable Systems

When analyzing the discussed formal model, it becomes evident that it is necessary to model all three aspects of dependable systems: the normal operation, the fault hypothesis, and the fault-tolerance mechanism. A strict separation of the related mechanisms supports a better maintainability and increases the reusability. However, most of the existing approaches do not support the modeling of all three aspects or mix these aspects.

10.4 Reliability and Safety Analysis

Dependability analysis techniques have been developed so as to evaluate the systems and correct the failures. Initially, reliability and/or availability were the most interesting attributes to be analyzed. For that, just the binary states (e.g., on or off) of the system and its components were considered. Therefore, Boolean methods such as reliability block diagrams, fault or success trees were adequate and sufficient. These classical methods are widespread. However, they provide a static view of the system only. As embedded systems grew rapidly, a dynamic view has been needed to analyze the dependability. Such a dynamic view can represent multiple states of a system changing over time. An example of a technique handling this behavior is Continuous Time Markov Chains approach.

The objective of the **reliability analysis** is to identify the kinds of system failures that are to be expected (i.e., qualitative analysis) or the distribution of the times-to-failure of a component, subsystem or system (i.e., quantitative analysis). The reliability analysis is performed during system design or operation to decide whether the reliability level of a system is acceptable or which parts of a system are particularily critical. Its results indicate how and which parts of the system should be improved.

In the following, we shortly describe selected, but typical reliability and safety analysis methods:

- Failure Modes, Effects and Criticality Analysis (FMECA)
- Fault Tree Analysis (FTA)
- Markov Chains
- Model-based Testing (MBT)

10.4.1 The FMECA Method

The **Failure Modes, Effects and Criticality Analysis** (FMECA) is basically a qualitative reliability analysis method that uses a static view of the system and/or its components. It analyzes potential failure modes within a system, classifies the severity, and determines the failure's effects on the system. It is widely used in the manufacturing industries in various phases of the product life cycle [17]. It also includes a criticality analysis that is used to chart the probability of failure modes against the severity of their consequences. The result highlights failure modes with relatively high probability and severity of consequences, allowing remedial effort to be directed where it will produce the greatest value [18].

FMECA is one of the first systematic approaches to failure analysis. It was developed in the 1950s for the use in U.S. military systems. It is put forward by international standards, in particular in SAE-ARP 5580 [19], IEC60812 [20], and BS 5760-5 [21].

Applying FMECA the system is split into subsystems. Within each subsystem the components and their relations are identified. Functional block diagrams are used to represent them. For each component a detailed FMECA worksheet (see Fig. 10.3) is specified. It includes:

– functions and operational modes;
– failure modes, their causes, and their detection methods;
– failures effects;
– failure rates and their severity; and
– a specification of risk reducing measures.

System:							Performed by:				
Ref. drawing no.:							Date:			Page: of	
Description of unit			Description of failure			Effect of failure					
Ref. no	Function	Operational mode	Failure mode	Failure cause or mechanism	Detection of failure	On the subsystem	On the system function	Failure rate	Severity ranking	Risk reducing measures	Comments

Fig. 10.3. FMECA worksheet

Typically, FMECA is integrated in the design process right from the beginning and updated during the development and maintenance. It is most often a bottom-up technique. FMECA does not handle dependencies between components, cannot handle systems with redundancy, and cannot cope with common cause failures or cascading failures. As single events are considered, the effects of sequences of events cannot be addressed. Furthermore, as FMECA has a focus on hardware component failures human errors and software errors cannot well be reflected.

On the other hand, FMECA is simple to apply. It requires, however, thorough knowledge of a system and its environment. FMECA can be tailored to meet specific industry or product needs. It helps to reveal weak points in the system structure during early phases of system design and by that, it can help to avoid expensive design changes. FMECA is very effective where system failures are caused by single components failures.

10.4.2 The Fault Tree Analysis Method

The **Fault Tree Analysis** (FTA) is used to show causes or combinations of causes that then lead to overall system failures. It is basically a quantitative reliability analysis method that uses a static view of a system.

A fault tree is a logic diagram that displays the interrelationships between a potential critical event in a system and the causes for this event. It analyzes combinations of causes using Boolean logic with and- and or-gates. The fault tree analysis (FTA) method was developed at Bell Telephone Laboratories [22]. It was extended by Boeing and became a part of the IEC 61025 standard [23].

FTA is used in the design phase to reveal *hidden* failures caused by underlying combinations of faults or errors. During system operation, it is used to identify potential hazardous combinations of component failure and operator or procedural faults. It is also used in combination with FMECA to analyze selected system parts.

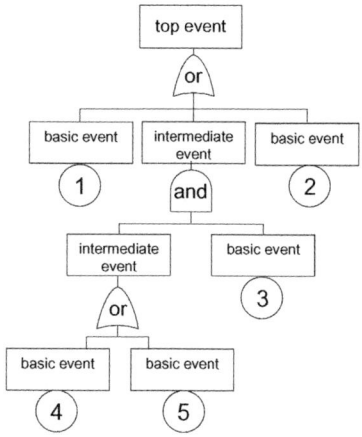

Fig. 10.4. A Fault Tree

A fault tree (see Fig. 10.4) is constructed following the procedure provided below:

(1) Select a top level event for analysis.
(2) Identify faults that could lead to the top level event and that represent an immediate, necessary, or sufficient cause that results in the upper event.
(3) Connect these fault events to the top event using logical and- or or-gates.
(4) Proceed level by level until all fault events have been described in an appropriate level of resolution.

Given a fault tree, the **minimal cut sets** can be determined: for a given event, the set of basic events that lead to this event are identified. In a qualitative analysis of a fault tree, the minimal cut sets give potential combinations of environmental factors, human errors, normal events, and component failures that may result in a critical event in the system.

For a quantitative analysis, failure rates for each basic event are assigned which are then cummulated to the probability of the top event (i.e., the unwanted incident) by assuming that all the basic event parts of the minimal cut set of the top event are independent and happen simultaneously.

Fault trees provide a static view on event combinations that may cause incidents. They cannot accurately model system dynamics. A fault tree is just a Boolean method (failure or success only). For the quantitative analysis, basic events are assumed to be statistically independent, so that the results are imprecise whenever this assumption does not hold.

Fault trees provide a clear picture of component failures and other events that may cause unwanted incidents. The graphical model is well known and fairly simple to explain. It forces users to understand the details of a system and to discover weaknesses at an early stage. It is able to handle common cause failures if component dependencies are well defined. It addresses redundant components in a system. Last but not least, the static nature of fault trees may be mitigated using scenario-based simulation of fault trees.

10.4.3 Markov Analysis

The **Markov chain** is a mathematical model for the random evolution of a memoryless system, for which the likelihood of a future state, at any given moment, depends only on the present system state and not on any past states. For reliability analysis Markov chains (also called Markov models) and their various flavors have been extensively used. Markov analysis enables a quantitative reliability analysis of the dynamic system behavior.

For Markov modeling the states of a system and transitions between them are considered. The system transitions are typically between a perfect state and a failure state. Transition probabilities define the degradation/failure rates and the repair rates. The Markov model has been developed by Andrej A. Markov [24]. It has been included in the standards IEC 61165 [25] and IEC 61508 [8].

A Markov model (see Fig. 10.5) is established according to the following procedure:

(1) Define all system states including failure states such as operation, degradation, maintenance, or repair.
(2) Define transitions between the states and assign failure and repair rates.
(3) Define initial state probabilities.

For large systems Markov models are often exceedingly large, complicated, and difficult to construct and validate. They suffer from the state space explosion problem. On the other hand, Markov models are able to handle systems that

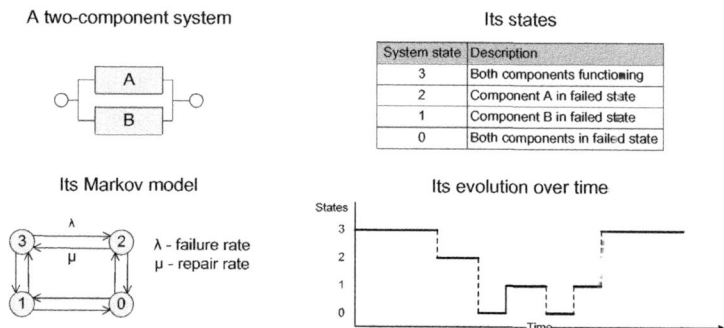

Fig. 10.5. Exemplified States Evolution over Time

exhibit strong dependencies between its components. System reconfiguration due to failures, repair, and switching strategies can easily be described. The analysis of a Markov model does not only give the probabilities for states, but also for sequences of events.

Although Markov models are a powerful and mathematically sound formalism for analysing system reliability, a Markov model is considered to be too low-level, which makes building a Markov model a tedious and error-prone task [26]. Hence this method is not yet widely used, despite the considerable benefits offered by Markov analysis [27].

10.4.4 Testing and Model-Based Testing

Testing is an analytic means for assessing the quality of systems [28, 29]. It "can never show the absence of failures" [30], but it aims at increasing the confidence that a system meets its specified behavior. Testing is an activity performed for improving the product quality by identifying defects and problems. It cannot be undertaken in isolation. Instead, in order to be in any way successful and efficient, it must be embedded in adequate system development process and have interfaces to the respective sub-processes.

Model-based Testing (MBT) relates to a process of test generation from a model of the system under test (SUT) by application of a number of sophisticated methods. It can be understood as the automation of black-box or white-box test design [29]. Several authors [31, 32, 33, 34, 35, 36] define MBT as testing in which test cases are derived in whole or in part from a model that describes some aspects of the SUT based on selected criteria in different contexts.

MBT allows tests to be linked directly to the SUT requirements, makes readability, understandability, and maintainability of tests easier. It helps to ensure a repeatable and scientific basis for testing and it may give good coverage of all the behaviors of the SUT [32]. Finally, it is a way to reduce the efforts and cost for testing [37].

The term MBT is widely used today with slightly different meanings. Surveys on selected MBT approaches are given in [38, 32, 29, 39, 40]. In the automotive industry MBT is used to describe all testing activities that are related to model-based development [41, 42]. To that end, the authors of [43, 44, 45, 46] define MBT as a test process that usually encompasses a combination of different test methods which utilize the executable system model as a source of information. Thus, the automotive viewpoint on MBT is rather process-oriented. A single testing technique is often not enough to provide an expected level of test coverage. Though, it strongly depends on the targeted coverage criteria, for example, white/box test criteria can be succesfully fulfilled with a single method. Relating to all the test dimensions different test approaches should be combined to complement each other (e.g., functional and structural). Then, analyzing testing activities throughout the entire test process, one can assume that if sufficient test coverage has been achieved on model level, the test cases can be reused for testing the control software generated from the model and the end-product unit within the framework of back-to-back tests [47]. With this practice, the functional equivalence between executable model, code, and product can be verified [41].

10.4.5 Summary: Reliability and Safety Analysis

This section reviewed reliability and safety analysis methods and provided details on FMECA, FTA, Markov models, and MBT. A comparison of these methods is given in Table 10.1.

Table 10.1. Comparison of Dependability Analysis Methods

Method	Modeling Concepts	Dynamic Modeling	Quantitative Evaluation
FMECA	Components and Failures	No	No
FTA	Events	No	Yes
Markov Models	States and Transitions	Yes	Yes
Test Methods	Any	Yes	Yes

Markov models and many testing methods allow to analyze system behavior dynamically. Though, Markov chains are not wide-spread basically because of their low abstraction level. Techniques described in Sec. 10.5, such as interaction-based models, AADL, or proprietary solutions, can be applied as complementary methods on a higher abstraction level.

10.5 Languages and Tool Support

After introducing the different methods for safety and reliability analysis, this section discusses selected examples for model-based development tools that target the area of dependable systems. Since tools based on the Unified Modeling

Language are already discussed in chapter *UML for Software Safety and Certication*, this section focuses on domain-specific approaches. Zougbi et al. pointed out that generic UML-based tools have the disadvantage of not covering all necessary aspects for modeling fault-tolerant real-time systems [48]. They also do not support adequate code generators supporting transformations from more sophisticated models than class diagrams and state charts [49]. The reason is mainly the lack of precise semantics of the UML models [50, 51]. The main advantage of domain-specific tools is the possibility to use restrictions (e.g., with respect to the model of computation) suitable for the intended domain. Therefore, it is possible to offer a better tool support, for example extensive code generation ability or formal verification. In the following, we discuss three concrete examples.

Within a project at the University of California in San Diego (UCSD), a taxonomy of software failures and interaction-based models for logical and deployment architectures were developed for the automotive domain [52]. Based on these models, a verification tool has been developed that allows the generation of models that can be feed into the SPIN model checker [53].

FTOS [54] is a model-driven tool developed at the Technical University München (TUM). It supports modeling of dependable systems and code generation for nonfunctional properties such us scheduling, communication within the distributed system, and fault-tolerance mechanisms. It is based on a meta code generation framework [55] and thus, supports expandability with respect to both the modeling language and code generation ability.

The third tool developed by LAAS-CNRS [56, 57] is based on Architecture Analysis and Design Language (AADL) [58]. The main contribution of this work is the definition of reusable fault-tolerance patterns that can be used at architectural level. These patterns can be instantiated and customized for a particular system. By transforming the AADL model into a stochastic model, dependability measures can be obtained.

10.5.1 Models

In the following, different approaches based on the applied models are discussed. Note, the terminology provided at the beginning of this chapter is used in the upcoming paragraphs independently of other variants proposed elsewhere (e.g., error and fault definitions).

Logical and Deployment Models. The first two approaches mentioned above offer two models to specify the application logic and hardware architecture (i.e., platform). In the UCSD approach, the **Logical Model** represents the Platform Independent Model (PIM) and the **Deployment Model** the Platform Specific Model (PSM) in the spirit of the Model Driven Architecture (MDA) [59]. The mapping is achieved by a **Mapping Model** as depicted in Fig. 10.6. This contrasts the approach of FTOS, where the hardware model is firstly defined and the resulting software model refers to this hardware model. This approach is motivated by the fact that the safety goals can only be reached when using the correct hardware architecture [8]. In AADL, interacting application components (e.g., processes, threads,

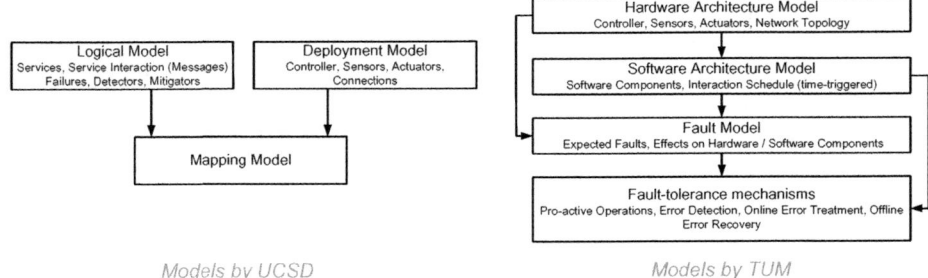

Fig. 10.6. Models and their Dependencies

subprograms, and platform components, such as processors, memory, buses) are specified as hierarchical collections in one model.

Another major difference is the model of computation. The design of an adequate model of computation can drastically leverage the implementation complexity of fault-tolerant embedded systems [60]. FTOS uses the concept of logical execution times [61] as a model of computation. This enhances the fault-tolerance mechanisms [62]. Within the software model, all interaction points between software components (i.e., actors) are specified with respect to time. An event-triggered execution can only be realized on basis of this time-triggered execution scheme. In contrast, the UCSD approach is based on a service-oriented architecture, where message-passing is used as a means for component interaction. The messages can be applied both for services executed on one controller and for services executed in the distributed system. This contrasts the approach in FTOS, where ports are used to realize the communication. AADL itself defines no concrete model of computation. Dynamic aspects are described by the selected operational model. This concept allows for the definition of different operational models for a system or a given system component to represent various system configurations and connection topologies.

Fault Models. All the reviewed approaches force the user to specify the fault hypothesis directly in the model. In the approach of the UCSD, a failure taxonomy allows for the description of the possible failures. The base failure taxonomy is depicted in Fig. 10.7. It can be augmented by domain specific failures. For each failure the cause can be specified and hardware/software, permanent/temporary, and unexpected/non occurrence behavior is distinguished. In addition, possible failure effects are categorized as *Nonhazardous*, *PotentiallyHazardous*, and *Hazardous*. The concrete effects and the affected components can be described in the logical model. The approach in AADL is similar. AADL error models allow for modeling component behavior in the presence of faults. Error models describe error states, error events, and error propagation. By the occurrence property, the arrival rate and probability of events and propagations can be specified.

In contrast to application specific errors, FTOS specifies a number of generic fault effects for each software/hardware component type in a distinct model.

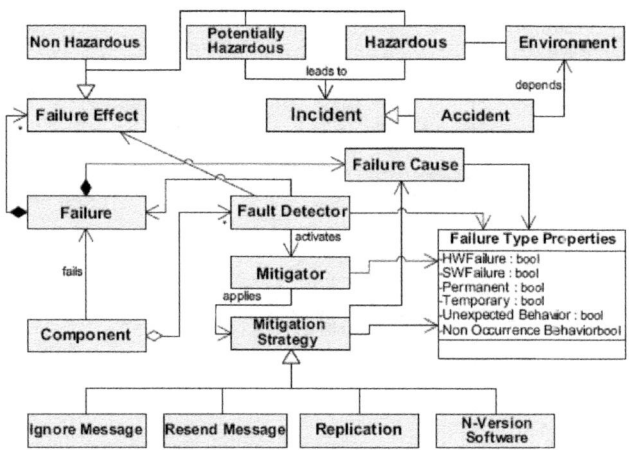

Fig. 10.7. Failure Taxonomy [52]

For network components for instance, FTOS defines seven different fault effects as suggested in the international standard IEC 61508 [8]: *DataCorruption, TimeDelay, DeletedTelegram, Repetition, InsertedTelegram, ResequencedTelegram, AddressingError*, and *Masquerade*. It is possible to specify which components might be affected by which fault effect and constrain the number of simultaneous faults. Because of generic fault effects, generic fault detectors can be applied. Code generation and verification are also supported [54].

Fault-Tolerance Mechanisms. The fault-tolerance mechanisms in the UCSD approach are specified in a similiar manner than the fault hypothesis within the logical model. For each failure effect a detector has to be specified. Detectors activate appropriate mitigators, similar to Arora's concept of Detectors and Correctors. Services can be used both as unmanaged service that defines system behavior without considering failures, and managed service that are equipped with detectors and mitigators. The services can be composed hierarchically allowing the fault-tolerance mechanisms to be applied at different levels of abstractions.

FTOS extends the concept of Arora and uses a separate model to specify the fault-tolerance mechanism. The system is split into fault containment units (FCU) and sets of components that can be affected by faults. In addition, relevant sets of fault configurations, and functions mapping each FCU to correct or false can be specified. At runtime, **tests** monitor one or more FCUs. If at least one associated test assumes the FCU to be faulty, the status of this FCU is set to false. Whenever the status of a FCU changes, the system determines the active fault configuration set. Changes of this set can trigger reactions (**error treatment**) by the system. Examples for error treatment operations are rollback operations or switches to a correct redundant unit. All error treatment operations are performed online. They can lead to the exclusion of erroneous components. The excluded components perform **recovery operations** offline

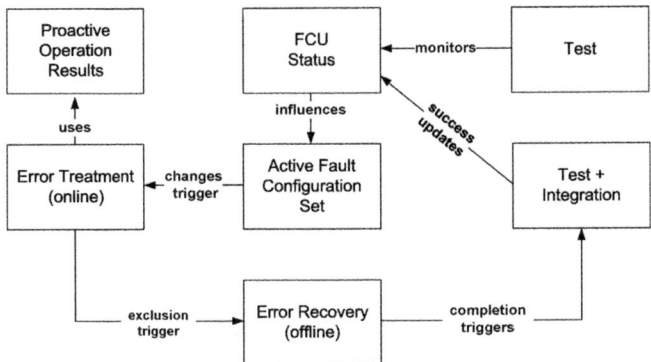

Fig. 10.8. Concept of Fault-Tolerance Mechanism

followed by tests to check the correctness of the repaired component. If successful, the component request the **integration** into the running system. The integration leads to a change of the active fault configuration set and triggers a new iteration of reactions. The relationship between different types is illustrated in Fig. 10.8.

For AADL, Rugina et al. propose different reusable fault-tolerance patterns at the architectural level (e.g., hot standby). The interaction between different components is specified within a dedicated pattern. The patterns should be customized for a concrete application, for example, by defining error detection strategies.

10.5.2 Implementations

All mentioned approaches have been tested in the context of different applications to point out the benefits of the model-based approach. UCSD used their approach to verify a central locking system in the automotive domain. The model was translated into another model that could be directly feed into the SPIN model checker. Based on the generated model one can verify the achievement of the safety goals or use the produced counter examples to refactor the system's architecture.

FTOS has been applied to prove the efficiency of the code generation and the possibility to cope with heterogeneous systems. One application is the classic "inverted pendulum" controlled by a triple-modular redundancy (TMR) system. Here, the generated code could be executed with a control response time of 2.5 milliseconds. Another application is the control of an elevator, consisting of a hot-standby system executing the control logic and five microcontrollers implementing the I/O functionality. Since the focus of FTOS is on the nonfunctional requirements, the code implementing the control functionality has to be provided manually. However, this gap can be closed by combining FTOS with existing tools for the development of control functionality [54].

The work of LAAS CNRS is used for dependability analysis. AADL models can be transformed to other models (e.g., stochastic Petri nets). In the context of a safety-critical subsystem of the French Air Traffic Control System, different architecture solutions were compared to evaluate their availability. The complete approach allows for a simple and fast evaluation of the design alternatives.

10.5.3 Summary: Language and Tool Support

Several tools for analysis and development of dependable systems are emerging. They differ in the number of covered aspects and in their application domain. For specific domains and application areas, these tools can facilitate the development/analysis process considerable. By using domain-specific concepts, such as a specific model of computation, or restricting the application purpose of the tool, these tools offer extensive code generation or analysis abilities. However, there are no generic tools available that can be used in the development process of arbitrary dependable systems.

10.6 Conclusion and Research Challenges

Within this chapter, we discussed the current state of the art with respect to model-based tools in the context of dependable systems. As dependability comprises very different aspects ranging from availability to maintainability. Here, safety and reliability were in focus. The chapter started with a definition of the relevant terms and concepts. Subsequently, in Sec. 10.3 we defined a formal model of a fault-tolerant system. This model can be used to discuss and compare different approaches and to get a good understanding of the concepts of fault-tolerant systems.

To illustrate the state of the art of model-based methods for reliability and safety analysis, Sec. 10.4 provided details on FMECA, FTA, Markov models, and MBT. A comparison of these methods can be found in Table 10.1.

Finally, Sec. 10.5 gave insight in some tools from academia that show the potential of model-based approaches with respect to formal verification and code generation. By using domain-specific models, the system, fault hypothesis, and fault-tolerance mechanisms can be specified. The presented tools allow the automatic synthesis of code or the translation into formal models that can be used as input for verification tools.

Regarding future research challenges, three main areas can be identified: model use throughout the whole development process, tool support for formal verification, and designing adequate fault models.

In currently available tools, different models of the system are used in each phase of the development process. Typically, these models can not be reused in the next phases, nor is there an automated transformation into adequate models. Especially in the area of dependable systems, where the developer has to provide a complete tracing from requirements to the resulting system, an integrated model-based development process with extensive tool support ranging from requirements analysis through code generation to validation would be tremendously beneficial. Some promising results have been provided by industry in the context of control systems [63, 40].

Furthermore, the integration of formal verification techniques is becoming more and more essential. The current state of the art with very complex mathematical models restricts the application to experts in formal verification. A promising approach is the automatic synthesis of these models out of domain-specific models that can be designed by non-experts. A major drawback is however that the counter examples are presented using the mathematical models. A translation back to original models is still an open research challenge.

Another major issue is the formal specification of the fault-hypothesis. Currently, this fault hypothesis is typically specified as a textual document that is not machine-readable and usually inconsistent or even contradictory. A formal model to specify the fault assumptions within the system model, which is also linked to the basic faults described in the certification guidelines, is only partially covered in some ongoing research (e.g., [54]).

Acknowledgements

We thank the anonymous referees for their valuable comments.

References

[1] Object Management Group: OMG Unified Modelling Language Specification. 2.1.2 edn. (November 2007)
[2] Arora, A., Gouda, M.: Closure and convergence: A foundation of fault-tolerant computing. IEEE Transactions on Software Engineering 19(11), 1015–1027 (1993)
[3] Arora, A., Kulkarni, S.S.: Detectors and correctors: A theory of fault-tolerance components. In: International Conference on Distributed Computing Systems, pp. 436–443 (1998)
[4] Laprie, J.C.: Dependable computing and fault-tolerance: Concepts and terminology. In: Proceedings of the 15th International Symposion on Fault Tolerant Computing Systems, pp. 2–11 (June 1985)
[5] Avizienis, A., Laprie, J.C., Randell, B.: Fundamental concepts of dependability. Technical report, LAAS-CNRS (April 2001)
[6] Department of Defense: Standard Practise for System Safety. MIL-STD-882D (2000)
[7] United Kingdom Ministry of Defence: Safety Management Requirements for Defence Systems. Def Stan 00-56 (2000)
[8] International Electrotechnical Commission: Functional safety of electrical/electronic/programmable electronic safety-related systems. IEC 61508 (2002)
[9] International Standards Organization: Quality management and quality assurance - Vocabulary. ISO 8402-1986 (1986)
[10] Pradhan, D.K.: Fault-Tolerant Computer System Design. Prentice-Hall, Englewood Cliffs (1996)
[11] Avizienis, A.: The four-universe information system model for the study of fault-tolerance. In: International Symposium on Fault-Tolerant Computing, Santa Monica, CA, vol. 12, pp. 6–13 (June 1982)
[12] Lee, P.A., Anderson, T.: Fault Tolerance: Principles and Practice. Springer, New York (1990)

[13] Powell, D., Chérèque, M., Drackley, D.: Fault-tolerance in delta-4. ACM SIGOPS Operating Systems Review 25(2), 122–125 (1991)
[14] Arora, A., Kulkarni, S.S.: Designing masking fault-tolerance via nonmasking fault-tolerance. IEEE Transactions on Software Engineering 24(6), 435–450 (1998)
[15] Kulkarni, S.S.: Component based design of fault-tolerance. PhD thesis, Ohio State University, Adviser-Anish Arora (1999)
[16] Randell, B., Lee, P., Treleaven, P.C.: Reliability issues in computing system design. ACM Computing Surveys 10(2), 123–165 (1978)
[17] Stamatis, D.H.: Failure Mode and Effect Analysis: FMEA from Theory to Execution. American Society for Quality (2003)
[18] Haimes, Y.Y.: Risk Modeling, Assessment, and Management. Wiley, Chichester (2005)
[19] Society of Automotive Engineers: Recommended Failure Modes and Effects Analysis (FMEA) Practices for Non-Automobile Applications. SAE ARP 5580 (2001)
[20] International Electrotechnical Commission: Analysis techniques for system reliability - Procedure for failure mode and effects analysis (FMEA). IEC 60812:2006 (2006)
[21] British Standards: Reliability of systems, equipment and components. Guide to the specification of dependability requirements. BS5760-4:2003 (2003)
[22] Ericson, C.: Fault Tree Analysis: A History. In: Proceedings of the 17th International System Safety Conference (1999)
[23] International Electrotechnical Commission: Fault Tree Analysis (FTA). IEC 61025 (1990)
[24] Markov, A.A.: In: Classical Text in Translation: An Example of Statistical Investigation of the Text Eugene Onegin Concerning the Connection of Samples in Chains. Science in Context. Cambridge Journals, 591–600 (2006)
[25] International Electrotechnical Commission: Application of Markov techniques. IEC 61165:2006 (2006)
[26] Boudali, H., Crouzen, P., Stoelinga, M.: Dynamic Fault Tree Analysis Using Input/Output Interactive Markov Chains. In: International Conference on Dependable Systems and Networks, pp. 708–717 (2007)
[27] Hausler, P.A., Linger, R.C., Trammell, C.J.: Adopting Cleanroom software engineering with a phased approach. IBM Syst. J. 33(1), 89–109 (1994)
[28] Wallmueller, E.: Software- Qualitätsmanagement in der Praxis. Hanser Verlag (2001) (in German)
[29] Utting, M., Legeard, B.: Practical Model-Based Testing: A Tools Approach. Morgan Kaufmann Publishers Inc., San Francisco (2006)
[30] Dijkstra, E.W.: Notes on Structured Programming. Circulated Privately (April 1970)
[31] Bernard, E., Legeard, B., Luck, X., Peureux, F.: Generation of test sequences from formal specifications: Gsm 11-11 standard case study. Softw. Pract. Exper. 34(10), 915–948 (2004)
[32] Utting, M.: Model-Based Testing. In: Proceedings of the Workshop on Verified Software: Theory, Tools, and Experiments, VSTTE 2005 (2005)
[33] Campbell, C., Grieskamp, W., Nachmanson, L., Schulte, W., Tillmann, N., Veanes, M.: Model-Based Testing of Object-Oriented Reactive Systems with Spec Explorer. Microsoft Research, MSR-TR-2005-59 (2005)
[34] Frantzen, L., Tretmans, J., Willemse, T.A.C.: A Symbolic Framework for Model-Based Testing. In: Havelund, K., Núñez, M., Roşu, G., Wolff, B. (eds.) FATES 2006 and RV 2006. LNCS, vol. 4262, pp. 40–54. Springer, Heidelberg (2006)

[35] Kamga, J., Herrmann, J., Joshi, P.: D-MINT Automotive Case Study. Deployment of Model-Based Technologies to Industrial Testing (D-MINT), ITEA2 Project, Deliverable 1.1 (2007)
[36] Tretmans, J.: Model based testing with labelled transition systems. In: Hierons, R.M., Bowen, J.P., Harman, M. (eds.) FORTEST 2008. LNCS, vol. 4949, pp. 1–38. Springer, Heidelberg (2008)
[37] Pretschner, A., Prenninger, W., Wagner, S., Kühnel, C., Baumgartner, M., Sostawa, B., Zölch, R., Stauner, T.: One evaluation of model-based testing and its automation. In: ICSE 2005: Proceedings of the 27th International Conference on Software Engineering, pp. 392–401. ACM, New York (2005)
[38] Broy, M., Jonsson, B., Katoen, J.P., Leucker, M., Pretschner, A.: Model-Based Testing of Reactive Systems. LNCS, vol. 3472. Springer, Heidelberg (2005)
[39] D-MINT Consortium: D-MINT Project - Deployment of Model-Based Technologies to Industrial Testing (2008), http://d-mint.org/ (last visited 01/05/09)
[40] Zander-Nowicka, J.: Model-based Testing of Real-Time Embedded Systems in the Automotive Domain. PhD thesis, Technical University Berlin (2009)
[41] Conrad, M., Fey, I., Sadeghipour, S.: Systematic model-based testing of embedded automotive software. Electr. Notes Theor. Comput. Sci. 111, 13–26 (2005)
[42] Bringmann, E., Krämer, A.: Model-based testing of automotive systems. In: ICST, pp. 485–493. IEEE Computer Society, Los Alamitos (2008)
[43] Rau, A.: Model-Based Development of Embedded Automotive Control Systems. PhD thesis, University of Tübingen (2002)
[44] Lamberg, K., Beine, M., Eschmann, M., Otterbach, R., Conrad, M., Fey, I.: Model-Based Testing of Embedded Automotive Software Using MTest. In: Proceedings of SAE World Congress, Detroit, US (2004); SAE technical paper 2004-01-1593
[45] Conrad, M.: Modell-Basierter Test Eingebetteter Software im Automobil: Auswahl und Beschreibung von Testszenarien. PhD thesis, Technical University Berlin (2004) (in German)
[46] Conrad, M.: A systematic approach to testing automotive control software. SAE Technical Paper Series, 2004210039, Detroit USA (2004)
[47] Wiesbrock, H.W., Conrad, M., Fey, I., Pohlheim, H.: Ein Neues Automatisiertes Auswerteverfahren für Regressions und Back-To-Back-Tests Eingebetteter Regelsysteme. Softwaretechnik-Trends 22(3), 22–27 (2002) (in German)
[48] Zoughbi, G., Briand, L.C., Labiche, Y.: A uml profile for developing airworthiness-compliant (rtca do-178b), safety-critical software. In: Engels, G., Opdyke, B., Schmidt, D.C., Weil, F. (eds.) MODELS 2007. LNCS, vol. 4735, pp. 574–588. Springer, Heidelberg (2007)
[49] Khan, M.U., Geihs, K., Gutbrodt, F., Gohner, P., Trauter, R.: Model-driven development of real-time systems with uml 2.0 and c. In: MBD-MOMPES 2006: Proceedings of the Fourth Workshop on Model-Based Development of Computer-Based Systems and Third International Workshop on Model-Based Methodologies for Pervasive and Embedded Software (MBD-MOMPES 2006), Washington, DC, USA, pp. 33–42. IEEE Computer Society, Los Alamitos (2006)
[50] Johnson, I., Snook, C., Edmunds, A., Butler, M.: Rigorous development of reusable, domain-specific components, for complex applications. In: CSDUML 2004 - 3rd International Workshop on Critical Systems Development with UML (2004)
[51] Bunse, C., Gross, H.G., Peper, C.: Applying a model-based approach for embedded system development. In: EUROMICRO 2007: Proceedings of the 33rd EUROMICRO Conference on Software Engineering and Advanced Applications (EUROMICRO 2007), Washington, DC, USA, pp. 121–128. IEEE Computer Society, Los Alamitos (2007)

[52] Ermagan, V., Krueger, I., Menarini, M., ichi Mizutani, J., Oguchi, K., Weir, D.: Towards model-based failure-management for automotive software. In: SEAS 2007: Proceedings of the 4th International Workshop on Software Engineering for Automotive Systems, Washington, DC, USA. IEEE Computer Society, Los Alamitos (2007)
[53] Holzmann, G.J.: The model checker spin. IEEE Trans. Software Eng. 23(5), 279–295 (1997)
[54] Buckl, C.: Model-Based Development of Fault-Tolerant Real-Time Systems. PhD thesis, TU München (October 2008)
[55] Stahl, T., Voelter, M.: Model-Driven Software Development. Technology, Engineering, Management, 1st edn. Wiley, Chichester (May 2006)
[56] Rugina, A.E., Feiler, P.H., Kanoun, K., Kaâniche, M.: Software dependability modeling using an industry-standard architecture description language. CoRR (2008)
[57] Rugina, A.E.: Dependability modeling and evaluation - From AADL to stochastic Petri nets. PhD thesis, LAAS CNRS (2007)
[58] International Society of Automotive Engineers: SAE Architecture Analysis and Design Language, AADL (November 2004)
[59] Miller, J., Mukerji, J.: MDA Guide. Object Management Group, Inc. (June 2003), Version 1.0.1, omg/03-06-01
[60] Wensley, J., Lamport, L., Goldberg, J., Green, M., Levitt, K., Melliar-Smith, P., Shostak, R., Weinstock, C.: Sift: Design and analysis of a fault-tolerant computer for aircraft control. Proceedings of the IEEE 66(10), 1240–1255 (1978)
[61] Henzinger, T.A.: Embedded software: Better models, better code. In: ICATPN, pp. 35–36 (2004)
[62] Buckl, C., Regensburger, M., Knoll, A., Schrott, G.: A model-based code generator in the context of safety-critical systems. In: Third Latin-American Symposium on Dependable Computing - Fast Abstracts Volume, pp. 3–4 (2007)
[63] Nicolescu, G., Mosterman, P.J. (eds.): Model-Based Design for Embedded Systems. CRC Press, Boca Raton (2009)

Part V

Approaches

11 The EAST-ADL Architecture Description Language for Automotive Embedded Software

Philippe Cuenot[1], Patrick Frey[2], Rolf Johansson[3], Henrik Lönn[4],
Yiannis Papadopoulos[5], Mark-Oliver Reiser[6], Anders Sandberg[7],
David Servat[8], Ramin Tavakoli Kolagari[4],
Martin Törngren[9], and Matthias Weber[10]

[1] Continental Automotive
[2] ETAS
[3] Mentor Graphics
[4] Volvo Technology
[5] University of Hull
[6] TU Berlin
[7] Mecel
[8] CEA LIST
[9] KTH
[10] Carmeq

Abstract. Current trends in automotive embedded systems focus on how to manage the increasing software content, with a strong emphasis on standardization of the embedded software structure. The management of engineering information remains a critical challenge in order to support development and other stages of the life-cycle. System modelling based on an Architecture Description Language (ADL) is a way to keep these assets within one information structure. This paper presents the EAST-ADL2 modelling language, developed in the ITEA EAST-EEA project and further enhanced in the ATESST project (www.atesst.org). EAST-ADL2 supports comprehensive model-based development of embedded systems and provides dedicated constructs to facilitate variability and product line management, requirements engineering, representation of functional as well as software/hardware solutions, and timing and safety analysis.

11.1 Introduction

Current trends in automotive software development focus to a large extent on how to manage the increasing software content. Hybrid vehicle control, active safety systems, diagnostics services, etc., all rely on embedded systems. The automotive industry faces the challenge of incorporating software and embedded systems engineering within traditional mechanical engineering enterprises. This challenge is addressed in many ways, including incorporation of new processes, tools, and the standardization of the embedded software structure. Software standardization is addressed in the AUTOSAR standardization initiative [1]. The AUTOSAR standard specifies how to model the software architecture and final implementation, but the requirements, functional content realized by this solution, and non-functional aspects such as support for safety analysis, are not covered [2].

EAST-ADL2 is an Architecture Description Language. As such it provides a basis for documenting and managing the various artefacts of an advanced embedded system (requirements, features, desired behaviours, software and hardware components), and their dependencies (refinement, allocation, composition, communication, etc.). Any modelling language is directed by the product aspects and process stages it intends to support. EAST-ADL2 is defined with the development of safety-related embedded control systems as a benchmark. EAST-ADL2 bridges the gap from vehicle content definition and early analysis via functional design to the implementation perspective and back to integration and acceptance testing up to vehicle-level. An early, high-level representation of the system can evolve seamlessly into the detailed specifications of the AUTOSAR language. In addition, EAST-ADL2 incorporates the following system development concerns:

- Modelling of requirements and verification/validation information,
- Feature modelling and support for product lines,
- Structural and behavioural modelling of functions and hardware entities in the context of distributed systems,
- Modelling of variability of the system design,
- Environment, i.e., plant model and adjacent systems, and
- Non-functional operational properties such as a definition of function timing and failure modes, supporting system level analysis.

The main role of EAST-ADL2 is that of providing an integrated system model. As such, EAST-ADL2 must address multiple aspects of a system [3] including:

- Documentation, in terms of an integrated system model.
- Communication, by providing predefined views as well as the information sufficient for generating a number of other views.
- Analysis of a complete embedded system through the description of system structure and properties. Special emphasis has been placed on modelling support for analysis of component interfaces, timing correctness and safety analysis.

EAST-ADL2 and AUTOSAR in concert provide means for efficient development and management of the complexity of automotive embedded systems from early analysis right down to implementation. Concepts from model based development and component based development reinforce one another [4].

The following sections briefly summarize the language capabilities of EAST-ADL2 and present an illustrative example of its use. We conclude by comparing EAST-ADL2 with related work and discuss ongoing activities in the development of the language.

11.2 Modeling and Analysis Capabilities of the EAST-ADL2

EAST-ADL2 is a domain-specific language specified through a metamodel and implemented/released as a UML2 profile.

The primary structural organization of EAST-ADL2 is the division of the model into different abstraction levels (see Fig. 11.1). On a high abstraction level, only the externally perceivable aspects of the embedded system are handled, while on a low abstraction level the implementation-specific solution is managed in an AUTOSAR-conforming software architecture. This ensures separation of concerns and provides means to trace between the solution and problem domains.

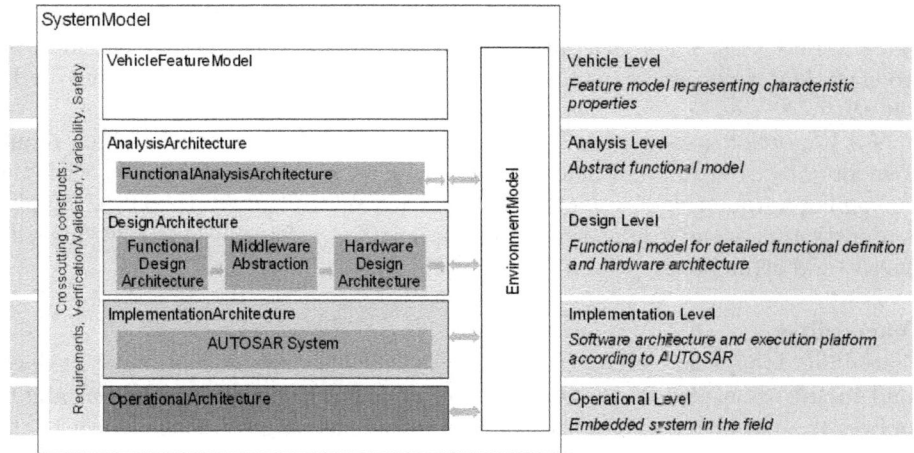

Fig. 11.1. EAST-ADL2 abstraction layers and relation to AUTOSAR. Cross-cutting concerns like requirements, V&V information, variability and safety/error modeling span all abstraction levels.

The abstraction levels implicitly represent different stages of an engineering process. However, the order in which a system is modelled could be top-down, middle-out, or bottom-up, in relation to the abstraction levels.

The structural organization of EAST-ADL2 is the backbone onto which additional modeling constructs are applied. Behavior, requirements, variability and safety aspects are examples of concepts that apply to the structural entities on several abstraction levels. A short summary follows below.

Behaviour

The goal of EAST-ADL2 with respect to behaviour is to define how model components (from different tools, in different modelling languages, or just representing code) are related to each other in order to capture behaviour and algorithms of the vehicle systems as well as the environment [5][6]. EAST-ADL enables structural components to refer to external behavior models, such as a Simulink model. The language also enables specification of triggering of blocks and precedence relations in their execution. The purpose of the behavioural definitions includes documentation, code generation and analysis, and the representation is chosen depending on the respective purpose.

Requirements

Requirements are captured in EAST-ADL2 according to the principles of SysML [7]: Requirements are separate entities that are associated to its target elements with a specific association, "ADLSatisfy".

Requirements are related to each other to support traceability between requirements. Typically, requirements on the higher abstraction levels of EAST-ADL2 are refined to more detailed requirements on lower abstraction levels.

Verification and Validation is supported through the concept of Verification & Validation Cases. These are linked to requirements and target entities, in order to show how a certain requirement is verified in the context of a specific model entity.

An important aspect of traceability is the possibility to follow which requirements are the results of safety concerns. This is needed to comply with the upcoming automotive standard for safety, ISO/WD 26262 [8]. EAST-ADL2 also supports this standard by providing support for safety case, safety integrity levels, and error propagation.

Variability

Variability is captured in EAST-ADL2 both in the feature models on vehicle level and in the architectures at analysis level and down. Feature model variability defines the permitted and expected variability regarding a certain aspect of the complete system. The idea is not to define how the system varies with respect to this aspect, but only that the system should exhibit such variability.

Variability on lower abstraction levels, on the other hand, defines how the feature variability is achieved. Variability mechanisms applied to the entities of EAST-ADL2 defines which of them are optional and under which circumstances they are included or excluded and the effect on structure and hierarchy. The mechanism is linked to the feature models such that variant choices in the feature model affects the variability resolution of the concrete architecture.

The variability management of EAST-ADL2 especially takes into account the automotive-specific challenges, e.g., management of a family of model ranges, different views on variability information (e.g., customer-related as opposed to development-related variability information), and extensibility of the variability management approach, e.g., for AUTOSAR modelling entities.

Error Modelling and Safety

State-of-the-art safety analysis techniques provide analysis support for deriving the causes and consequences of errors, based on a representation of the dependencies between system components. EAST-ADL2 provides means to manage the safety-related information together with the engineering information in a systematic way. An error model is defined with a structure that may be independent of the nominal system architecture. This way, the error model may capture errors and error propagation at the level of detail and according to a structure that is appropriate for the safety analysis at hand.

Further, it is possible to trace from the final implementation back to safety-related design choices and the applicable hazards. The requirement constructs for traceability and explicit support for safety cases according to the Goal Structure Notation [9] are relevant parts of EAST-ADL2 in this context.

11.3 A Small Case Study

To explain how the model is organized, an electric steering column lock function will be used. This is a security function for preventing any steering wheel movement without an authorized key. Traditional solutions for locking a steering column use the position of physical starter key as the authentication and unlocking mechanism. The introduction of immobilizers improved vehicle security by allowing advanced cryptography for authentication control prior to engine start. With a keyless engine start solution the steering lock also needs to be realized by the embedded system: a mechanical lock placed on the steering column is the actuation element, and a control unit reads the immobilizer transponder code and vehicle state and controls the mechanical lock accordingly.

11.3.1 Vehicle Features: Vehicle Level

To document what the embedded system provides to the user and other external stakeholders, a feature model is used. The feature model can be used as an entry point to related requirements, use cases, and other constructs. The feature model can be used to expose what the system provides and how a product line is organized in terms of available options and dependencies between options.

The steering column lock can be represented by a feature tree according to Figure 11.2. Steering column lock may be mechanical or electronic, and the electronic version may be based on a key or be key-less. Top level requirements are linked to each feature (not shown in Figure 11.2).

11.3.2 Abstract Functional Description: Analysis Level

The vehicle features are realized at the Analysis level by abstract functions ("ADLFunction") and devices that interact with the vehicle environment ("FunctionalDevice"). The Analysis level captures the principal interfaces and behaviour of the embedded system without design details or decisions on implementation technology.

The "Functional Analysis Architecture" for the example is sketched in Figure 11.2. The "ECL_Function" is the primary controller. It requires certain inputs including vehicle speed, engine status, the key position and more. The

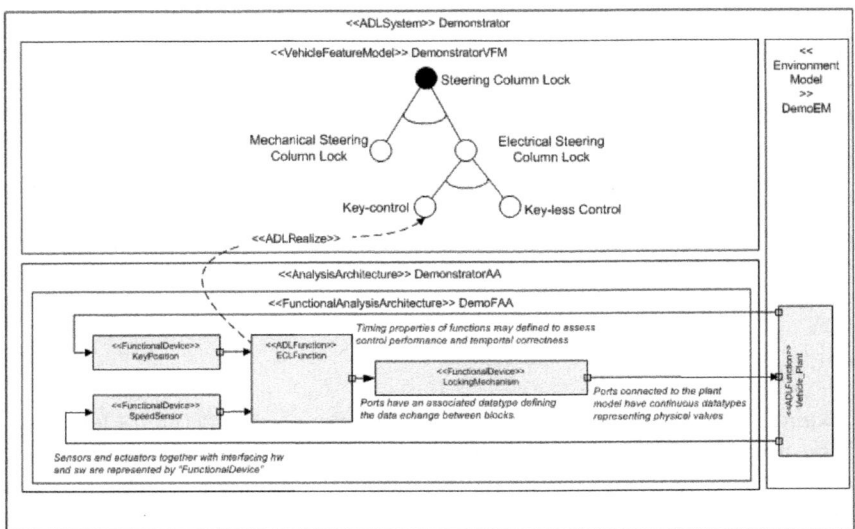

Fig. 11.2. The feature model and abstract functional representation of the Steering Column Lock

"ECL_Function" outputs signals to the ECL actuator. Signals for exchange with the environment are continuous while signals within the embedded system are typically modelled as discrete. All sensors and actuators are modelled as FunctionalDevices with ports connecting to the physical environment. For those familiar with dynamic simulation environments such as MATLAB/Simulink, this is a similar view, allowing mixed discrete/continuous signals, and implicit time-marching.

11.3.3 Concrete Functional Description: Design Level

On the Design level, models are refined with more implementation-oriented aspects that allow a subsequent software decomposition of the functional architecture. While our functional analysis architecture above did not differentiate between application software, middleware and hardware, the functional design architecture now separates these areas of the system implementation. To distribute the systems functionality among these areas already constitutes an important design decision.

The abstract interface elements on analysis level ("FunctionalDevices") are realized by hardware elements such as sensors, actuators and amplifiers, and the software parts for signal transformation ("LocalDeviceManager"). Middleware abstraction projects the platform services and functionality (OS, AUTOSAR Basic software, etc.) to the functional level. The hardware architecture is introduced in parallel to capture the hardware entities as abstract elements (e.g. I/O, sensor, actuator, power, Electronic Control Units (ECU), electrical wiring including communication buses) describing the topology of the electronic architecture of the systems. Design level allows preliminary allocation of functional entities to ECUs and provides the basis for verification either by simulation or analysis techniques such as timing and dependability analysis.

The DesignArchitecture contains three parts, as shown in Figure 11.1, representing the application software, execution platform and hardware respectively. To show how they are related, a part of the example related to vehicle speed is shown in Figure 11.3. The Hardware Design Architecture can be seen as a circuit diagram of the system. The "StalkEcu" with its connected speed sensor is represented with its wiring. In addition, a transfer function of hardware devices can be specified to capture sensor and hardware characteristics as well as other behaviour of the hardware architecture.

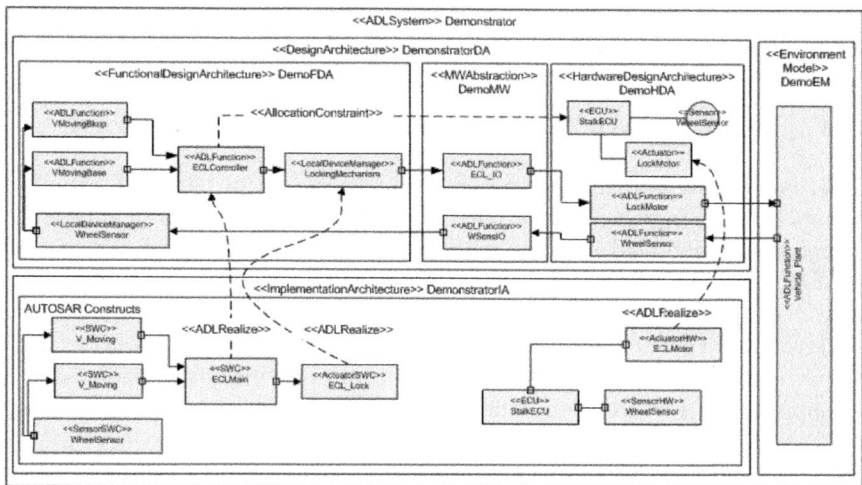

Fig. 11.3. Parts of design architecture and implementation architecture with realization and allocation relations. Note that only selected model entities and relations are shown.

The Middleware Abstraction contains a representation of the software platform that the applications rely on. In the example, the driver/interface software for the speed sensor is represented. The "WSensIO" represent the platform functionality that provides pulse rate of the wheel sensor.

The Functional Design Architecture contains the functionality that is subsequently realized by application software. The "WheelSensor" LocalDeviceManager translates to wheel speed according to the characteristics of the wheel sensor in use. The abstract function "ECLFunction" on analysis level is realized by three separate functions, two of which represent a redundant decision on whether the vehicle is moving.

11.3.4 Software Architecture: Implementation Level

System implementation in software is not represented by EAST-ADL2 entities, since this is the scope of AUTOSAR. However, the AUTOSAR entities are part of the system model to support traceability. Figure 11.3 also shows the AUTOSAR model and its relation to the EAST-ADL2 functions. As a realistic example would be too complex, only one-to-one mappings between AUTOSAR and EAST-ADL2 entities are shown. Readers not familiar with AUTOSAR may ignore the details and consider the shown entities as parts of the software and hardware architecture, respectively: The AUTOSAR software architecture typically shows a different structure than the functional architecture on design level. The purpose of the AUTOSAR hardware entities is to capture details necessary for the correct configuration of software. EAST-ADL2 provides a more abstract view of the hardware architecture, with a functional description of hardware elements and support for early assessment of feasibility of the system realization.

11.4 Related Work, Conclusions and Further Work

Model based development for embedded systems, and in particular automotive systems can be supported in various ways. The AADL is a modelling language dedicated to embedded systems with its roots in the aerospace domain. Compared to the EAST-ADL2 and AUTOSAR combination it covers parts of this scope. However, because of its overlap with AUTOSAR on the software architecture level, and the lack of complementary abstraction levels it does not provide an appropriate structural framework for automotive systems development. Also, the support for feature modelling, requirements and variability is unique for EAST-ADL2.

SysML and MARTE are UML profiles that augment plain UML with constructs for systems engineering and embedded real-time systems modelling, respectively. Both approaches, and even plain UML are useful tools in automotive development and EAST-ADL2 has integrated some of these concepts, for example requirement concepts from SysML and timing constructs from MARTE. But the abstraction levels and tailored model structure as well as complementary constructs of EAST-ADL2 adds a framework that both supports the modelling needs and guides modelling in a way that improves model exchange and understanding between stakeholders.

Off-the-shelf tools like SCADE, ASCET, Simulink, etc. all support model based development with analysis and synthesis to various degrees. Our conclusion, however, is that no single tool will be used for an entire vehicle development project, but model integration is necessary. EAST-ADL2 supports this aspect by allowing external representation of behaviour and concepts for integration with requirements management tools.

Another effort with large impact on automotive systems is the safety standard developed by the ISO working group on functional safety for road vehicles (ISO TC 22/SC 3/WG 16), ISO/WD 26262. The standard calls for rigorous development methods and requires documentation that shows that adequate measures are taken to achieve safety. EAST-ADL2 provides a framework that makes this possible and includes dedicated constructs for safety assessment and documentation.

Having had the opportunity to define this architecture description language in parallel with the dynamic phases of the definitions of AUTOSAR and ISO/WD 26262, EAST-ADL2 has a good potential to become a de facto standard as it fits well with the major critical needs of the automotive industry of today.

The language has received further momentum from its deployment in several industrial research projects that claim the central role of EAST-ADL2 in their work. Among others, the ADAMS project (www.adams-project.org) leads the dissemination of MARTE, where AUTOSAR and EAST-ADL2 are of primary importance for the automotive domain. Another example is the EDONA project (http://www.edona.org) in which EAST-ADL2 is a cornerstone of an integrated tool suite for the automotive domain. Finally, TIMMO (www.timmo.org) defines a methodology and representation of timing aspects in automotive embedded systems, where EAST-ADL2 together with AUTOSAR is the basis.

To enable a wide spread use of EAST-ADL2, its UML2 implementation is released as a public UML2 profile. The profile is supported in the open-source UML modeller Papyrus which can be downloaded on the www.atesst.org or the www.papyrusuml.org websites. In the ongoing European research project ATESST2, the EAST-ADL2 is currently extended in several areas:

- Modelling concepts for requirements and verification and validation are extended to support e.g. views on requirements and product line support.
- Timing modelling extensions, as being developed in the TIMMO project, will be integrated into the EAST-ADL2.
- Various aspects of native behavior descriptions are being further investigated for potential inclusion including explicit support for modes of operation and representation of continuous-time behavior as part of environment models.
- Variability mechanisms are used to choose between different behavioural representations.
- Variability concepts are extended to support product-line oriented manufacturer-supplier exchange.
- New dependability and cost modelling concepts are being developed to support multi-objective optimisation of system models, in conjunction with tools such as HiP-HOPS [10] that provide such advanced capabilities. The aim of optimization is to automatically evolve models that do not necessarily meet dependability requirements (e.g. safety, reliability or availability) to designs that fulfil such requirements with minimal costs [11]. Optimization can be done via exploration of potential design spaces using meta-heuristics such as genetic algorithms. The specification of design alternatives and variant sub-architectures the combinations of which define the potential design space can be described in EAST-ADL2 by using the variability constructs of the language. This work pushes the boundaries of the state-of-the-art in this area, as no modelling language provides support for such unique capabilities in design.

- Specialised plug-ins, based on the UML profile of the language, are being developed to achieve practical integration of EAST-ADL2 with existing tools. For example, data exchange will be supported with a plug-in for the RIF [12] requirement interchange format. As another example behavioural simulation, safety analysis and optimisation of models will be supported with plug-ins for Hip-HOPS and MATLAB/Simulink..

In addition, a methodology for the EAST-ADL2 will be developed, that explains the use and the interrelation of the different modelling concepts on the different abstraction levels during system specification and design as well as integration and testing.

Acknowldegements

The authors acknowledge the financial support provided by the European Commission through the Project ATESST2 (call FP7-ICT-2007-2, grant agreement number 224442).

References

[1] AUTOSAR Development Partnership: AUTOSAR Development Partnership (2007), http://www.autosar.org
[2] Sangiovanni-Vincentelli, A., Di Natale, M.: Embedded system design for automotive applications. Computer 40(10), 42–51 (2007)
[3] Törngren, M., Chen, D.J., Malvius, D., Axelsson, J.: Model based development of automotive embedded systems. In: Handbook on Automotive Embedded Systems. Taylor and Francis CRC Press - Series: Industrial Information Technology (invited) (forthcoming 2008), ISBN=9780849380266
[4] Törngren, M., Chen, D.J., Crnkovic, I.: Component-based vs. model-based development: A comparison in the context of vehicular embedded systems. In: EUROMICRO-SEAA, pp. 432–441 (2005)
[5] ATESST consortium: Report on behavioral modeling within east-adl2, d3.2 deliverable. Technical report (December 2007), http://www.atesst.org/
[6] Sjöstedt, C.J., Shi, J., Törngren, M., Servat, D., Chen, D., Ahlsten, V., Lönn, H.: Mapping Simulink to UML in the Design of Embedded Systems: Investigating Scenarios and Structural and Behavioral Mapping. In: OMER4 Post-Proceedings (2008)
[7] SysML Partners: Systems Modeling Language (SysML) open source specification project, http://www.sysml.org
[8] International Organization for Standardization: ISO Working Draft 26262 Baseline 10 (2007)
[9] Kelley, T.P.: Arguing Safety - A Systematic Approach to Managing Safety Cases. PhD thesis, University of York (1998)
[10] Papadopoulos, Y., McDermid, J.A.: Hierarchically performed hazard origin and propagation studies. In: Felici, M., Kanoun, K., Pasquini, A. (eds.) SAFECOMP 1999. LNCS, vol. 1698, pp. 139–152. Springer, Heidelberg (1999)
[11] Papadopoulos, Y., Grante, C.: Evolving car designs using model-based automated safety analysis and optimisation techniques. J. Syst. Softw. 76(1), 77–89 (2005)

[12] HIS: Specification Requirements Interchange Format (RIF), version 1.1a (2007)
[13] OMG: Uml profile for modeling and analysis of real-time and embedded systems (marte), beta1, omg document number: ptc/07-08-04 (August 2007)
[14] The Motor Industry Software Reliability Association (MISRA): Development guidelines for vehicle based software (1994)
[15] International Electrotechnical Commission: Functional safety of electrical/electronic/ programmable electronic safety-related systems - part 0: Functional safety and iec 61508 (2005)
[16] Törner, F., Chen, D.J., Johansson, R., Lönn, H., Törngren, M.: Supporting an automotive safety case through systematic model based development - the east-adl2 approach. In: SAE World Congress (2008), SAE paper number 2008-01-0127

12 Fujaba4Eclipse Real-Time Tool Suite

Claudia Priesterjahn, Matthias Tichy, Stefan Henkler,
Martin Hirsch, and Wilhelm Schäfer

Software Engineering Group, Department of Computer Science,
University of Paderborn, Paderborn, Germany
{cpr,mtt,shenkler,mahirsch,wilhelm}@uni-paderborn.de

Abstract. The Fujaba Real-Time Tool Suite supports modeling and verification of software in mechatronic or embedded systems. It also addresses the specification of advanced systems which reconfigure part of their structure and behavior at runtime. The Fujaba Real-Time Tool Suite requires a rigorous development process concerning the use of the different (partially refined) UML diagrams. All diagrams have a formally and well-defined semantics which allow to check models for given safety properties. Further, the tool suite provides a tight integration with software tools used by control engineers like CaMEL-View and Matlab to enable the simulation of production code of a complete system.

12.1 Introduction

Fujaba is an Open Source UML CASE tool project which was kicked off by the software engineering group at the University of Paderborn in 1997. Current major contributors to Fujaba are research groups at the University of Paderborn, the University of Kassel, the Technical University of Darmstadt, the Hasso-Plattner Institute at the University of Potsdam, the University of Bayreuth, the Technical University of Dresden and the University of Antwerp. Minor contributions come from a number of other places like Tampere and Victoria. In 2002, Fujaba has been redesigned and became the Fujaba Tool Suite with a plug-in architecture allowing developers to add functionality easily while retaining full control over their contributions. There are different Fujaba tool suites available, consisting of the Fujaba Core with different sets of plug-ins, each of which supports modeling and analysis for different domains.

One of the above mentioned tool suites is the Fujaba Real-Time Tool Suite which supports modeling and verification of software in mechatronic or embedded systems. The Fujaba Real-Time Tool Suite requires a rigorous development process concerning the use of the different (partially refined) UML diagrams. All diagrams have a formally and well-defined semantics which allow to check models for given safety properties. Further, the tool suite provides a tight integration with software tools used by control engineers like CaMEL-View and Matlab and a transformation from domain-spanning models of the early development phases to domain-specific models of the Fujaba Real-Time Tool Suite.

In 2008, this tool suite received an IBM Real-Time Innovation Award. In addition, another FUJABA tool suite (supporting the teaching of object-oriented concepts in undergraduate education) was acknowledged by an IBM Eclipse Innovation Grant in 2004.

In this paper, we present an overview of the features and the corresponding development process of the Fujaba4Eclipse Real-Time Tool Suite[1]. The whole approach is called MUML (cf. [1]). We further show how the MUML was employed to develop the software of a prototype of a new type of public transport system, i.e. a non trivial case study.

12.2 Features

The software of mechatronic systems is characterized by hard real-time constraints and the integration of controllers to control the dynamics of mechanical components. Hence, our approach supports the modeling and formal verification of so-called hybrid systems and the specification of timed behavior. As formal verification techniques like model checking suffer from the state space explosion problem, we developed a modular and compositional verification approach (cf. [2]). Code synthesis (for C++ and Java real-time) takes the specified real-time requirements into account such that the code exhibits the specified time constraints [3].

In more detail, the structure (architecture) of the system is specified by software components (as well as the relevant parts of the physics of the system, e.g. for the control engineers) and connectors between them using a slightly refined and formally defined UML 2.0 component model. The behavior of each port to port connection between components is specified by so-called **real-time coordination patterns**. Using an extended version of timed automata called **Real-time Statecharts**, which includes a number of additional syntactical constructs, the pattern specifications define, besides the particular communication protocols, all related required time constraints like invariants, guards, worst case execution times (WCET), and deadlines. The coordination patterns consist of different roles, which correspond to a particular port's behavior. Each pattern is individually verifiable concerning safety properties, specified using ATCTL, using the model checker UPPAAL.

The complete behavior of **components**, consisting of a number of ports as depicted in Figure 12.1(a), is automatically composed of all port roles of corresponding patterns. The user just sets a few parameters like eliminating nondeterministic choices in the pattern definition, i.e. synchronizing all behavior using internal events which do not affect the external behavior. A well defined refinement relation, which is checked by the tool, guarantees that the already verified properties still hold after composition without the need to check them again. The only check remaining is to make sure that the composed automaton does not include any deadlocks [2].

---

[1] The Fujaba4Eclipse Real-Time Tool Suite is available for download at http://wwwcs.uni-paderborn.de/cs/fujaba/projects/realtime/index.html

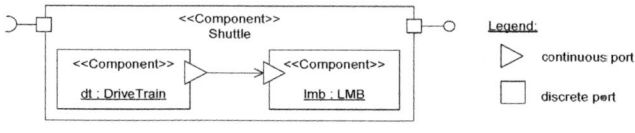

(a) Component Diagram

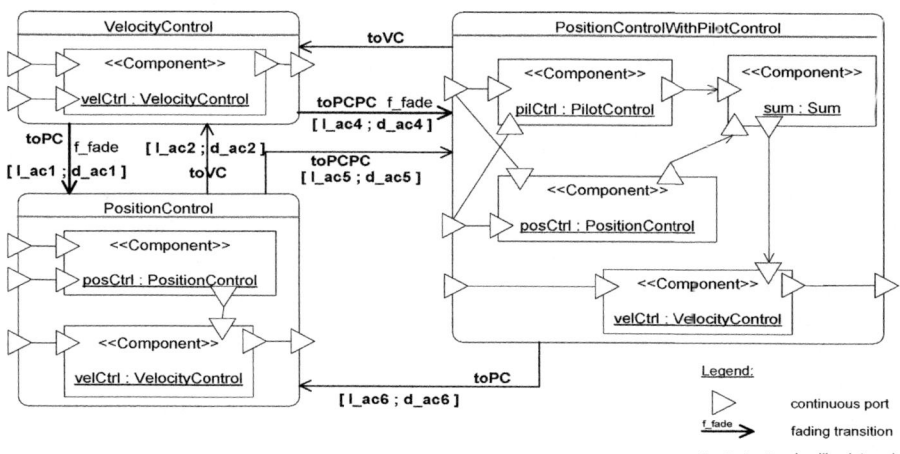

(b) Hybrid Reconfiguration Chart

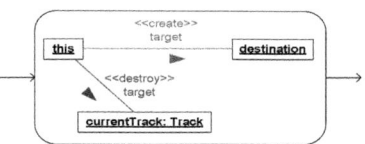

(c) Part of a Story Diagram

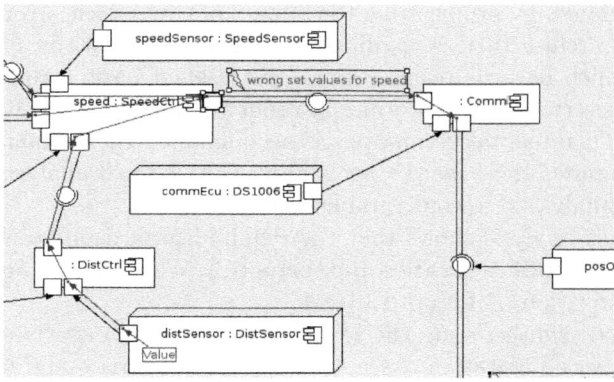

(d) Qualitative Hazard Analysis

Fig. 12.1. Development with the Fujaba4Eclipse Real-Time Tool Suite

To integrate **controllers** into the component model, they are embedded into hierarchical component structures. **Hybrid reconfiguration charts** (s. Figure 12.1(b)) are used to specify the different controller modes. The provided embedding concept enables the specification and modular verification of reconfiguration, i.e. the activation and deactivation of software as well as hardware components, across multiple components. Simple consistency checks ensure again that the verified real-time constraints of the coordination patterns are still valid in spite of the embedding. Thus, a verification of the whole system is not necessary, because the verification results of the individual patterns and components hold for the complete system [4].

Advanced systems may change their structure during runtime, i.e. a part of or even a complete component may be replaced or removed. These structural changes which may correspond e.g. to a removal of a pattern or the addition of a component, are also specified at design time. As a usual infinite state space has to be specified, we employ a grammar-based formalism in the sense of a generator definition. In more detail, a graph transformation system defines all valid and usually infinitely many system configurations by a finite set of rules. The correctness of such a rules set concerning safety properties and reachability of only valid system configurations is automatically verifiable [5] and [6, 7, 8]. Timing constraints of the execution of rules may also be specified such that the real time constraints of a system reconfiguration are expressable and analyzable on the model level as well.

We employ a special formalism to define the transformation rules called Story Diagrams [9] as depicted in Figure 12.1(c). We have shown in [10] that this formalism supports the computation of worst case execution times such that the time constraints defined on the model level correspond to the real execution time on a particular implementation platform. This approach "only" assumes that the implementation platform guarantees time constraints for basic operations like adding or deleting a graph node or edge resp.

The tool suite further supports a **hazard analysis** [11]. That analysis identifies random faults by propagating the impact of component errors through the whole system architecture. A qualitative analysis, as shown in Figure 12.1(d), determines which hazards result from a given set of basic errors (bottom up) or which basic errors have to occur in order to make a given hazard happen (top down). This qualitative analysis is accompanied by a quantitative analysis which computes the hazard's probability. The hazard analysis furthermore supports the analysis of reconfigurable systems.

If the hazard analysis shows that the required hazard probability is not satisfied, we apply **fault tolerance patterns** [12]. We again use Story Diagrams for their specification (s. Figure 12.1(c)).

Input to code synthesis are the hybrid reconfiguration charts and the graph rules defining reconfiguration. As not all system properties and the whole system behavior can be checked on the model level, the resulting code is executed using an advanced simulation system, which is partly based on the integration of a commercially available control engineering tool. The key point of this approach is that the simulated code is the same as the production code which is driving the real system, thus avoiding variations of behavior in simulation and implementation.

12.3 Case Study: RailCab

In terms of ecological values, public transport by bus or railway is deemed superior to individual transport by car. Unfortunately, individual transport clearly provides more flexibility and comfort for the passenger. The RailCab project[2] was founded at the University of Paderborn in 1998 in order to develop a new railway system that features the advantages of both techniques in terms of cost and fuel efficiency as well as flexibility and comfort. The novel system is characterized by autonomous vehicles operating on demand instead of trains being determined to a fixed schedule. RailCabs exhibit self-adaptive properties and operate in a safety-critical domain. Consequently, they provide an excellent case study for the Fujaba4Eclipse Real-Time Tool Suite.

We used the Fujaba4Eclipse Real-Time Tool Suite in different scenarios in the RailCab project. These scenarios include the active steering, the suspension/tilt [13] and the air gap adjustment system [14].

One particular problem is to reduce the energy consumption due to air resistance by coordinating the autonomously operating RailCabs in such a way that they build **convoys** whenever possible. Such convoys are built on-demand and require a small distance between the different RailCabs such that a high reduction of energy consumption is achieved. The convoy operation is clearly safety-critical. It requires a rigorous development approach for the real-time coordination between the RailCabs as well as for the integration of feedback controllers.

The first step is to **specify the structure and behavior** of the system [15]. The structure reflects typically the relevant parts of the physics. In our case study, we specify a RailCab component which embeds further components like a drive component. Like the RailCab component, the drive component could be composed of multiple other component instances. This leads to an architectural description of the RailCab, consisting of multiple layers. The coordination behavior of a convoy is specified by a real-time coordination pattern. We further embed the controllers of the RailCab – the distance and velocity controllers – into the internal component behavior and specify the reconfiguration between the different controller modes.

After building the model, we **verify** it applying two different techniques. First, we verify for the convoy scenario the real-time protocol behavior as well as the internal real-time component behavior through model checking, thus ensuring the safety critical property that the RailCabs will not collide [15]. Therefore, we check that all RailCabs are driving in convoy mode simultaneously. Second, we check structural properties as the correct instantiation of patterns and the consistent reconfiguration. For the first property we check, whether RailCabs

---

[2] http://www.railcab.de/en/index.html

driving on consecutive tracks apply the appropriate real-time coordination pattern to ensure they keep enough distance between each other. For consistency in reconfiguration, we check the correct embedding of controllers [3].

For **hazard analysis**, we consider the hazard of a RailCab driving in convoy mode at a wrong speed, which might result in collision. Therefore, a value failure on the output of the speed control is specified. Then, we employ the top down analysis to determine the errors that result in the hazard, e.g. wrong values from the speed sensor. The next step is to compute the hazard's probability by quantitative analysis. Thereafter, we determine the error's propagation paths leading to the hazard (see Figure 12.1(d)) by bottom up analysis thereby obtaining improvement points in the system's architecture [12].

After obtaining the final model, the tool **generates source code** that integrates the continuous and the discrete behavior. Then, we validate it through **simulation** that uses the same code as the final implementation, thus avoiding variation of behavior in simulation and implementation.

12.4 Conclusions and Future Work

We applied the Fujaba4Eclipse Real-Time Tool Suite successfully in the development of the RailCab's software. Up to now, we focused mainly on the forward engineering of the software. We are currently complementing our approach by a reverse engineering part [16]. This enables the integration of legacy software into our rigorous development approach. Due to space constraints, a discussion of related work is only contained in the cited papers.

References

[1] Burmester, S., Tichy, M., Giese, H.: Modeling Reconfigurable Mechatronic Systems with Mechatronic UML. In: Aßmann, U. (ed.) Proc. of Model Driven Architecture: Foundations and Applications (MDAFA 2004), Linköping, Sweden, pp. 155–169 (June 2004)

[2] Giese, H., Tichy, M., Burmester, S., Schäfer, W., Flake, S.: Towards the Compositional Verification of Real-Time UML Designs. In: Proc. of the 9th European software engineering conference held jointly with 11th ACM SIGSOFT international symposium on Foundations of software engineering (ESEC/FSE-11), pp. 38–47 (September 2003)

[3] Burmester, S., Giese, H., Henkler, S., Hirsch, M., Tichy, M., Gambuzza, A., Müch, E., Vöcking, H.: Tool support for developing advanced mechatronic systems: Integrating the fujaba real-time tool suite with camel-view. In: Proc. of the 29th International Conference on Software Engineering (ICSE), Minneapolis, Minnesota, USA, pp. 801–804. IEEE Computer Society Press, Los Alamitos (May 2007)

[4] Giese, H., Burmester, S., Schäfer, W., Oberschelp, O.: Modular Design and Verification of Component-Based Mechatronic Systems with Online-Reconfiguration. In: Proc. of 12th ACM SIGSOFT Foundations of Software Engineering 2004 (FSE 2004), Newport Beach, USA, pp. 179–188 (November 2004)

[5] Becker, B., Beyer, D., Giese, H., Klein, F., Schilling, D.: Symbolic Invariant Verification for Systems with Dynamic Structural Adaptation. In: Proc. of the 28th International Conference on Software Engineering (ICSE), Shanghai, China, pp. 72–81. ACM Press, New York (2006)

[6] Burmester, S., Giese, H.: Visual Integration of UML 2.0 and Block Diagrams for Flexible Reconfiguration in Mechatronic UML. In: Proc. of the IEEE Symposium on Visual Languages and Human-Centric Computing (VL/HCC 2005), Dallas, Texas, USA, pp. 109–116. IEEE Computer Society Press, Los Alamitos (September 2005)

[7] Tichy, M., Henkler, S., Holtmann, J., Oberthür, S.: Component story diagrams: A transformation language for component structures in mechatronic systems. In: Postproc. of the 4th Workshop on Object-oriented Modeling of Embedded Real-Time Systems (OMER 4), Paderborn, Germany (2008)

[8] Hirsch, M., Henkler, S., Giese, H.: Modeling Collaborations with Dynamic Structural Adaptation in Mechatronic UML. In: Proc. of the ICSE 2008 Workshop on Software Engineering for Adaptive and Self-Managing Systems (SEAMS 2008), Leipzig, Germany, pp. 33–40. ACM Press, New York (May 2008)

[9] Fischer, T., Niere, J., Torunski, L., Zündorf, A.: Story diagrams: A new graph rewrite language based on the unified modeling language. In: Ehrig, H., Engels, G., Kreowski, H.-J., Rozenberg, G. (eds.) TAGT 1998. LNCS, vol. 1764, pp. 296–309. Springer, Heidelberg (2000)

[10] Burmester, S., Giese, H., Seibel, A., Tichy, M.: Worst-case execution time optimization of story patterns for hard real-time systems. In: Proc. of the 3rd International Fujaba Days 2005, Paderborn, Germany, pp. 71–78 (September 2005)

[11] Giese, H., Tichy, M.: Component-Based Hazard Analysis: Optimal Designs, Product Lines, and Online-Reconfiguration. In: Górski, J. (ed) SAFECOMP 2006. LNCS, vol. 4166, pp. 156–169. Springer, Heidelberg (2006)

[12] Tichy, M., Henkler, S., Meyer, M., von Detten, M.: Safety of component-based systems: Analysis and improvement using fujaba4eclipse. In: Companion Proceedings of the 30th International Conference on Software Engineering (ICSE), Leipzig, Germany, pp. 1–2 (May 2008)

[13] Burmester, S., Giese, H., Oberschelp, O.: Hybrid UML Components for the Design of Complex Self-optimizing Mechatronic Systems. In: Braz, J., Araújo, H., Vieira, A., Encarnacao, B. (eds.) Informatics in Control, Automation and Robotics I, Springer, Heidelberg (March 2006)

[14] Henkler, S., Hirsch, M., Kahl, S., Schmidt, A.: Development of self-optimizing systems: Domain-spanning and domain-specific models exemplified by an air gap adjustment system for autonomous vehicles. In: ASME International Design Engineering Technical Conferences and Computers and Information in Engineering Conference, New York, USA, ASME, August 3-6, pp. 1–11 (September 2008)

[15] Burmester, S., Giese, H., Hirsch, M., Schilling, D., Tichy, M.: The fujaba realtime tool suite: Model-driven development of safety-critical, real-time systems. In: Proc. of the 27th International Conference on Software Engineering (ICSE), St. Louis, Missouri, USA, pp. 670–671. ACM Press, New York (May 2005)

[16] Giese, H., Henkler, S., Hirsch, M.: Combining Compositional Formal Verification and Testing for Correct Legacy Component Integration in Mechatronic UML. In: de Lemos, R., Di Giandomenico, F., Gacek, C., Muccini, H., Vieira, M. (eds.) Architecting Dependable Systems V. LNCS, vol. 5135, pp. 248–272. Springer, Heidelberg (2008)

13 AutoFocus 3 - A Scientific Tool Prototype for Model-Based Development of Component-Based, Reactive, Distributed Systems

Florian Hölzl and Martin Feilkas

Institut für Informatik
Technische Universität München
D-85748 Garching, Germany
{hoelzlf,feilkas}@in.tum.de

Abstract. We give an introduction of the AUTOFOCUS 3 tool[1], which allows component-based modeling of reactive, distributed systems and provides validation and verification mechanisms for these models. Furthermore, AUTOFOCUS 3 includes descriptions of specific technical platforms and deployments. The modeling language is based on precise semantics including the notion of time and allows for a refinement-based methodology for the development of reactive systems, typically found in user-accessible embedded realtime-systems.

13.1 Introduction

FOCUS is a general theory providing a model of computation based on the notion of streams and stream processing functions [1]. It is suitable to describe models for distributed, reactive systems. Based on this mathematical semantic foundation, we have developed a CASE tool, named AUTOFOCUS 3, to allow for graphical description of systems according to this model of computation. While FOCUS allows different techniques to build formal specifications of component-based, distributed systems, AUTOFOCUS 3 only uses some of these techniques as shown in the following. Furthermore, FOCUS allows to use different models of time expressed through the notion of streams: in particular untimed, timed and time-synchronous streams. AUTOFOCUS 3 is based on the time-synchronous notion of streams, which corresponds to a discrete notion of time based on globally synchronized clocks. FOCUS targets at precise description of applications on a logical level. Time is divided into logical ticks and logical components interact synchronously with each other in this setting.

In order to develop real distributed systems, the application must be executed on real hardware. Thus, during the development of the system, it is convenient to have several levels of abstraction and different development views. While early

---

[1] http://af3.in.tum.de/ provides a set of tutorials and screen shots of the current released version.

requirements are captured in natural language to allow flexibility, later levels use more formal specifications and finally add technical details. We believe that such a multi-layered set of models is the only way to cope with the complexity of today's systems by applying a strict separation of concerns during the phases of the development process.

AUTOFOCUS 3, as presented here, currently covers the lower-most levels of abstraction, namely the logical system architecture and the technical architecture, which provides the application execution environment. Higher levels of abstraction and requirements oriented models have been studied in earlier versions [2]. Our goal is to provide a prototypical tool implementation that clearly distinguishes between the models of different levels of abstraction and provides support for methods to combine these models into a description of the system under development along the complete development process. Larger case studies showing the complete methodology from requirements to deployment have been published: [3] presents a case study from the field of business systems, while [4] presents a case study from the automotive domain.

13.2 Capabilities of AutoFocus 3

This section briefly describes the modeling techniques of the two levels of abstraction currently supported by the AUTOFOCUS 3 tool: the logical architecture and its mapping onto a hardware/software execution platform. The logical architecture describes application specific components, while the topology describes the execution environment. For embedded systems the latter is usually a set of distributed control units and communication busses. Finally, the deployment model describes the mapping from application components onto execution and communication units.

13.2.1 Logical Architecture

The logical architecture defines a model of the system under development from an abstract point of view. The system's functionality is described independently of the concrete hardware/software environment and also independently of the concrete distribution of (parts of) the system on these resources.

The system consists of a set of communicating components, each having its own behavior specification, which may be stateful or stateless. Components exchange pieces of data in the form of typed messages. The semantic foundation assumes a global, discrete notion of time, e.g. the components are synchronized to a global clock.

Component Architecture
In AUTOFOCUS 3 the system's model is described as a set of communicating components. Each component has a defined interface (e.g its black-box view) and an implementation (e.g. its white-box behavior). The interface consists of a set of communication ports. A port is either an input port or an output port. It is

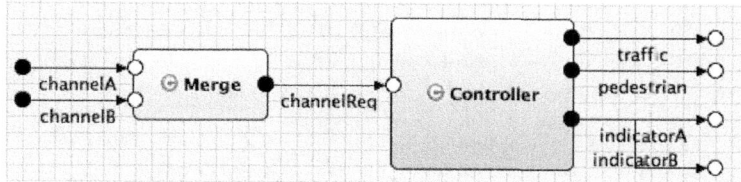

Fig. 13.1. A simple pedestrian crossing traffic light system

identified by its name and it has a defined type, thus describing which messages can be sent or received via this port.

Components can exchange data by sending messages through output ports and receiving messages via input ports. Communication paths are described by channels. A channel connects an output port to some input port, thus describing the sender/receiver relation. The data type of both ports must be compatible, of course. Under certain conditions, a component is allowed to send messages to itself, i.e. the model contains a feedback channel or more general a feedback loop. Output ports allow multi-cast messages, while input ports only allow a single incoming channel. From the logical point of view, channels transmit messages instantaneously.

Fig. 13.1 shows an example of a pedestrian traffic lights system, which consist of a controller for the application behavior and a merge component that merges button signal from both sides of the road. Note that AutoFocus 3 provides a hierarchical structuring of components in order to deal with larger systems in an easily comprehensible manner.

Causality and Time

AutoFocus 3 component networks are executed synchronously based on a discrete notion of time and a global clock. In this setting a component belongs to one of two classes. A *strong* causal component (the blue 'Controller'-component in Fig. 13.1) has a reaction delay of at least one logical time tick which means that the current output cannot be influenced by the current input values. A *weak* causal component (marked yellow in AutoFocus 3, c.f. 'Merge' in Fig. 13.1) may produce an output, which depends on the current input, e.g. the component's reaction is instantaneous. From the semantics point of view, networks consisting of strong causal components are always well-defined, e.g. for the recursive equation system induced by the channel connections unique fixed-points always exist. Component networks including weak causal components are also well-defined under the constraint that no weak-causal cycles exist, i.e. no weak causal component may send a signal that would be fed back to itself in the current time tick.

Stateful Behavior

To define stateful component behavior, we use a simple input / output automaton model. The automaton consists of a set of control states, a set of data state variables and a state transition function. One of the control states is defined to be

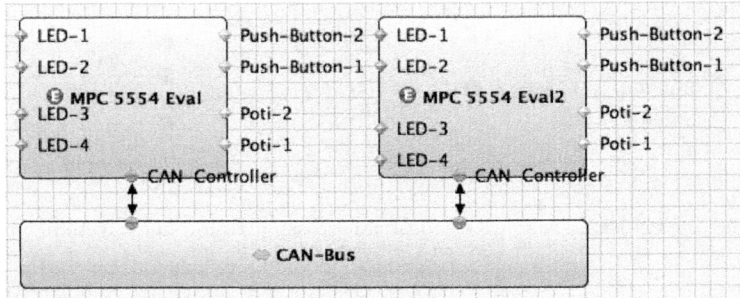

Fig. 13.2. Two ECU example topology for automotive lab hardware

the initial state of the component, while each data state variable has also a defined initial value. The state transition function is defined as a mapping from the current state, the current input values, and the current data state variable values to output values and subsequent data state variable values. A single transition has a source control state and a target control state, defines a set of input patterns, a set of preconditions over data state variables and variables bound in input patterns, and characterizes the output patterns and successor data state variable values.

Stateless Behavior
Time and again, some component has a relatively simple behavior like prioritizing certain input values or making a pre-computation. These components might not need data and control states at all. For this reason AUTOFOCUS 3 provides a simple tabular behavior specification that gives a (possibly non-deterministic) mapping from input patterns to output patterns.

Validation and verification
AUTOFOCUS 3 supports techniques to verify the logical architecture early in the development process, such as automatic test case generation and model checking. [4] presents the application of model checking techniques to verify the logical architecture. Automatic test case generation (from a separate test model) has also been applied in this case study to ensure the functional correctness of the system.

13.2.2 Technical Architecture

The topology architecture describes the execution environment of the system by means of execution control units (ECU) and busses. Embedded systems can observe their environment through sensors and influence it via acutators, which are connected to some ECU (like I/O devices in classical computers) or directly on some bus.

Fig. 13.2 shows an example of a topology with two ECUs connected to a common CAN bus. Each ECU provides a set of hardware ports: some LEDs, push buttons and potentiometers. These ECUs are part of our automotive lab[2]

---

[2] http://www4.in.tum.de/lehre/automotivelab/

demonstration hardware which is actively used in academic and industrial case studies and students' education [4].

Deployment
Having described the logical architecture and the execution environment, these two views of the system must be related to each other. In particular, each logical component has to be mapped onto some execution resource. Furthermore, logical signals need to be mapped to hardware ports, e.g. I/O devices or bus messages.

For the given example system, we could define a distributed deployment by assigning the `Merge` component to one ECU and the `Controller` component to the other. Since the `Merge` sends a signal to the `Controller` the connecting channel `channelReq` is automatically deployed on the connecting CAN bus. For the remaining signals, in particular the input signals of `Merge` and the output signals of `Controller`, we use the push buttons and LEDs, respectively.

Code Generation
Having completed the deployment by assigning components to ECUs and signals to hardware ports and bus messages, we can build our system to be run on the real hardware. Practically, this means to use all of the information of the model and produce C code from it, which can be compiled and flashed onto the demonstration hardware. Most of this task can be automated by suitable code generators or at least supported by the tool. Currently, this area is the most active part of our tool development activities.

13.3 Conclusion

We have given a compact introduction to the core features of AUTOFOCUS 3. We have shown two system model layers: the logical architecture describing the system application by means of communicating components and the technical architecture describing the execution environment of the application by means of electronic control units, sensors, activators, and busses. We have shown a deployment model, which describes the mapping of system components to these execution resources, thus relating the abstraction levels to each other. Finally, we have obtained a complete model to be used for automatic code generation.

Of course, the model provided here is very simplistic. In particular, concerning hardware and software resources of the execution environments, we have not treated issues like precise timing and task scheduling, bus message identification and mappings of data values to bus messages. Further work must be done in this direction, possibly also including upcoming hardware techniques like multi-core processors.

AUTOFOCUS 3 is currently implemented on the Eclipse platform[3]. Since we also develop other CASE-oriented tools, e.g. with different semantic foundations or different target domains, like machine engineering, we have built a common infrastructure, which allows a modular architecture. The deployment extension

---
[3] http://www.eclipse.org/

presented here makes heavy use of this modularity and extensibility. In detail, the topology and deployment features can be extended in specific ways: here, we have shown our automotive lab extension, which provides a event-triggered execution environment, as an example. Other extensions might include time-triggered architectures based on, e.g., FlexRay and OSEKtime. [5] already presents the formally verified mapping between the logical architecture and a time-triggered execution platform using the AUTOFOCUS task model, which is closely related to the model of computation of AUTOFOCUS 3.

We believe that rigorous division of the engineering process into different levels of abstraction with suitable models and precise semantics is a fundamental step towards model-based software engineering, in particular for embedded systems. We also believe that appropriate relations between these abstraction levels is vital and must be well understood from the methodological point of view, and of course supported by suitable tools. We have presented first steps towards this vision by the example of AUTOFOCUS 3.

Acknowledgements

We are especially grateful to Bernhard Schätz for providing continuous discussions, in particular on the language semantics and deployment questions, to Benjamin Hummel for his great work in building large parts of the tool infrastructure, and to Wolfgang Schwitzer for his work on the deployment extension of AUTOFOCUS 3.

References

[1] Broy, M., Stølen, K.: Specification and Development of Interactive Systems: Focus on Streams, Interfaces and Refinement. Springer, Heidelberg (2001)
[2] Geisberger, E., Grünbauer, J., Schätz: Interdisciplinary requirements-analysis using the model-based rm tool autoraid. In: Automotive Requirements Engineering (AURE 2006) Workshop at IEEE Intl. RE Conf. (2006)
[3] Broy, M., Fox, J., Hölzl, F., Koss, D., Kuhrmann, M., Meisinger, M., Penzenstadler, B., Rittmann, S., Schätz, B., Spichkova, M., Wild, D.: Service-oriented modeling of cocome with focus and autofocus. In: The Common Component Modeling Example: Comparing Software Component Models, pp. 177–206. Springer, Heidelberg (2008)
[4] Feilkas, M., Fleischmann, A., Hölzl, F., Pfaller, C., Rittmann, S., Scheidemann, K., Spichkova, M., Trachtenherz, D.: A top-down methodology for the development of automotive software. Technical Report TUM-I0902 Technical Report, Technische Universität München (2009)
[5] Botaschanjan, J., Broy, M., Gruler, A., Harhurin, A., Knapp, S., Kof, L., Paul, W., Spichkova, M.: On the correctness of upper layers of automotive systems. Formal Aspects of Computing 20(6), 637–662 (2006)

14 MATE - A Model Analysis and Transformation Environment for MATLAB Simulink

Elodie Legros[1], Wilhelm Schäfer[2], Andy Schürr[1], and Ingo Stürmer[3]

[1] Technische Universität Darmstadt, Real-Time Systems Lab
{legros,schuerr}@es.tu-darmstadt.de
[2] University of Paderborn, Software Engineering Group
wilhelm@uni-paderborn.de
[3] Model Engineering Solutions, Berlin
stuermer@model-engineers.com

Abstract. In the automotive industry, the model driven development of software is generally based on the use of the tool MATLAB Simulink. Huge catalogues with hundreds of modeling guidelines have already been developed to increase the quality of models and ensure the safety and reliability of the generated code. In this paper, we present the MATLAB Simulink Analysis and Transformation Environment (MATE), a tool using metamodeling techniques and visual graph transformations to automate the analysis and correction of models according to these guidelines. The MATE approach is illustrated by a typical example, and compared to other classical approaches for model analysis.

14.1 Introduction

Nowadays, model-based development is common practice within a wide range of automotive embedded software development projects. In model-based development, de facto standard modeling and simulation tools such as MATLAB Simulink are used for specifying, designing, implementing, and checking the functionality of new controller functions. The quality and efficiency of the software are strongly dependent upon the quality of the model used for code generation. For that purpose, modeling guidelines such as the MathWorks Automotive Advisory Board (MAAB) guidelines [1] have been defined and are usually adopted. There are tools available to help the modeler check a model according to these guidelines, such as the Simulink Model Advisor (part of the Simulink toolbox), the Model Examiner [2] or MINT [3]. They assist the developer in reporting violations of block settings, model configurations, or modeling styles that do not comply with such guidelines. However, for huge controller models, this can add up to a few hundreds, or even a few thousands, violations that must be corrected manually by the modeler. That is a cumbersome, complex, and expensive task. An in-house case study at Daimler detected in a model more than 2,000 guideline violations. After closer inspection, it was estimated that tools such as MATE could fix automatically up to 45% (900) of these 2,000 rule violations, and approximately 40% with user feedback. Only 8% required a manual correction. Finally, 4% remained undefined, i.e. the modeler had to determine whether the reported violations really infringed upon a modeling guideline (Cf. [4] for details).

For that reason we developed the MATLAB Simulink Analysis and Transformation Environment (MATE)[1]. This environment supports the detection of previously specified guideline violations. In addition to the Model Advisor or MINT, the Model Examiner and MATE also provide the possibilities of model repair operations. While the Model Examiner supports only static repair operations, i.e. without performing a model transformation (refactoring), MATE provides also layout improvements and modeling pattern instantiations. Apart from that, MATE is capable to perform high-level model analysis operations, such as control flow and data flow analysis.

14.2 Approach

Analysis as well as refactoring of MATLAB models demands full access to MATLAB's model repository. This requires an intimate knowledge of the API written in M-Script, a proprietary script language. Because both the used language and the tool's API evolved over many years, learning how to program reliable model checks and transformations using this approach costs time and efforts. It requires an intimate knowledge of the MATLAB API. MATE overcomes this problem by providing a layer of uniform API adapters on top of which visual graph queries can be developed in a more human friendly manner and on a considerably higher level of abstraction. Another benefit of the graph transformation is the possibility to express model transformations, and, hence, to define repair actions in an abstract way.

MATE provides two ways to analyze and repair models: *online* and *offline*. The corresponding architecture is represented in Fig. 14.1. The *online* modus enables an interactive analysis of a model within MATLAB. The communication between MATE and Simulink is realized through an Online-Adapter which supports all required read-and-write operations. The MATE-Tool represents the Java application controlling the execution of the analysis and transformation operations. On the other hand, the *offline* modus works on an efficiently processable main memory representation of the data representing the models to be checked and is generally used for complex analyses and corrections. The models are exported in their proprietary *mdl*-format, and imported in the Model Repository through the Reader/Writer as instances of the Simulink metamodel represented in Fig. 14.2. The module SDM Transformation - SDM Analysis defined in both parts (online and offline) of the MATE architecture represents the specifications of model analysis and correction.

---

[1] MATE project: *Model Engineering Solution* and *DaimlerChrysler* in partnership with the universities of Paderborn, Siegen, Kassel and the TU Darmstadt.

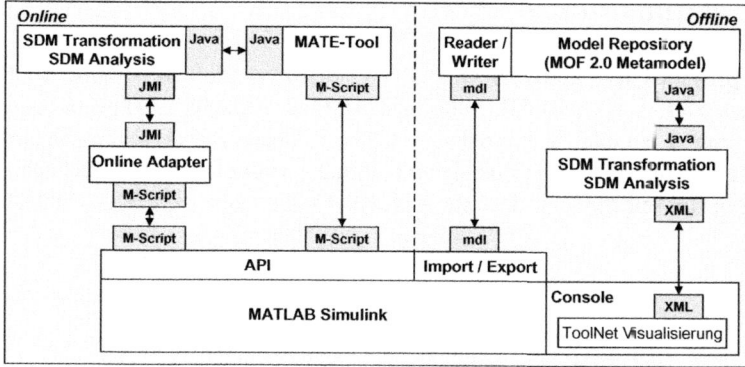

Fig. 14.1. MATE Architecture

Inside MATE, guideline specifications are based on a MOF 2.0 [5] compliant metamodel of Simulink. This metamodel is specified using the meta-CASE-Tool MOFLON presented in chapter 16. The real metamodel is very large and complicated. Therefore, only a very small version is presented in Fig. 14.2. The simplified metamodel consists of a single package Simulink containing only the most important classes. A Simulink model is a System that may contain a hierarchy of Subsystems with Blocks as leafs. Blocks are the atomic processing units. They are connected to each other by connecting their Outports and Inports via Lines.

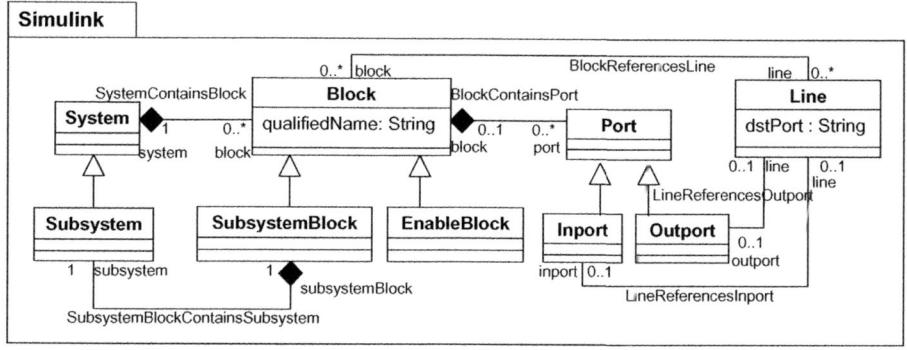

Fig. 14.2. Very simplified Simulink metamodel

This metamodel acts as graph schema for the specification of graph transformation rules. The technique of graph transformations is on the one hand applied for the detection of incorrect models as well as, on the other hand, for the (semi)automatic repair of identified errors. MATE uses the visual graph transformation approach of Story Driven Modeling (SDM) supported by the graph transformation tool Fujaba [6] and our plug-in MOFLON [7]. The next section gives an example of guideline specification with SDM. For more details about SDM, please refer to [8].

14.3 Application

MATE provides the analysis and correction of models according to modeling guidelines such as the MAAB guidelines [1]. One of the MAAB guideline concerns the naming of Enable Port blocks (jc_0281, "Naming of Trigger Port block and Enable Port block" [1]) The Enable block's name matches the name of the corresponding enable signal of the regarded subsystem (Cf. Fig. 14.3).

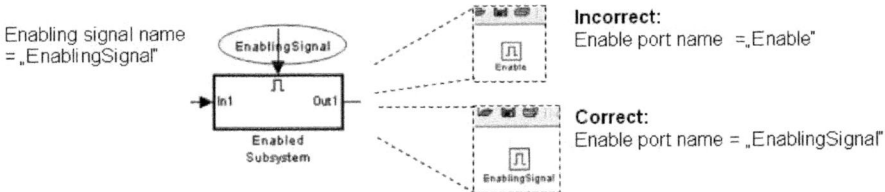

Fig. 14.3. Naming of Enable Port Block

Such a guideline can be implemented in different ways. It is state of the art that MATLAB modeling guidelines are implemented using the imperative scripting language M-Script. The M-Script specification of the guideline presented above would be the following:

```
function f_block_h = guideline_2(system, cmd_s)
  top_h = get_param(bdroot,'Handle');
  f_block_h = [];
  subsys = get_param(get_param(find_system(top_h,'BlockType','EnablePort'),'Parent'),
                                                     'Handle');
  for k=1:length(subsys)
    subsys_handle = get_param(subsys{k},'Handle');
    porth = get_param(subsys{k},'PortHandles');
    enable_port_name = get_param(porth.Enable,'Name');
    enableh = find_system(subsys{k},'SearchDepth',1,'BlockType','EnablePort');
    enable_block_name = get_param(enableh,'Name');
    if ~(strcmp(enable_port_name, enable_block_name))
       f_block_h = [f_block_h;subsys_handle];
    end
  end % for
end % function
```

In fact, the implementation of model guidelines with M-Script is nothing else than traversing graph structures and implementing graph pattern matching operations with an imperative language. Thus, implementing guidelines with M-Script is rather a task of programming skills and detailed API knowledge than a task of a conceptual and well structured conversion of an informal description into a formal one.

Since modeling guidelines represent constraints on model elements or relations between model elements which have to be respected, the OMG's logic-based language OCL [9] can be used for a formal description of rules like the modeling guideline described above:

```
context SubsystemBlock inv:
if self.containedBlock ->exists(b:Block j b.oclIsTypeOf(EnableBlock))
```

```
then self.containingSubsystemBlock.incomingLine ->select( line j line.dstPort = "enable")
    ->collect(qualifiedName)
 -> intersection (self.containedBlock ->select(b:Block j b.oclIsTypeOf(EnableBlock))
    ->collect(qualifiedName))
 -> notEmpty()
endif
```

Though, OCL is not very well-suited for the specification of complex patterns, where we have to navigate along different paths through a model and to compare their results. Even worse, OCL is not able to define most modeling guidelines about naming conventions since the available operations on strings and characters are pretty poor. In the same way, guidelines requiring complex arithmetic operations cannot be defined using OCL.

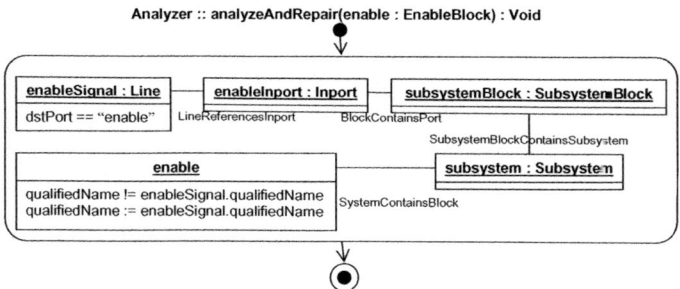

Fig. 14.4. Guideline specification with SDM

Therefore, we considered another approach when developing MATE, namely the visual graph transformation approach. To define these graph transformations, we use the visual SDM syntax. Fig. 14.4 simultaneously checks and fixes violations of the guideline presented above. It matches any occurrence of a pattern, where an EnableBlock and an EnableSignal object, which belong to the same Subsystem, do not have the same qualifiedName attribute. The line with the ":="-operator inside the enable object rectangle assigns the name of the matched enableSignal to the regarded enable object.

This specification shows how visual graph transformations facilitate the search for more complex object/link patterns. Moreover, while OCL only defines model checking operations, the SDM graph transformations provide incorporated repair actions. In case of guidelines that cannot be translated into graph transformation (e.g. naming conventions), Java code can be embedded within a story diagram by using so-called *Java Statements*, whereas OCL proposes no alternative for guidelines that cannot be specified as OCL constraints. A more detailed comparison between the SDM approach and other classical approaches for the analysis and correction of models, as well as the M-Script and OCL specification of the guideline described forehand, can be found in [10].

14.4 Conclusion

The adoption of modeling guidelines for the design of automotive controller models is important. MATE supports the developer in analyzing MATLAB Simulink models and automatically or interactively transforming such models into guideline-compliant ones. This tool is based on the use of visual graph transformations to specify the guidelines. The advantages of graph transformation are the higher level of abstraction, and the possibility to define not only check but also repair actions. The used language SDM is a visual graph transformation language. The visual aspect facilitates the specification of patterns to be matched as well as the navigation through a model to find information to be checked, i.e. tasks which are frequently needed when specifying modeling guidelines.

Although the SDM graph transformations are pretty well-suited, they are not equipped for handling *all sorts* of modeling guidelines. Moreover, many guidelines require the definition of quite similar transformation rules. Therefore, we are just extending the SDM syntax with generic and reflective features to increase the reusability and expressiveness of story diagrams [11]. Since MATLAB Simulink is widely used in the automotive industry, MATE should not remain a pure research prototype but be developed according to the needs of the industry. Therefore, we are integrating MATE within the Model Examiner developed by *Model Engineering Solution* [2] so that it can be used and, hence, evaluated in the context of concrete industrial use cases.

References

[1] MathWorks Automotive Advisory Board Homepage, http://www.mathworks.com/industries/auto/maab.html
[2] Model Examiner Homepage, http://www.model-engineers.com/our-products/model-examiner.html
[3] Mint Homepage, http://www.ricardo.com/engineeringservices/controlelectronics.aspx?page=mint
[4] Stürmer, I., Kreuz, I., Schäfer, W., Schürr, A.: Enhanced simulink and stateflow model transformation: The MATE approach. In: Proc. of MathWorks Automotive Conference (MAC 2007), June 19-20, Dearborn (MI), USA (2007)
[5] Object Management Group: Meta Object Facility (MOF) 2.0 Core Specification. ptc/03-10-04 (2003)
[6] Fujaba Homepage, http://www.fujaba.de
[7] MOFLON Homepage, http://www.moflon.org
[8] Zündorf, A.: Rigorous Object Oriented Software Development. University of Paderborn, Habilitation Thesis (2001)
[9] OCL Specification, http://www.omg.org/docs/ptc/03-10-14.pdf
[10] Amelunxen, C., Legros, E., Schürr, A.: Checking and Enforcement of Modeling Guidelines with Graph Transformations. In: Schürr, A., Nagl, M., Zündorf, A. (eds.) Proceedings of the Third International Symposium on Applications of Graph Transformations with Industrial Relevance (October 2007)
[11] Amelunxen, C., Legros, E., Schürr, A.: Generic and Reflective Graph Transformations for the Checking and Enforcement of Modeling Guidelines. In: Proc. of the Visual Languages and Human-Centric Computing (VL/HCC 2008), pp. 211–218 (September 2008)

15 Benefits of System Simulation for Automotive Applications

Oliver Niggemann[1], Anne Geburzi[2], and Joachim Stroop[2]

[1] Institute for Industrial IT, University of Applied Sciences, Lemgo, Germany
[2] dSPACE GmbH, Technologiepark 25, 33100 Paderborn

Abstract. The automotive industry faces the challenge of handling increasingly complex software systems in modern vehicles. The solution may be twofold: (*i*) A model-based development paradigm on the system level, using standards such as AUTOSAR, (*ii*) the usage of such models for a seamless testing and quality assurance process, using simulations and (reusable) tests.

This paper describes, from an automotive industry perspective, system models and their advantages for manufacturers and suppliers. The main focus is on tooling for offline system simulations and on the introduction of such solutions in standard development processes. For this, different industrial simulation and testing scenarios are outlined; starting with single software component tests and ending with virtual integrations. For each phase of the development process, chances and problems are discussed.

15.1 System Models

System models—a buzz word which has been spreading through journals and conferences for years, has reached industry. But it is more than just a buzz word: System models will improve the development of ECUs. Developers will use system models to formalize the design process for distributed ECU systems—a process which currently is getting out of hand (see [1], [2]). System models will also help to increase the rate of software reuse and to test complex systems earlier.

But for many developers, system models will first of all cause additional work. Another model to be created. Another model to keep updated. Another tool that must be used to edit such models. So why are many manufacturers currently introducing these models in their development processes?

The electronic system of most modern vehicles has reached a complexity that rivals almost any other known embedded system. Vehicles comprise up to 80 different ECUs that are connected via several different communication buses. ECUs often contain 1 MB of software (ROM) which may be built of more than 200 different software modules. And many vehicle functionalities such as braking or lane keeping are not implemented by one single ECU but rely on several communicating ECUs, i.e. they are implemented as distributed software systems.

In addition to these technical issues, there are other constraints that make the design of electronic vehicle systems one of the most difficult challenges known to computer science: Most ECUs must work under hard real-time constraints. And often safety-related constraints must be considered—after all, we all trust ABS and ESP systems to work under harsh conditions and at high velocities. Furthermore, ECUs are not implemented by the car manufacturers themselves but by many different ECU suppliers, making the integration process of ECU networks rather difficult. And finally, hard cost constraints apply to these systems; no customer wants to pay for invisible vehicle electronics.

Like models used in traditional engineering disciplines, system models are a way to formalize these complex design processes (see [3]). An improved planning process will mean fewer integration problems later: (i) Work is shared in an elaborate way between developers and suppliers. Each working party gets a formal description of interfaces and other applicable constraints. (ii) Software can be designed in a way that makes reuse as easy as possible. The more software is reused, the fewer new implementations are needed and the fewer problems will occur. (iii) Formal models can be analyzed or simulated in early stages, helping developers to find errors as soon as possible.

In this paper, the focus will be on the simulation of system models to support early error identification. Because a significant percentage of errors are due to distributed software–25% according to [4]–we will especially describe the simulation and testing of such systems.

15.1.1 State of the Art and AUTOSAR

It is important to remember that model-based development has been used in engineering fields such as control, civil, or electric engineering for much longer than in computer science. This becomes evident by the success of tools such as The Mathwork's MATLAB ®, Simulink®, and Stateflow® development tools which are commonplace in almost all application development processes. These solutions use graphical models to represent control algorithms or physical systems. The integration of these existing tools into new tool chains is crucial for the success of new development paradigms such as system models.

The term "system model" is not defined formally and is therefore used ambiguously. This paper describes the implementation of system models (including their simulation) within a commercial tool that aims at the software production process in automotive industry. Such tools nowadays normally follow the AUTOSAR ([5]) standard. I.e. while different model-based system development approaches have been used in the past (see e.g. [6, 7, 1, 2, 8, 9]), AUTOSAR has lead to a standardization for the modeling of embedded production software in the vehicle. Such system models comprise the following parts.

Application Software Models: In the automotive community, these models are currently the main novelty and the main point of controversy. They are used to model several aspects of the software:

- Software component models refer to reusable application software modules. AUTOSAR uses an AtomicSoftwareComponentType for this.
- Composition models define the hierarchical interconnection of application software components based on the structure of the software architecture. In AUTOSAR this corresponds to the CompositionType of the Software Component Template, in UML [10] one could use the Component Diagram, or with SysML [11] the Internal Block Diagram. Similar approaches exist in EAST-ADL [12, 13].

Hardware Topology Models: Current automotive hardware topologies normally comprise (i) electronic control units (ECUs), (ii) buses such as CAN, LIN, FlexRay and (iii) I/O such as DIO, ADC, or PWM. AUTOSAR has defined the ECU Resource Template to describe these hardware elements. SysML provides the Block Definition Diagram to express such static system structures.

Platform Models: To use hardware elements additional software modules or configuration settings are needed. E.g. an ECU would need an operating system, I/O relies on I/O drivers, and communication buses must be configured by defining communication matrices. Here mainly two types of models are used:

- Interfaces and configuration settings of ECU basic software such as operating system, COM stack or I/O drivers are modeled. Standards such as OSEK's OIL format or AUTOSAR's ECU Configuration Template are used in the automotive domain.
- Besides essential parameters such as bus type and bus speed, buses and bus controlers are mainly configured by communication matrices. Important standards here are ASAM's FIBEX or AUTOSAR's system template.

Deployment Models: Another important type of information is how application software is mapped onto the hardware topology and how it is connected to basic software:

- Mapping of software components onto ECUs. E.g. in UML deployment diagrams are used for this. AUTOSAR uses the system template.
- Connection of software components to basic software modules such as I/O drivers (e.g. analog/digital converters) or to COM stacks (e.g. for a CAN bus).
- Mapping of communication between software components onto bus signals, e.g. AUTOSAR uses the system template to map signals contained in a port of an AUTOSAR software component onto a bus message.

Behavior Models: These are models of the behavior of software components, e.g. given as C code, Simulink or TargetLink models. AUTOSAR allows the integration of C code for software components—this code could be generated by code generators from graphical models such as Simulink.

Plant Models: Plant models describe the environment of the ECU network, i.e. the vehicle, the driver, road conditions, etc. Such models are often given as

C code, Modelica models or as Simulink models. In AUTOSAR and most other system model formalisms such models are not part of the standard itself; only sensors and actuators can be modeled.

So far, no standard exists that covers all these aspects. For example, AUTOSAR [5] does not include plant and behavior models.

SystemDesk—the dSPACE system modeling and simulation tool—offers an integrated modeling environment for automotive systems (see figure 15.1). While being mainly based on AUTOSAR, it extends the standard by interfacing with function and plant behavior models (mainly Simulink) to offer a holistic and complete solution. As the next section will outline, this is necessary to make the system models executable.

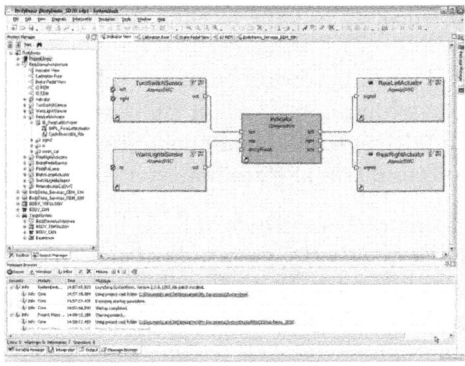

Fig. 15.1. Modeling Software Architectures with SystemDesk

While formal software and system models (e.g. according to AUTOSAR) are now rapidly becoming part of a mainstream automotive software development process, the question remains open what types of more abstract models are needed in earlier phases. Several research projects (e.g. compare EAST's Vehicle Feature Models in [8] or Mobilsoft's Services and Interaction Models in [14]) have proposed a hierarchical decomposition of a vehicle's functionality – sometimes combined with a model of the vehicle variants. Other approaches also provide more (e.g. [15]) or less formal (e.g. [11]) model types for early system designs and requirements.

15.2 System Simulation

Companies have introduced model-based development processes (e.g. with Simulink) to speed up algorithm development, to ensure feasible documentation of the solution and to leverage automatic code generation methods. But model-based design has a useful by-product: The ability to simulate and to test the algorithm design in early stages, e.g. on the developer's PC. A similar development can already be seen with system models.

System models are called *executable* if they can be simulated. Simulation scenarios and thus test cases might cover both offline simulations on PCs or real-time simulations on designated hardware platforms, e.g. hardware-in-the-loop (HIL) simulations. First experiences with dSPACE's SystemDesk tools can be found in [16] and [17].

But AUTOSAR system models are in most cases not executable. This is due to their focus on the structural and integration aspects of systems. However, when behavior descriptions for software and hardware modules are added, system models can be used for simulation purposes (see also [18]). For example, AUTOSAR is not executable because it lacks a way of describing application and plant behavior, but by adding behavior models, i.e. for software components, AUTOSAR models are made executable. Typical behavior models are Simulink, IBM's Rhapsody, or plain C code.

Executable system models provide the chance to simulate aspects that traditional, behavior-focused models can either not describe at all, or only with significant effort. Examples are (see also [16]):

- Effects caused by bus communication such as signal delays or signal synchronization
- Timing effects such as task scheduling, execution times of functions, or periodic communication tasks
- Mode management functionality such as system boot-up behavior or initialization and system sleeping/wake-up behavior
- Errors and problems on the system level such as bus faults (e.g. CRC errors), jitter effects in task executions, or driver-related faults.

From a theoretical perspective, traditional behavior-based modeling environments can also be used to simulate these aspects. However, since behavior-based modeling environments provide no first-level way of describing typical system aspects, so this approach involves a high workload and in many cases high maintenance costs.

For example, the scheduling behavior of an ECU can be modeled and simulated using finite state machine models such as Stateflow. However, this effectively requires manual reimplementation of operating system behavior. System models, on the other hand, provide formalisms to describe scheduling configurations. That is, the simulation task is easier when high-level system models are used instead of traditional behavior-based modeling formalisms.

SystemDesk implements this idea systematically. During system design, SystemDesk tries to keep the model executable on condition that the user provides suitable behavior models (see [19]). SystemDesk follows the paradigm of requiring as little simulation-specific work as possible. In most cases, users just have to provide the test cases and can then simulate the system model.

Figure 15.2 shows a typical SystemDesk simulation: Some variables are simulated and plotted over time. Detailed descriptions of SystemDesk's simulation capabilities can be found in section 15.3.

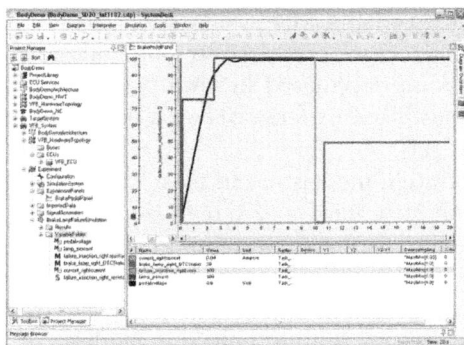

Fig. 15.2. Simulating a System Model with SystemDesk

15.3 Applications of System Simulation

System simulation can be used in different stages of the development cycle. Please note that the (AUTOSAR) system models are enriched during the development phases. While the first scenarios rely on some parts of the system models only (e.g. models for software components only), later scenarios need fully defined system models.

15.3.1 Specification Verification

In early design stages, manufacturers decompose their electronic system into interconnected functions or software components—often disregarding ECUs and communication buses. These software components are normally not implemented yet; instead their behavior is only roughly specified (e.g. using Stateflow models). But even in these early stages, developers want to verify whether the function specifications work correctly and whether the interplay of local component specifications produces the wanted overall behavior. For this, the network of components (and their behavior) must be simulated in a suitable execution order—with or without plant models. SystemDesk offers both scenarios in its virtual functional bus (VFB) simulation mode (VFB is a term coined by AUTOSAR).

15.3.2 Software Component Tests

SystemDesk provides the capability of simulating a single (AUTOSAR) software component. This includes stimulating input signals, and measuring outputs and internal variables. Software components such as AUTOSAR rely on other ECU software modules such as the operating system, mode management, diagnostic services or I/O drivers. SystemDesk provides an ECU-like simulation framework which allows the execution of production software components in an offline simulation on the developer's PC. Another use case is the verification of software components implemented by third-party software suppliers: ECU integrators are interested in testing these components without looking at the behavior definition models (e.g. Simulink/TargetLink models). In many cases, for reasons of intellectual property protection, such components only exist in the form of object files.

15.3.3 ECU Tests

Later on in the production process, ECU integrators (normally ECU suppliers) want to verify the correct behavior of ECUs. At this stage, application software components have been implemented. In order to simulate whole ECUs, not only the application software components but also the ECU platform must be simulated. SystemDesk takes basic software configurations of the production ECUs and uses them to simulate the application software on a PC as realistically as possible. The SystemDesk simulation uses emulations of the basic software models for this: An AUTOSAR operating system is emulated for the production ECU under development, and the NVRAM and fault memory are also reproduced on the PC. A practical example, the verification of ECU diagnosis algorithms within SystemDesk at Daimler AG, is presented in [20].

15.3.4 Virtual Integration

Manufacturers need to verify whether the interconnected ECUs work together correctly. Traditionally such tests are done using hardware-in-the-loop (HIL) simulations. SystemDesk offers virtual integration using an offline simulation. For this, besides the ECU simulation from section 15.3.3, a simulation of the communication buses is needed. Also plant models are used to simulate the vehicle and specific driving maneuvers. A major challenge for the introduction of such virtual HIL simulations is the usage of the same test cases and models for the offline and the HIL simulation respectively. First experiences at AUDI can be found in [17].

15.4 Summary

In this paper, the authors tried to show that system models are not only a new buzz word but really offer advantages for the automotive industry. System models enable users to get an overall view of vehicle software systems. When behavior descriptions for software modules and plants are added, system models become executable and can be used for simulation purposes. Accordingly, system models allow the verification and simulation of the E/E architecture in early stages of the development process of electronic control units. Thus it is possible to perform HIL-like test scenarios in an offline simulation on the PC.

The offline simulation of system models therefore supports manufacturers and suppliers in their task of detecting design errors in distributed systems, problems caused by bus communication, etc. In view of the increasing complexity of software in modern vehicles and the resulting increase in the number of possible faults, this will prove to be a significant development step.

References

[1] Kraft, D., Lapp, A., Schirmer, J.: Elektrik/Elektronik-Architektur - Die Herausforderung für die Automobilindustrie. In: VDI Berichte Nr. 1789 (2003)
[2] Steiner, P., Schmidt, F.: Anforderungen und Architektur zukünftiger Karosserieelektroniksysteme. In: VDI Berichte Nr. 1789 (2003)
[3] Niggemann, O., Stroop, J.: Models for Model's Sake. In: ICSE Experience Track on Automotive Systems (2008)
[4] McKinsey, Company: Managing innovations on the road. In: Automotive Electronics (2005)
[5] AUTOSAR 3.1 Specifications, http://www.autosar.org
[6] Eppinger, K., Berentroth, L.: Plattform versus Flexibilitaet: Die Siemens VDO EMS 2 Plattform Architektur. In: VDI Berichte Nr. 1789 (2003)
[7] Hietl, H., Streit, W.: Integration komplexer Elektroniksysteme am Beispiel des neuen A8. In: VDI Berichte Nr. 1789 (2003)
[8] Voget, S., de Boer, G., Heidl, P., Adis, F., Virnich, U.: Analyse fahrzeugweiter Softwarekonzepte im Rahmen des europäischen Förderprojekts ITEA-EAST/EEA. In: AUTOREG, Wiesloch, Germany (2004)
[9] Wolf, F.: Integrationsverfahren für Softwaresysteme im Antriebsstrang. In: VDI Berichte Nr. 1789 (2003)
[10] OMG: Unified Modeling Language, UML (2003), http://www.omg.org
[11] SysML, http://www.sysml.org
[12] EAST-EEA, http://www.east-eea.net
[13] Lönn, H., Saxena, T., Nolin, M., Törngren, M.: FAR EAST: Modeling an Automotive Software Architecture Using the EAST ADL. In: ICSE Workshop on Software Engineering for Automotive Systems (SEAS). IEE (2004)
[14] Wild, D., Fleischmann, A., Hartmann, J., Pfaller, C., Raapl, M., Rittmann, S.: An Architecture-Centric Approach towards the Construction of Dependable Automotive Software. In: SAE World Congress (2006)
[15] Hartmann, J., Rittmann, S., Wild, D., Scholz, P.: Formal Incremental Requirements Specification of Service-oriented Automotive Software Systems. In: Proceedings of the Second IEEE International Symposium on Service-Oriented System Engineering, SOSE 2006 (2006)
[16] Thiessen, C., Geburzi, A., Lamberg, K., Niggemann, O.: Übertragung von HIL-Tests in die Offline-Simulation. In: 2nd AutoTest Technical Conference (2008)
[17] Stichling, D., Niggemann, O., Schmidt, K., Reichelt, S., Maleuda, M.: Durchgängige Systemtests von der virtuellen Integration bis zum Verbundtest. ATZ elektronik (November 2008)
[18] Otterbach, R., Niggemann, O., Stroop, J., Thümmler, A., Kiffmeier, U.: System Verification throughout the Development Cycle. ATZ Automobiltechnische Zeitschrift (April 2007)
[19] Nawratil, P., Niggemann, O.: Der kleine grüne Pfeil ... Hanser automotive (July 2008)
[20] Adam, V., Kohlweyer, M., Balzer, H., Nawratil, P.: Diagnoseverifikation in frühen Entwicklungsphasen. Elektronik automotive (December 2008)

16 Development of Tool Extensions with MOFLON

Ingo Weisemöller[2], Felix Klar[1], and Andy Schürr[1]

[1] Fachgebiet Echtzeitsysteme
Technische Universität Darmstadt, Germany
{klar,schuerr}@es.tu-darmstadt.de
[2] RWTH Aachen University, Germany
http://www.se-rwth.de

Abstract. The increasing complexity of embedded systems is accompanied by an increasing number and complexity of models, modeling languages and tools in the development process. This results in a need for appropriate tool support at the metamodel level. Besides the necessity to develop new languages and tools, there is also a large demand for extensions to existing tools as well as for integration frameworks. Such frameworks ensure consistency between data that is distributed over several tools. In this chapter, we present MOFLON, a metamodeling tool primarily focused on tool extension and integration. It adopts several standards such as MOF 2.0 and JMI. It also supports story driven modeling as a means of describing on-model transformations as well as a combination of MOF QVT and triple graph grammars for model-to-model transformations and integration. We present a typical application of these features to tools used in the development of embedded systems.

16.1 Introduction

Because the number of software development processes, especially for embedded systems, has rapidly increased recently, the number of modeling languages and commercial off-the-shelf (COTS) modeling tools has increased as well. Therefore, documents created with these modeling tools are also becoming harder to manage and maintain. These documents and models may be difficult to understand and to develop further. Therefore, modeling guidelines are a wide spread approach to improve readability and maintainability of these documents. Such guidelines may also enforce properties of the model that are necessarily required for automatic processes, such as code generation. Data spread across several documents may be redundant and needs to be kept consistent. These documents are usually developed with different COTS tools. Most of such tools neither provide proper interfaces to couple them with one another, and they do not provide a way to define domain specific rules for data consistency between several documents. Thus, alignment and adjustment of this data is usually performed manually, which results in considerable efforts and costs.

Since new tools are not usually an option in ongoing processes, tool extensions are a more adequate way to enforce modeling guidelines and to ensure consistency between several models. The metamodeling tool MOFLON is focused on efficient development of such extensions. We use MOFLON to develop tool adapters that comply to the Meta Object Facility (MOF) [1] and to the Java Metadata Interface (JMI) [2], and thus provide standardized access to model data. Based on these adapters, we use model transformations to describe rules for analysis and semi-automatic repair of models according to guidelines. MOFLON also allows to define model-to-model transformations and consistency rules in a declarative notation based on MOF Query/View/Transformation (QVT) [3].

The remainder of this chapter is outlined as follows: In Section 16.2 we describe the core features and briefly introduce the standards adopted by MOFLON. Section 16.3 provides an overview of usage scenarios for MOFLON, and in Section 16.4 we give a short summary and present some ideas for future versions of MOFLON.

16.2 History and Overview of Features

The development of MOFLON began in 2002. Based on code generated by the MOF Model Compiler (MOMoC) [4] from a simplified version of the MOF metamodel, we developed a graphical editor as a plugin for the UML tool Fujaba [5]. Besides this editor, the graph transformation environment of Fujaba was reused for model transformations. This step required a refactoring of the existing environment in order to make it work on an abstract metamodel interface, which could be implemented by plugins. Having completed these steps successfully, we released MOFLON 1.0 in December 2006.

More recent versions of MOFLON introduced an editor and code generator for model-to-model transformation rules based on triple graph grammars (TGGs) (MOFLON 1.1, July 2007), a compiler for the Object Constraint Language (OCL) [1] based on the Dresden OCL toolkit [6] (MOFLON 1.2, December 2007) and modularization concepts for model-to-model transformations (MOFLON 1.3, December 2008).

16.2.1 MOF Editor and Code Generation for MOF Models

MOFLON adopts the MOF 2.0 standard [1] by the Object Management Group (OMG). MOF compliant metamodels describe the abstract syntax of modeling languages in a notation based on UML class diagrams. MOFLON supports the complete MOF (CMOF); in comparison to its subset essential MOF (EMOF), which is, for instance, supported by the Eclipse Modeling Framework [7], CMOF has much more sophisticated association and modularization concepts, which are substantial for metamodeling in the large. Constraints can be added to metamodels using the Object Constraint Language (OCL) [1] in MOFLON.

The code generated from metamodels by MOFLON complies to the JMI standard by Sun. This defines tailored interfaces, which are specific to the respective metamodel, and reflective interfaces, which provide generic access to model and metamodel data. Our mapping from MOF 2.0 to JMI is an extension of the JMI mapping for MOF 1.4 defined by the OMG. Because JMI does not describe an event mechanism, MOFLON metamodels implement the interface of Netbeans' metadata repository (MDR) [8] for events.

16.2.2 Additional Frontends

Besides the graphical MOF editor, MOFLON provides import modules for several other frontends. UML models can be imported from Rational Rose or Sparx Systems Enterprise Architect. For Enterprise Architect, there is also a plugin [9] that introduces MOF diagrams, provides a toolbox for editing MOF models, and performs checks on these models to ensure they can be imported and used for code generation in MOFLON. The import from UML tools is based on the XML Metadata Interchange (XMI) standard. Because many tools have their own extensions to or interpretations of XMI, one can run XSL Transformations on the XMI data before the import. This results in low efforts to develop import modules for further tools. Currently, we are also working on a textual frontend.

16.2.3 Model Transformations

Since one of our core areas of application is model analysis and repair, MOFLON can be used to describe rules and constraints for this. We make extensive use of OCL constraints, pattern matching and model transformations for model analysis and repair. MOFLON uses the transformation engine provided by Fujaba, with a set of code generation templates that has been adopted to MOF and JMI.

Model transformations in MOFLON are described in story diagrams [10], which are a combination of UML activity diagrams and an adopted version of collaboration diagrams. The control flow of a transformation is specified in an activity diagram. Inside each activity, pattern matching and replacement is described in an extended collaboration diagram. Chapter 14 of this book gives an example of model transformations with MOFLON.

16.2.4 Triple Graph Grammar Editor

Model-to-model transformations can be specified using the MOFLON triple graph grammar editor. TGGs [11] are a formal transformation language that allows to relate model elements with each other. TGGs specify bidirectional model-to-model transformations in a declarative manner. TGG rules can be translated into operational transformation rules. These can be used to perform forward and backward transformations as well as consistency checks on related models. TGGs are closely related to the model transformation standard QVT [12]. However, since QVT is not based on a formal foundation and, therefore, also suffers from a lack of precision, we decided to base our transformation implementation on TGGs, which have formally and precisely defined semantics.

16.3 Usage Scenarios

Extensions to COTS tools, which we develop with MOFLON, typically perform analysis and repair tasks on single models, or they keep data across several tools consistent. A combination of both kinds of extensions is possible.

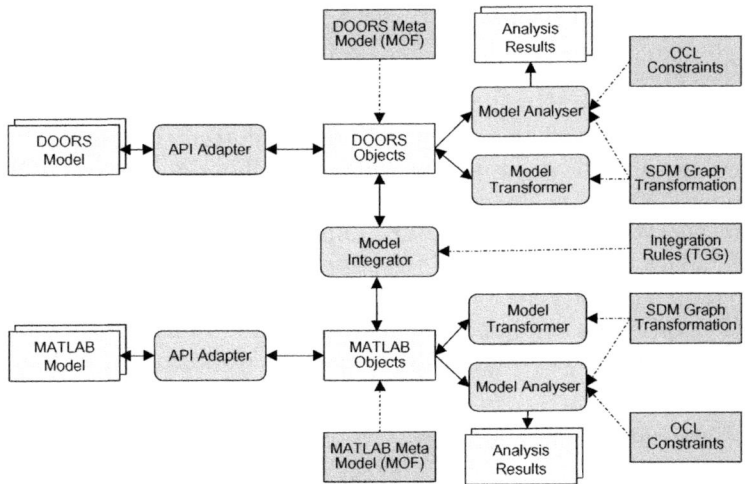

Fig. 16.1. Integration Scenario Including Model Analysis and Repair

Figure 16.1 provides an overview of such a combination. It shows the integration of the requirements engineering tool DOORS with the systems modeling environment MATLAB/Simulink. Adapters provide standardized interfaces to the data in each tool, i.e. the adapter provides JMI compliant objects to all other components. This is, for instance, required for the model transformation rules to work properly. For both the DOORS and the MATLAB data, there is a model analyzer and transformer, which take OCL constraints and model transformation rules as input and apply them to the models. Moreover, there is a model integrator, which applies TGG rules to keep data between the tools consistent.

16.3.1 Tool Adapters

The code generated by MOFLON for model transformations requires a JMI compliant metamodel to run. In order to perform analysis and repair actions on models in tools, we need a JMI compliant interface to this data. We use MOFLON to describe the API and data structure of the tool in a metamodel, and to generate the interfaces and a substantial part of the adapter implementation with a customized set of templates. As an example, Figure 14.2 shows the metamodel of the modeling and simulation tool MATLAB/Simulink.

Since adapters use calls to the proprietary tool API, a part of it needs to be written manually. An evaluation based on the MATLAB/Simulink adapter has shown that about 95% of the adapter (measured in lines of code) can be generated. This includes the interfaces and most of the implementation of the reflective methods, whereas calls to the tool API must be implemented manually. Further increment of this percentage will be possible, if some API calls like setting attribute values in model elements are generated with tool specific templates.

16.3.2 Model Analysis and Repair

With the JMI compliant tool adapter, one can perform model analyses and repairs, which are implemented by means of OCL constraints and model transformations. Minor repairs may be performed automatically, but more complex actions require a user to choose one of several possible repair actions. Analyses and repairs with MOFLON, especially on MATLAB/Simulink models are discussed in detail in chapter 14 of this book.

16.3.3 Integration Framework

Integration rules specified in the TGG editor can be translated to operational graph analysis and transformation rules by MOFLON. Figure 16.2 provides a more detailed view of the integration between DOORS and MATLAB/Simulink models.

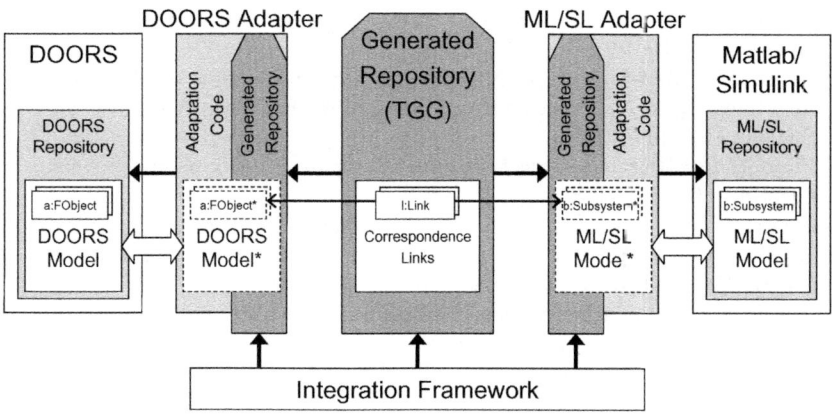

Fig. 16.2. Integration between DOORS and MATLAB/Simulink [13]

Access to the tool repositories is provided by the JMI adapters. The integration framework applies the TGG rules to the models. For instance, it may ensure that for every use case in DOORS, which is specified in a so called formal object (FObject in the figure), a corresponding subsystem must implement this use case in the MATLAB/Simulink model.

16.4 Conclusions and Future Work

The metamodeling tool MOFLON is designed for the rapid development of tool extensions rather than for developing tools from scratch. It includes editors and code generators for MOF compliant metamodels, OCL constraints, endogenous and exogenous transformations. Typical areas of application are model analysis and repair as well as model-to-model consistency checking and integration.

Future versions of MOFLON will provide enhanced possibilities to use commercial or open source tools for metamodel and transformation editing as well as more sophisticated modularization concepts for metamodeling in the large [14].

Acknowledgments

We would like to thank Tobias Rötschke, Alexander Königs and Carsten Amelunxen, who have initiated the MOFLON project and contributed a lot to it.

References

[1] OMG, Inc.: Catalog of OMG Modeling and Metadata Specifications (November 2008),
http://www.omg.org/technology/documents/modeling_spec_catalog.htm
[2] Dirckze, R.: Java Metadata Interface (JMI) Specification, v1.0 (June 2002)
[3] Amelunxen, C., Königs, A., Rötschke, T., Schürr, A.: MOFLON: A Standard-Compliant Metamodeling Framework with Graph Transformations. In: Rensink, A., Warmer, J. (eds.) ECMDA-FA 2006. LNCS, vol. 4066, pp. 361–375. Springer, Heidelberg (2006)
[4] Bichler, L.: Tool Support for Generating Implementations of MOF-based Modeling Languages. In: Proceedings of the Third OOPSLA Workshop on Domain-Specific Modeling (2003)
[5] Zündorf, A.: Rigorous Object Oriented Software Development. University of Paderborn (2002),
http://www.se.eecs.uni-kassel.de/fileadmin/se/publications/Zuen02.pdf
[6] Loecher, S., Ocke, S.: A Metamodel-Based OCL-Compiler for UML and MOF. Electr. Notes Theor. Comput. Sci. 102, 43–61 (2004)
[7] The Eclipse Foundation: Eclipse Modeling – EMF – Home (2008),
http://www.eclipse.org/modeling/emf/
[8] netbeans.org: Metadata Repository (MDR) Project Home (2008),
http://mdr.netbeans.org/
[9] Patzina, S.: Anpassung eines UML-Modellierungswerkzeuges für die Metamodellierung domänenspezifischer Sprachen. Master's thesis, TU Darmstadt (2008)
[10] Amelunxen, C., Rötschke, T., Schürr, A.: Graph Transformations with MOF 2.0. In: Giese, H., Zündorf, A. (eds.) Proc. 3rd International Fujaba Days 2005, vol. tr-ri-05-259, pp. 25–31. Universität Paderborn (September 2005)
[11] Schürr, A.: Specification of Graph Translators with Triple Graph Grammars. In: Mayr, E.W., Schmidt, G., Tinhofer, G. (eds.) WG 1994. LNCS, vol. 903, pp. 151–163. Springer, Heidelberg (1995)

[12] Königs, A.: Model Integration and Transformation - A Triple Graph Grammar-based QVT Implementation. PhD thesis, Technische Universität Darmstadt (2009)
[13] Amelunxen, C., Klar, F., Königs, A., Rötschke, T., Schürr, A.: Metamodel-based Tool Integration with MOFLON. In: 30th International Conference on Software Engineering, pp. 807–810. ACM Press, New York (2008) (Formal Research Demonstration)
[14] Weisemöller, I., Schürr, A.: Formal Definition of MOF 2.0 Metamodel Components and Composition. In: Czarnecki, K., Ober, I., Bruel, J.-M., Uhl, A., Völter, M. (eds.) MODELS 2008. LNCS, vol. 5301, pp. 386–400. Springer, Heidelberg (2008)

17 Towards Model-Based Engineering of Self-configuring Embedded Systems

DeJiu Chen, Martin Törngren, Magnus Persson,
Lei Feng, and Tahir Naseer Qureshi

Mechatronics Lab, Department of Machine Design,
Royal Institute of Technology (KTH), Stockholm, Sweden
{chen,martin,magnper,leifeng,tnqu}@md.kth.se

Abstract. In self-configuring embedded systems, upgrades, attachment of devices, relocation of applications and adjustment of performance parameters can be carried out during run-time for the purposes of information/function integration, maintenance, performance, resource efficiency, and robustness. We describe a model-based engineering approach to support the development of such systems. Essential ingredients include a combined usage of a system model, simulation combined with a number of formal techniques, and run-time models used as a basis for on-line decision making, with the overall goal to ascertain flexible, yet dependable, system behavior.

17.1 Introduction

Distributed embedded systems are traditionally configured at deployment time, i.e. tasks are statically allocated to nodes and settings are fixed for the entire life time of the product. It is usually not easy to upgrade the software or add new functions after deployment time.

In a *self-configurable* system, on the other hand, it would be possible to upgrade the software (or parts thereof), add new modules (hardware and software) to the system, and change system properties, e.g. periodicity of tasks and on which nodes they run, either at run-time or when the system is in an idle mode. We use the following definition:

> An embedded system is *self-configuring* if it is able to autonomously adapt to changing environmental conditions or internal status, by altering its structure, behavior, and data to meet its functionality and quality requirements.

Self-configuration relies on the system's awareness of its internal and external status, its built-in knowledge about system variability and rules for inferring and planning configuration changes, and appropriate run-time mechanisms for performing dynamic configuration management.

There are many reasons why the needs for dynamic and self-configuring systems are increasing, including cost-efficient and reliable field maintenance and upgrades of software, integration of external resources for information and/or functionality sharing. A typical example of this is the infotainment area in automotive systems, where new devices and software have much shorter lifecycle compared to the rest of the vehicle.

Many embedded systems are moreover exposed to a varying number of events and applications running at the same time, and with varying resource requirements over time. The applications are often heterogeneous in nature, e.g. event- and clock triggering, and tasks of different criticality. Taking this into account causes a further need for adaptation of the application behavior, configuration, and resource deployment at run-time in order to optimize system performance and improve system robustness.

New model-based development approaches and software platforms are required to meet these needs[1]. In this chapter, we describe work in the FP6 project DySCAS[2], which has developed a middleware architecture for dynamically configurable automotive embedded systems. The architecture itself is described in [3, 4, 5] and is out of scope for this chapter. The reader is especially referred to [3] for elaboration on the middleware architecture, the corresponding processes and more details on the methodology described in the following.

In this brief chapter, we focus on summarizing and synthesizing methodological experiences. We describe a suggested model-based approach to developing a middleware architecture for self-configuring automotive embedded systems (Section 17.2), together with a case study on concrete methodology actually used in the DySCAS project (Section 17.3).

17.2 Capabilities

Self-configuring embedded systems highlight the use of software as a means to provide advanced features that allow transparent handling of uncertainties, such as changing internal status and/or environment conditions. At the same time, such features often involve complex decisions, behavior synchronization, platform interactions and error handling, while requiring additional computation and communication resources. It is clear that a configuration change, if not designed and performed properly, would introduce faults of various types (i.e., primary, secondary, and command faults [6]) or cause violations of the intended architecture (i.e., architectural erosion [7]). Therefore, the DySCAS approach emphasizes the importance of systematic reasoning and decision making, and predictable behavior in performing dynamic configuration management for the reasons of performance, dependability, verification and validation.

The DySCAS reference architecture provides a domain-specific strategy and approach for introducing context-aware, self-managing dynamic configuration and QoS behavior into automotive embedded systems through middleware. To support the architecture design, a model-based methodology has been developed, illustrated in Fig. 17.1. The methodology distinguishes the platform-neutral aspect of architecture design from the platform-specific aspects, and adopts an incremental approach including the following stages:

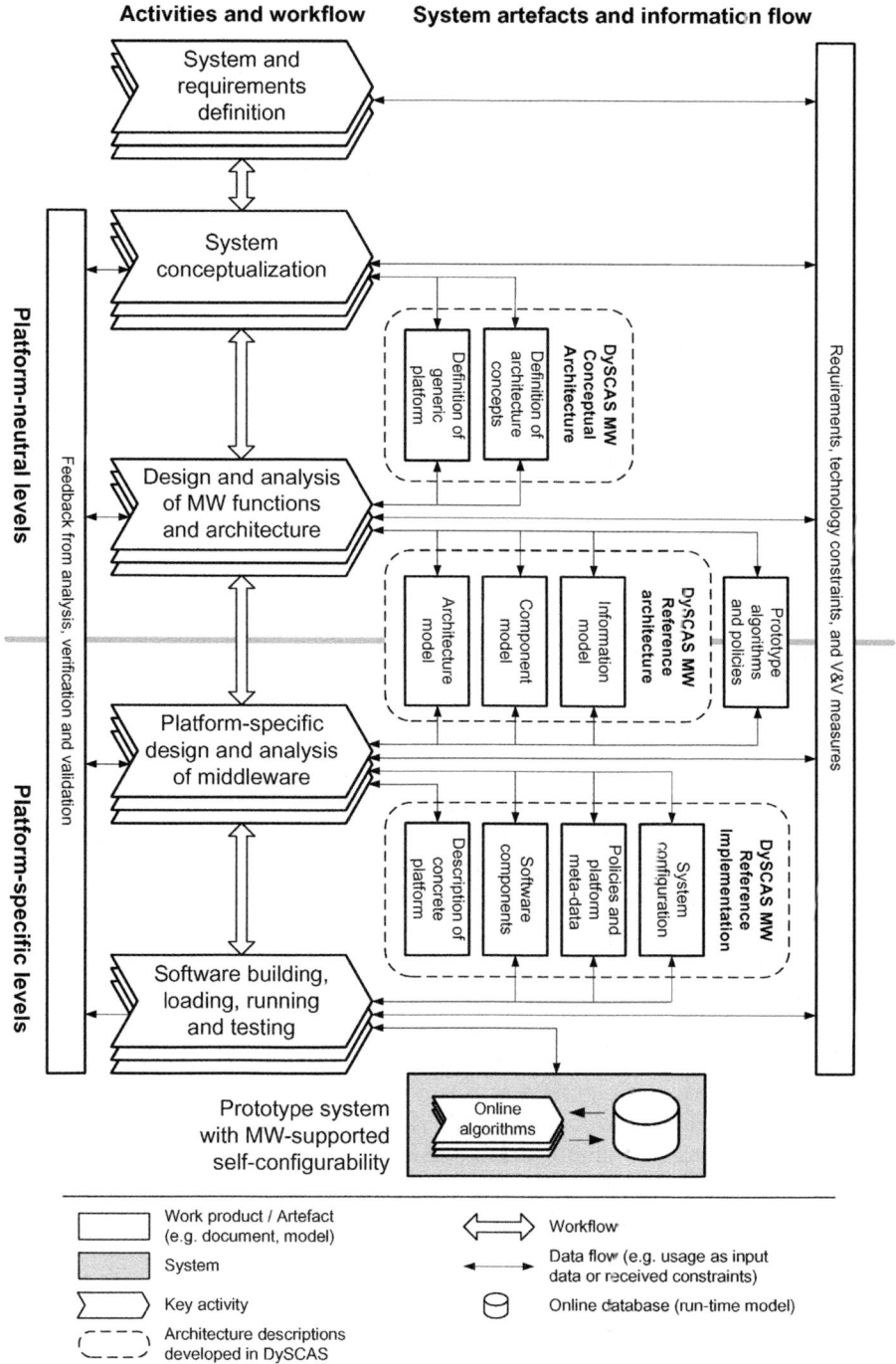

Fig. 17.1. Outline of the adopted design methodology of middleware architecture for self-configurable systems

- System and requirements definition.
- Architecture conceptualization and analysis.
- Design and analysis of MW functions and architecture (on generic platforms).
- platform-specific design and analysis of MW functions and architecture.
- Software building, loading, running, and testing.

Each architecture design stage is characterized by an iterative loop of requirement specification and refinement, design and verification, involving the following major activities:

- Assessing available information, including the requirements and technology constraints.
- Performing quality-oriented design and defining effectiveness measures for architecture evaluation, in respect to performance and dependability.
- Peforming structural design to specify system components and their data, functionality, properties, composition and connections.
- Performing behavioral design to specify modes and transitions, actions and control flow, interactions and synchronizations, and other related properties.
- Evaluating alternative solutions, describing refined/derived constraints and requirements, and repeating the structure and behavioral design for any unsatisfactory solutions.

There are various concerns of performance and dependability, e.g. in regards to logic and behavior, timing and synchronization, error detection and handling. In the DySCAS methodology, such concerns have been treated incrementally throughout the design in separate stages. At the platform-neural architecture design stages, the fundamental paradigms of communication and computation have been decided using a generic platform definition. At this level, the design focuses on the structuring and behavioral semantics of the middleware services, the implementation strategy in regards to partitioning, allocation, and scheduling, and the necessary platform support, that together promote the decision determinism, behavior and timing predictability, and delimitation of error propagation. The solutions at these stages have been captured and specified within the DySCAS component model and architecture model [3]. For example, each DySCAS middleware component is a self-contained and active object with its own thread-of-control and hierarchically defined behavior, triggered by signals. The inter-component communication is based on asynchronous message passing with well-defined buffering queues. To avoid undesired behavior during execution (e.g., deadlock, overflow and starvation), with the purpose of simplifying verification, a component always reads and removes all messages from all its input message queues, and carries out the computation using these messages. After the computation, the component writes messages to its output message queues if necessary. The strategy of implementation, as a part of the component and architecture models, is that the middleware services, as well as the application and system tasks subjected to dynamic configuration, should have hierarchical task scheduling support.However, the concrete definition of particular scheduling solutions are left to the platform-specific design stages, where a specific platform is chosen for the implementation.

While the architecture design solutions are captured in UML, a combined usage of simulation and formal analysis techniques, supported by model transformations, has been employed to support advanced architecture verification and validation in regards to the logic, algorithmic and execution behavior, and performance/resource usage. To address safety and robustness concerns, a preliminary hazard analysis through FFA (Functional Failure Analysis) and FMEA (Failure Modes and Effects Analysis)has also been performed for the envisioned self-configurable automotive systems [8].

Compared to traditional statically configured systems, one key difference of the DySCAS approach is the support for middleware users to systematically design and embedded various design information, quality concerns, and predefined variability for dynamic configuration decisions in the forms of meta-data. For example, each application component in DySCAS is represented not only by the executable code itself, but also by associated meta-data describing the conditions and contracts under which the component can operate and be adaptive. The covered concerns include for example its dependencies on other components, restrictions on hardware binding and amount of resources, and its QoS regions. One fundamental concept is variability, which refers to the specification of possible variations that a particular system configuration can have. Traditionally, variability is used for the feature configuration of product lines and dealt with off-line [9]. In DySCAS, the concept is extended to many more configuration items, levels of abstractions, and product lifecycle concerns. Based on such meta-data, the middleware provides dynamic configuration management support by monitoring run-time system conditions and then reacting to such conditions according to some a priori defined control policies/rules. It is assumed that all meta-information and configuration management policies/rules that provide the middleware-based online configuration management are derived from corresponding off-line function and architecture design, verification and validation activities. The issues of particular concern include impacts of changes on the overall system functionality, end-to-end performance, and dependability.

Some further explanations of key artefacts in the methodology are as follows.

- The conceptual architecture outlines the middleware services and components, boundaries, and environments. It also provides a generic platform definition characterizing some common features or concerns of hardware architecture and system support.
- The information model of the DySCAS reference architecture provides a framework and a set of predefined data types for specifying system configuration and configuration variability at architectural levels, embedding such meta-information for run-time configuration management and QoS control, capturing environmental conditions and system internal status, and describing expected dynamic configuration tasks and activities.
- The component model of the DySCAS reference architecture provides a common means for packaging configuration management data and control

functions, exploiting a policy-based mechanism for the implementation and maintenance of reasoning and decision algorithms, and controlling fundamental service behaviour in regards to execution, interaction and synchronization.
- The architecture model of the DySCAS reference architecture provides a layered data hierarchy and control strategy for identifying, structuring, and deploying various middleware services. Together they provide distributed run-time configuration management.

As a complement, we found it useful to apply safety analysis techniques. While the intended middleware system is not primarily intended to be used in safety critical applications, the use of such analysis techniques promotes system robustness, by stimulating and aiding reasoning about critical failure modes, fault hypotheses and causes of component failures. The platform-neutral design has also been refined into a concrete implementation on the specfic platforms. For example, the componentization, communication and execution details will depend on the exact real-time operating system and communication protocols available on the platform. Much of the analysis can also be repeated and refined at different levels of abstraction. Some prototype on-line configuration algorithms were tested through simulation and verified formally.[10]

17.3 Case Study

The case study focuses on a few examples on how models have practically been used within the DySCAS project.

17.3.1 Architecture Modelling with UML

The DySCAS architecture has been modelled in UML[11], using the MagicDraw UML tool[12]. This model is presented in [3]. See Fig. 17.2 for an overview.

For middleware implementers using the DySCAS architecture framework, suitable modeling languages and tools (based on UML or domain-specific) will be needed to document the exact interpretation of the DySCAS concept that has

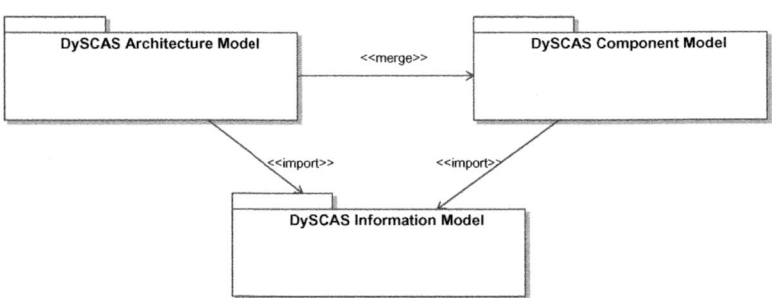

Fig. 17.2. A package overview of the DySCAS Reference Architecture in UML

been used. Further, implementers of hardware devices and software applications intended to run on DySCAS systems will have to use compatible languages and tools to capture relevant meta-data such as limitations, dependencies and other requirements the devices/applications have, and which they put on their environment.

17.3.2 Verification and Validation through Analysis

Several model-based approaches to V&V have been used in the development of DySCAS, both to supply early design feedback and to perform quality control.

As a high-level design support tool, logic simulation of the middleware using the SimEvents[13]/Stateflow/Simulink/Matlab environment has been performed [14]. These simulations have provided feedback for the architecture development. Results include a specification of an example model transformation from the UML-based specification of the middleware architecture to the simulation environment, simulation guidelines and conceptual evaluation of the DySCAS reference architecture.

Several different variations of on-line reconfiguration algorithms were developed [10]. The development was supported with simulation covering applications, middleware and their software/hardware realization using the TrueTime[15] toolbox for Simulink. The level of abstraction supports evaluation of algorithms and timing properties. Formal verification (i.e. mathematical proofs of their properties) on the algorithms was also performed.

To further introduce the reconfiguration problem: Assume a set of embedded computers, with limited memory and CPU resources, connected by a network; a set of application tasks with specified resource usage; timing requirements; eligible computers on which tasks can run, for each task a list of required and provided services; and the overall benefit metrics for the system, based on each task's *significance*. Further assume that a task can only execute when enough resources are available and all required services are present in the system.

The challenge then is to find an algorithm to allocate tasks to computers, such that all allocated tasks are schedulable and provide maximal benefit to the system. Events such as hardware failures, addition of new resources, and workload disturbances may trigger self-reconfiguration. This problem is provably NP-hard, and hence intractable for online configuration. To enable quick system response, a suite of efficient algorithms to find acceptable solutions (rather than optimal) has been proposed. They are capable of online configuration management and adaptive resource management, both in polynomial time of the number of tasks and computers of the network[10].

Further, a safety analysis model was developed, using the FFA (Functional Failure Analysis) and FMEA (Failure Modes and Effects Analysis) methods to find sensitive failure modes of the middleware itself.[8] Such types of verification are vital to be able to guarantee system properties such as availability.

17.3.3 Run-Time Models

During run-time, in-memory models are used to keep track of the current execution status of the system. This is examplified by DyLite[16], a partial reference implementation of DySCAS focusing on QoS and reconfiguration. Using a model of the applications' behavior, describing resource usage over time based on an automata formalism. The automata states correspond to different QoS modes, and along with each mode, maximum resource usage levels for CPU time, network communication and memory usage are documented. This information is used to predict the maximum resource utilization of the applications.

Additionally, applications explicitly switch between different modes by calling a function from the DyLite API. As parameters, the application gives the set of modes that it wants to switch to. In return, after considering system constraints, the middleware returns which one of these modes will actually be used. The middleware may in certain cases also block the application, even indefinately (i.e. in the case of overload situations). Since the DyLite middleware is able to predict future resource usage, it is able to give resource availability guarantees to applications; either the application is granted the requested resources, or it won't run at all.

17.4 Conclusions and Future Work

In this chapter, we have presented a model-based development methodology for middleware for dynamically self-configuring distributed embedded systems, based on experiences in the DySCAS project. There are several interesting threads of ongoing and desired work. We plan to perform validation on the DyLite implementation to validate the system behavior models, thus closing the loop from models to reality. While all the aspects of the outlined methodology were covered in the DySCAS project, many tool-related issues require further work, e.g. to automate the model transformations we have defined. Finally, further work is required to elaborate the developed methodology and architecture, in particular in order to support systematic analysis and synthesis, where repeating all verification at each level will no longer be required.

Acknowledgement

As a part of DySCAS project, this work is funded by the 6th framework program of the European Commission. Project number: FP6-IST-2006-034904.

References

[1] Törngren, M., Chen, D., Malvius, D., Axelsson, J.: Model based development of automotive embedded systems. In: Automotive Embedded Systems Handbook. Industrial Information Technology. Taylor and Francis, CRC Press (2008)

[2] DySCAS, http://www.dyscas.org

[3] DySCAS Consortium: D2.3 DySCAS system specification. Technical report (2008)
[4] Anthony, R., Ekelin, C.: Policy-driven self-management for an automotive middleware. In: Proceedings of First International Workshop on Policy-Based Autonomic Computing (PBAC 2007), at the Fourth IEEE International Conference on Autonomic Computing, in Jacksonville, Florida, USA, June 11-15 (2007)
[5] Chen, D., Anthony, R., Persson, M., Scholle, D., Friesen, V., de Boer, G., Rettberg, A., Ekelin, C.: An architectural approach to autonomics and self-management of automotive embedded electronic systems. In: Proceedings of the 4th European Congress Embedded Real-Time Software (ERTS 2008), Toulouse, France, January 29-February 1 (2008)
[6] Leveson, N.G.: Safeware: System safety and computers. Addison-Wesley Publishing Company, Reading (1995)
[7] Perry, D.E., Wolf, A.L.: Foundations for the study of software architecture. ACM SIGSOFT Software Engineering Notes 17(4), 40–52 (1992)
[8] Feng, L., Törngren, M., Chen, D.: Safety analysis of dynamically self-configuring automotive systems. Technical report, Mechatronics Lab, Department of Machine Design, KTH, Stockholm (2008)
[9] Cuenot, P., Chen, D., Gérard, S., Lönn, H., Reiser, O., Servat, D., Kolagari, R.T., Törngren, M., Weber, M.: Improving dependability by using an architecture description language. In: de Lemos, R., Gacek, C., Romanovsky, A. (eds.) Architecting Dependable Systems IV. LNCS, vol. 4615, pp. 39–65. Springer, Heidelberg (2007)
[10] Feng, L., Chen, D., Törngren, M.: Self configuration of dependent tasks for dynamically reconfigurable automotive embedded systems. In: Proceedings of 47th IEEE Conference on Decision and Control, Cancún, Mexico, December 9-11 (2008)
[11] UML, http://www.uml.org
[12] MagicDraw, http://www.magicdraw.com
[13] SimEvents, http://www.mathworks.com/products/simevents/
[14] Naseer Qureshi, T., Chen, D., Törngren, M., Feng, L., Persson, M.: Experiences in simulating a dynamically self-configuring middleware: A case study of DySCAS. Technical report, Mechatronics Lab, Department of Machine Design, KTH, Stockholm (2008)
[15] TrueTime, http://www.control.lth.se/truetime/
[16] Persson, M., García, J., Feng, L., Naseer Qureshi, T., Chen, D., Törngren, M.: DyLite: Design, implementation and experiences. Technical report, Mechatronics Lab, Department of Machine Design, KTH, Stockholm (2009)

18 Representation of Automotive Software Description Means in ASCET

Ulrich Freund

ETAS GmbH,
Borsigstrasse 14, 70469 Stuttgart, Germany
{ulrich.freund}@etas.com
http://www.etas.com

18.1 Introduction

Embedded automotive real-time software is developed according to the V-Cycle. The control engineers start with the so-called function development where they specify the control-algorithm. This control-algorithm transforms input signals to output signal reflecting also the state variables and parameters. A software engineer partitions the control-algorithm to executable software components, which are then transformed to C-code with a code-generator. The software engineer also integrates the generated code in a μ-Controller (μC) of an ECU. Afterwards, the ECU is integrated into a vehicle network by a software engineer too. The integrated control-algorithm is then tuned by a control engineer. This V-Cycle is shown in figure 18.1. From a software engineering point of view, the design artifact created in the V-Cycle is the embedded control software. The actual values of the control-algorithm's parameters, i.e. the calibration data, is seen as part of the control software. However, the first and the last design step in the V-cycle are not done by software engineers, but by control engineers. As a result, neither pure control engineering nor pure software engineering descriptions appear

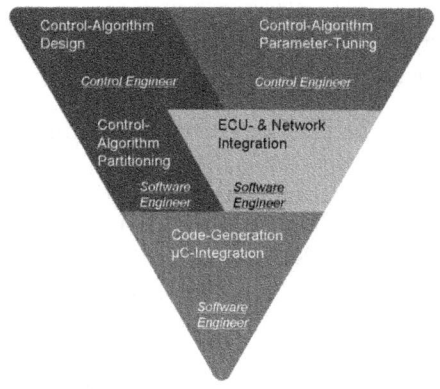

Fig. 18.1. The V-Cycle for Embedded Automotive Software Development

to be appropriate as sole description means. Applying dedicated descriptions at each design step is a valid approach, but the transition from control to software engineering requires dedicated attention because two different ways of thinking intersect. At this design step, an appropriate description well apprehended by control and software engineers can avoid a lot of misunderstandings.

18.2 Overview of Design Means for Automotive Software Design

18.2.1 Description Means for Control Engineering

Control engineers describe the control-algorithm by means of signal flow and expect the implemented ECU-software to expose its calibration parameters for trimming. Thus, the control engineer has to specify the following items:

- Data-Flow-Diagrams: The most prominent design approach stems from control engineering where components containing control-algorithms are coupled in a data-flow manner. When coupled to a physical model of a plant, the control-algorithm can be executed in a closed-loop simulation running typically on an engineering workstation. As a rule, hierarchies in the data-flow diagram will be flattened before the simulation starts.
- Control-Algorithm Parameters: A control-algorithm consists of a structural part, e.g. PID-controller, and a parameter part, e.g. P, Tv, and Ti. While analytical parameter design techniques typically require a relatively exact dynamic model of the plant, which is not always available, automotive engineers tend to trim the vehicles on a test-track in driving experiments. The more parameters the structure of the control algorithm provides, the more flexibility has the calibration engineer on the test-track for trimming.

18.2.2 Description Means for Software Engineering

Software engineers transform control-algorithm to concurrent embedded software, requiring the specification of:

- Software-Components and Tasks: Creating reusable control engineering components is part of the software engineering. A well-proved concept in software engineering is the encapsulation of algorithms in classes by means of methods and the instantiation of the class in another algorithm, which will be called by a task. Tasks interact with communication means like queues and remote procedure calls, whose mutual exclusive access is encapsulated in system-functions. Proxy-concepts relieve a component programmer from knowing all system resources by name.
- Synchronous-Reactive Design: A different approach in software engineering is synchronous-reactive design. Besides the clock-concept of triggering computational blocks in an appropriate order and a zero time assumption on

execution time for blocks an important element of synchronous-reactive systems is the name-matching approach for signals. In practice, this means that a signal has one sender (writer) and n receivers (readers). The signal is thus becoming a global variable. An algorithm developer has to know the global name of the signal for receiving and sending.
- Automotive Software Design Practice: Due to the static nature of embedded automotive software prohibiting dynamic creation of control-engineering objects, the concept of a global signal space is used in the automotive industry for more than three decades. In actual vehicles there are typically around thousand signals available on several buses while electronic control units (ECUs) e.g. for engine-management work on more than thousand signals internally. At the vehicle manufacturer as well as at the supplier there are dedicated departments established for maintaining the global signal pool.
- Integer Arithmetic: Since a considerable number of μCs do not support floating point operations, a control-algorithm has to be implemented in fixed point arithmetic [1]. This implies that a control-engineering signal has an integer representation in bits. To achieve a higher resolution, there is no one-to-one mapping of the physical value to its integer value, but a conversion formula determines the physical limits of a signal and distributes the values between the limits to the available bits in the integer variable. Arithmetic operations on integer represented signal have to take into account the conversion formula applied on the signal. As a result, additional adaptive operations need to be applied to the original operation.
- Memory Layout of μCs: Electronic Control Units employ one or more(μCs). These μCs have different kinds of memory available. For example, parameters change their value only during the trimming phase and are therefore located in a flash-memory. This section might be changed for series production to ROM. Of course, the variables representing state-variables in the structural part of the control-algorithms have to be located in a RAM section. But the RAM might be available internally on the μC or externally in the ECU. As a result, the embedded software realizing a control-algorithm will be distributed over several parts of the μC- and ECU-memory. ROM has also several flavors. There might be pure ROM, flash or EEPROM. Similar to RAM all kinds of ROM might be internal of a μC or external in the ECU.

18.3 Integration of the Design Approaches in ASCET

The embedded automotive design tool ASCET [2] combines the design approaches described above by executable models and transforms the model μC-production C-Code via code generation.

Control-Algorithms are constructed in ASCET by means of classes and modules. While classes represent sequential encapsulation, modules are real-time components working on a global signal space.

18.3.1 Classes

Similar to the class encapsulation approach in an object-oriented programming language the ASCET classes consist of attributes and methods. However attributes can be classified in simple and complex variables or parameters. Complex variables are instances of other classes.

Methods have return values and in, inout, and out arguments. Methods have a collection of sequentially ordered statements. Statements assign the results of expressions to variables. Expressions may involve arithmetic operations, variables, parameters, arguments and method calls to instances of other classes. Statements can be constructed graphically or textually. There are statements for iterations and conditional execution. ASCET classes do not support inheritance. Recursions can be modeled but are not recommended in a safety critical environment. Methods can be called iteratively.

18.3.2 Modules

A module consists of send- and receive-messages. Messages represent signals in a global name space. When modules are grouped in an ASCET system, the messages are mapped by name thus creating a global message name space in the ECU. Modules provide so called processes reading messages, performing computations, and writing messages. Processes are mapped to tasks in a sequential manner. A process at position n will be executed when its operating system task is triggered and the n-1 preceding processes also mapped to the task are finished.

Besides representing an ECU global signal-name space, messages serve as inter-process communication mechanism between tasks. Depending on the mapping of processes to tasks, the task properties (e.g. preemptive), and the task schedule message access and buffering will be protected appropriately.

Modules can contain instances of classes. A typical execution of a process starts by reading data from receive messages and calling methods. When calling a method, data read from the messages will be written to arguments of the method while result of a method call will be written to send messages. The called method might call methods of complex attributes. Since modules contain messages representing signals on a global name space, modules can only be instantiated once.

Figure 18.2 shows a module running a data-flow graph for a throttle controller. One can see at the left hand side the receive messages *current-position* and *target-pos* while on the right hand side there is the send message *new-position*. In the data-flow from the receive- to the send-messages one can see that every variable is annotated with a tag of the process it runs, in this example the processes *normal* and *out*. The method compute of the instantiated PID class is also called by the process *normal*. Parameters (*e.g. T1, Td, Ti, Pgain*) as described in section 18.2.1 have a different color and shape. When the process *out* calls the PID's method *out* the return value will be written to the send message *new-position*. It is up to the ECU integrator to assign the processes *normal* and *out* in alphabetical order to one task or to assign the processes to one task in a different order or even to two different tasks.

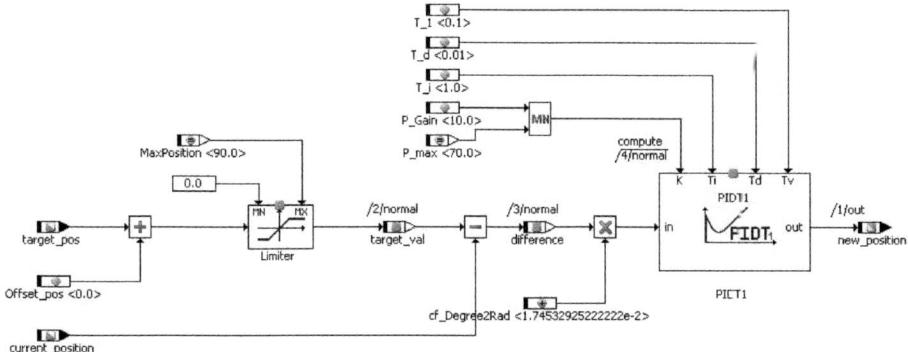

Fig. 18.2. ASCET Module for a Throttle-Controller

18.3.3 Model-Types

ASCET elements like messages, variables, parameters, and method arguments are typed by so-called model-types. The model-types are *cont* representing continuous values, *signed discrete* and *unsigned discrete* for representing the respective integer values and *logic* for logical values. In a refinement step, the model types are replaced by platform types like real64 or uint8.

18.3.4 Tasks

An ASCET system represents a control-algorithm constructed of modules. These modules communicate via messages. Messages are read or written in processes when they are executed by tasks. As written in section 18.3.2, processes of modules are assigned to tasks and messages constitute interprocess communication buffers. Publication [3] shows that an appropriate scheduling strategy of tasks running processes with message access keeps the synchronous reactive hypothesis.

Depending on the stage of development, the ASCET system contains, besides the modules of the control-algorithm, either I/O-modules for accessing μC-registers or continuous time blocks representing a plant model. The processes of the I/O-modules will be mapped to tasks while the numerical solvers of the continuous time blocks represent tasks on their own.

18.3.5 Implementations: Integer Arithmetic and Memory Section

Depending on the actual development step the ASCET model types are transformed to float or integer C-variables. When variables or parameters of continuous type are transformed to an integer, one has to specify a conversion formula. During code-generation this formula will be evaluated and original operators defined for the model type of variable will be concatenated with additional adaptive operations.

When code will be generated for a μC, one has to specify a memory section for every variable and parameter. From a C-programming point of view, the only difference between a variable and a parameter is that the parameter will be located in a kind of ROM.

18.3.6 Codegeneration Approach

All attributes of modules and classes in an ASCET system are encapsulated during code-generation in C-structs. The attached implementation w.r.t. memory allocation will be reflected during code-generation. Methods and processes are transformed to C-functions which are of type void in case of a process or to C-functions with arguments, instance handle and return value when the method belongs to a class which is multiple instantiated in the ASCET system. Since messages serve interprocess-communication, they are generated together with the operating system and managed in a different way. The final variable layout depends on the buffering technique and the access protection, which themselves depend on the OS-layout. Messages have a unique name in an ASCET system.

18.4 Conclusion

Development of embedded automotive software requires implies a control-engineering as well as a software-engineering view on the very same design artifact. ASCET combines both views by an object-based real-time representation of control-algorithms. These representations are transformed to embedded C-code by means of code-generators.

References

[1] Schäuffele, J., Zurawka, T.: Automotive Software Engineering. Vieweg Verlag, Wiesbaden (2003)
[2] ETAS GmbH Stuttgart: ASCET User-Manual (2008)
[3] Baleani, M.: Correct-by-construction transformations across design environments for model-based embedded software development. In: Proceedings of DATE, München (2005)

19 Papyrus: A UML2 Tool for Domain-Specific Language Modeling

Sébastien Gérard[1], Cédric Dumoulin[2], Patrick Tessier[1], and Bran Selic[3]

[1] CEA LIST, Laboratory of Model Driven Engineering for Embedded Systems (LISE), Boîte courrier 65, Gif sur Yvette Cedex, F-91191 France
{Sebastien.Gerard,Patrick.Tessier}@cea.fr
[2] LIFL and INRIA-Lille Nord Europe, University of Lille, France
Cedric.Dumoulin@lifl.fr
[3] Malina Software Corp., Nepean, Ontario, Canada
Selic@acm.org

Abstract. This chapter outlines Papyrus, a tool for graphical modeling of UML2 applications. It is an open-source project, designed as an Eclipse component, and based on the existing EMF-based realization of the UML2 meta-model. The goal of this open-source project is twofold. First, it is a complete, efficient, robust, and methodologically agnostic implementation of a UML2 tool to both industry and academia. Second, it is an open and flexible facility for defining and utilizing domain-specific modeling languages using a very advanced implementation of the UML profile concept.

Keywords: UML2, UML profile, DSL, MDD, MDE, Modeling and Eclipse.

19.1 Introduction

As part of its Model-Driven Architecture (MDA) initiative, the Object Management Group (OMG, http://www.omg.org) – an international consortium representing numerous industrial and academic institutions – has provided a comprehensive series of standardized technology recommendations in support of model-based development of both software and systems in general. These cover core facilities such as meta-modeling, model transformations, and general-purpose and domain-specific modeling languages. A key component in the latter category is UML (the Unified Modeling Language, [1]) which has emerged as the most widely used modeling language in both industry and academia.

A number of tools supporting UML are available from a variety of sources. However, these are generally proprietary solutions whose capabilities and market availability are controlled by their respective vendors. This can be problematic for industrial users, who may require highly-specific tool capabilities as well as long-term support which, from a vendor perspective, often extend beyond the point of commercial viability. Consequently, some industrial enterprises are seeking open-source solutions for their UML tools. Of course, the fact that open-source solutions are often less costly is another influencing factor, although it is typically not the primary motivation for selecting them, since there is always a cost involved in integrating such solutions. Indeed, industrial developers using open-source solutions need time and money to account for debugging or developing new features in their projects.

A similar situation is encountered in research which typically depends on open source tools, as most proprietary products are too constraining and inflexible to allow optimized implementation of new ideas and prototypes.

Therefore, the Eclipse platform along with its Model Development Tools (MDT, http://www.eclipse.org/modeling/mdt) subproject is the environment of choice for developing open-source tools for modeling. Indeed, MDT focuses on big "M" modeling within the Modeling project of Eclipse. To achieve this goal, MDT aims at providing: (a) an implementation of industry standard meta-models, and (b) exemplary tools for developing models based on those meta-models. The UML2 meta-model implementation was its first component (also known as the "UML2 Component", http://wiki.eclipse.org/MDT-UML2). This component has become the *de facto* standard implementation of the UML2 meta-model (note that it is also the basis for the UML2 tool suites provided by IBM). Based on the UML2 component, several open-source tools emerged providing facilities for graphical modeling within Eclipse. In early 2008, three of the main community initiatives developing such a tool —MOSKitt (http://www.moskitt.org/eng/), Papyrus (http://www.papyrusuml.org) and TOPCASED (http://www.topcased.org) — decided to merge their efforts and provide a joint contribution to the Eclipse MDT. This new graphical editor, named as Papyrus (http://wiki.eclipse.org/MDT/Papyrus-Proposal), was accepted by the Eclipse's Project Management Committee in August 2008, and the first code delivered in November 2008: http://wiki.eclipse.org/MDT/Papyrus.

The following section outlines the architecture of the Papyrus tool and its main capabilities in terms of UML2 graphical modeling and openness for customization. Section 19.3 provides some snapshots of UML2 diagrams created with Papyrus illustrating its graphical capabilities. Finally, section 19.4 gives some conclusions and a roadmap for the tool.

19.2 Capabilities

As explained previously, Papyrus is a graphical editing tool for UML2. It is Eclipse-based and it uses the Eclipse Graphical Modeling Framework (GMF, http://www.eclipse.org/modeling/gmf/). This section first provides an overview of Papyrus capabilities and its architecture, and then expands on two of its distinguishing features: Subsection 19.2.3, describes the general Papyrus capabilities for graphical UML2 modeling, while subsection 19.2.4 focuses on the abilities of the tool to be customized for domain-specific needs through the use of UML profiles to define domain-specific modeling languages (DSLs).

19.2.1 Overview

Papyrus is a tool consisting of several editors, mainly graphical editors but also completed with other editors such as textual-based and tree-based editors. All these editors allow simultaneous viewing of multiple diagrams of a given UML model. Modifying an element in one of the diagrams is immediately reflected in others diagrams showing this element. Papyrus is integrated in Eclipse as a single editor linked to one UML 2 model. Papyrus provides a main view, showing model diagrams, and additional views including an outline view, a property view, and a bird's-eye view. The multiple diagrams are managed by Papyrus rather than by Eclipse. Model diagrams can be arranged in tabbed views, and several tabbed views can be arranged side by side (left, right, top and bottom). Such views can be created, re-sized, moved, and deleted by dragging them to the desired position.

Papyrus is highly customizable and allows adding new diagram types developed using any Eclipse-compatible technology (GEF, GMF, EMF Tree Editors, ...). This is achieved through a diagram plug-in mechanism. In fact, even the default diagrams use this mechanism, allowing their easy replacement if desired.

19.2.2 Global Architecture and Design Tenets

As shown in Fig. 19.1, the Papyrus top-level architecture consists of two main parts: a core providing the common services used by the various diagram editors, and a part containing the various model editors.

One of the main functions of the core component of Papyrus is to enable collaboration of the different editors regardless of their specific implementation technology (e.g., EMF or GMF). The main services provided by the papyrus core are:

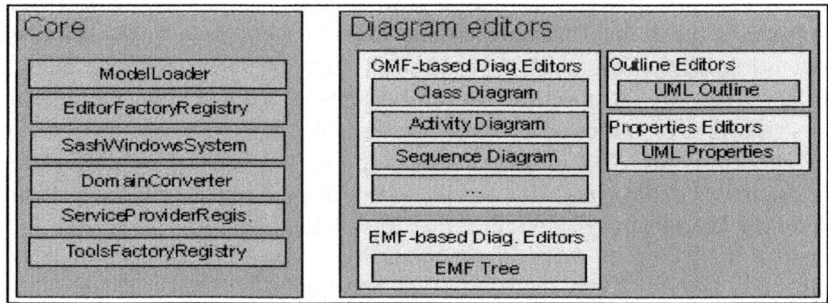

Fig. 19.1. This figure sketches the architecture model of the papyrus tool

- *A sash windows system.* This provides the ability to have multiple diagrams opened simultaneously, and to arrange them as desired.
- *Model life-cycle support.* This allows loading and saving of models.
- *Pluggable diagram editor factories.* A diagram factory is used to create a particular kind of diagram in Papyrus. For example, there is a factory for

class diagrams. The plug-in mechanism allows registering such factory in the core, enabling thereby the management of existing diagrams via the sash window system.
- *Pluggable tools factories.* A tool factory allows creation of a tool that will interact with the model. Examples of such tools are: code generators, transformations engines, model checkers, and refactoring facilities.

The Papyrus core is not linked to a particular technology other than EMF, allowing use of various different technologies. The core also provides facilities that enables diagrams and tools using the same implementation technology (e.g., GMF) to share classes.

In Papyrus, the UML meta-model and the graphical models are at the heart of the proposed architecture: diagram editors and other external tools can be used to define and modify any models, but they can also be used to observe them and react accordingly. Thus, diagram editors and external tools can be independent while still reacting to each other's actions.

Both the property view and the outline editors are also independent model editors in Papyrus. This means that they can interact with the model like any other diagram editors, providing hence another interesting and useful point of view on the model.

Several diagram editors have been developed and integrated successfully in Papyrus, using different technologies like GEF, GMF, EMF Tree Editors, SWT.

The UML part of Papyrus consists of diagrams partly generated with GMF, and of a property view editor also generated with a dedicated framework. Section 19.3 provides some snapshots of UML diagrams edited with Papyrus.

19.2.3 UML2 Graphical Modeling Capabilities

Papyrus allows supporting all the diagram types defined in UML2. Hence, a Papyrus user can build UML2 models that strictly conform to the standardized specification of the language. Papyrus can then be used to check against the UML specification: if something cannot be represented in Papyrus, this generally means that it is not allowed by the specification.

Papyrus provides a lot of functions in support of UML 2 modeling. The following list is not exhaustive (due to space limitation), but is still a representative sample of the advanced modeling features of Papyrus:

- *A tools palette.* Each diagram type has its own palette of tools allowing creation of UML elements in the diagram.
- *A property view.* This allows the editing of any property of a UML element as well as any of its graphical properties. These properties are organized by categories for ease of use. There is also an 'expert' category, allowing access to all the properties defined in the UML meta-model.
- *Contextual accelerator actions.* When the cursor is moved to an area where some actions are available, these actions are shown in a popup near the cursor. An action can then be selected directly without going to the palette.

– *Contextual text editors enabling syntax highlight, completion and content assist.* When a text-based model element is accessed, e.g., a property, Papyrus helps by providing a list of possible values, such as a list of existing types, or existing cardinalities. Papyrus also validates the text according to its grammar (if the latter is defined). This functionality is customizable: one can provide one`s own implementation (e.g., a specific text editor and validation rules) for specific needs.
– *OCL constraint specification and checking.* OCL constraints can be specified for each UML element at the model or at the meta-model level, such as in a profile definition. Furthermore, these constraints can be checked against the current model.
– *Model import.* Papyrus can import models, profiles or elements from other files. Imported elements can be used in the model and edited the same as any other elements.

Figure 19.2 illustrates the default perspective of the Papyrus tool: The diagram editor is in the top right of the figure, with the corresponding palette to its right. At the bottom right is the properties view editor. This editor is used to edit, and modify (if required), the properties of the element selected in the diagram editor or in the project outline editor. The latter is located in the bottom left part of the figure. Finally, in the top left part of the figure is the file system navigator, which enables access to different existing projects.

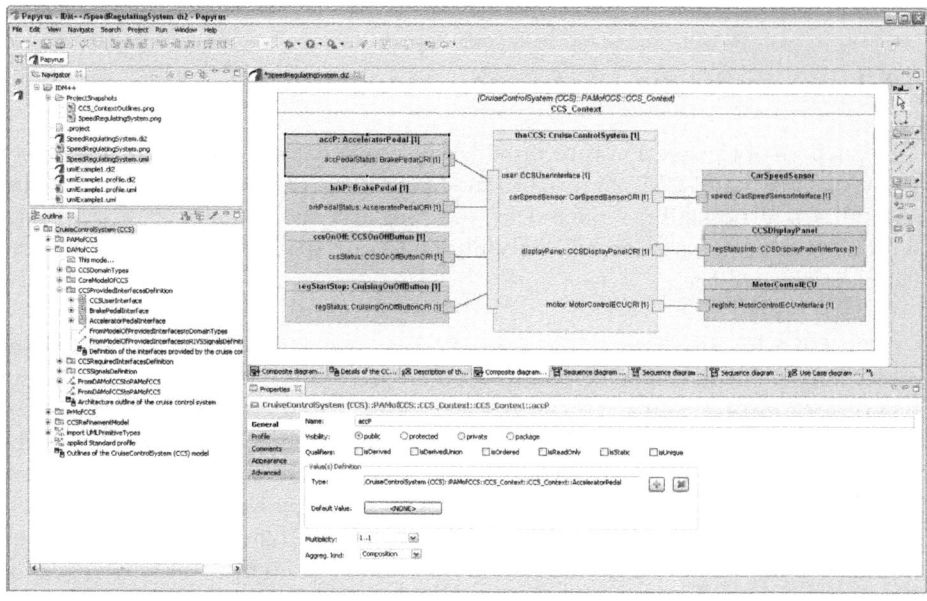

Fig. 19.2. Default perspective of Papyrus

19.2.4 Building DSL Tools Profiling the UML2

In accordance with its primary goal of realizing the UML2 specification, Papyrus provides extensive support for UML profiles. It includes all the facilities for defining and applying UML profiles in a very rich and efficient manner. It also provides powerful tool customization capabilities similar to DSML-like (Domain Specific Modeling Language) meta-tools. The main intent here is to enable the profile plug-in of Papyrus to dynamically drive its customization as defined by a profile. This means that when applying a profile, the tool may adapt its visual rendering and GUI to fit the specific domain supported by that profile. Of course, if the profile is unapplied later, the tool resumes to its previous configuration.

When designing a UML2 profile, it may be necessary to customize one or more existing UML2 diagram editors. For that purpose, Papyrus supports customization of existing editors with the added capability of extending such customizations by adding new tools relevant to the stereotypes defined in the UML profile. For example, the SysML requirements diagram editor is designed as a customization of the classical UML2 class diagram editor with additional features for direct manipulation of all the concepts defined in the SysML requirements diagram (see example show in section 19.3).

Finally, when embedding a profile within an Eclipse plug-in, a designer can also provide a specific properties view that will simplify (i.e., make more user friendly) the manipulation of the stereotypes and their related properties. The outline editor and the menu of the tools can also be customized to fit domain-specific concerns appropriate to the profile.

19.3 Case Study

The purpose of this section is to provide the reader some sense of the facilities of Papyrus in terms of UML2 graphical modeling. It is, of course, infeasible to give the full set of possibilities offered by the tool in a short summary article such as this. Hence, this section will only illustrate one specific papyrus editor—the class diagram editor.

The following two figures represent a class association between two classes and a multipoint-dependency relationship linking three classes (two sources and one target) respectively.

The Class diagram example is very illustrative of the Papyrus tenets in terms of its implementation goal: 99,9% of the specification without imposing any methodological constraints or assumptions. This is very important because it is common in many tools that some particular form of UML2 usage is not supported because of hard-coded implementation choices. The intent of basic Papyrus is to not favour any specific methodology and thereby offer an unhampered access to the full extent of UML2. Furthermore, it is intended to give methodologists and users the ability to construct specific customizations of the tool to support methodologies that are best suited to given profiles. That is, when applied to a

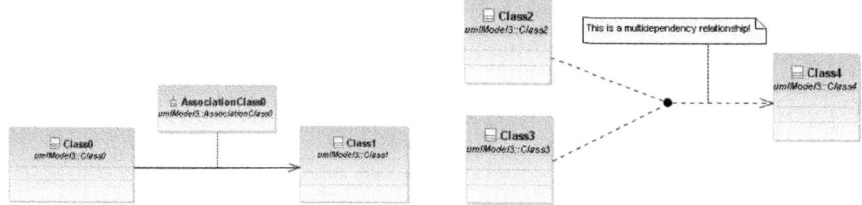

(a) Example of class association modeling

(b) Example of multipoint dependency relationship modeling

user model, such a profile also reconfigures the GUI and behavior of the Papyrus tool to fit a domain-specific methodology.

The following figure illustrates the palette customization feature of Papyrus. This palette provides both standard UML tool capabilities (called "UML Links" and "UML Elements" respectively) as well as an extended set of domain-specific tool capabilities required for the EAST_ADL2.0 language profile. (EAST-ADL2.0 is an architectural description language, intended for modeling automotive embedded systems (http://www.atesst.org/)).

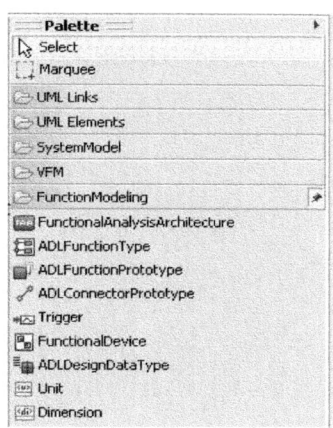

Fig. 19.3. Example of customized palette

19.4 Conclusions and Future Work

This article introduces the Papyrus tool, an Eclipse-based graphical editor for UML2 modeling. This open-source application was devised with two principal objectives in mind. First, it aims at implementing the complete UML specification (currently in its 2.2 revision), enabling thereby its potential use as the reference implementation of the OMG standard. Furthermore, it was designed as a highly scalable and robust tool for supporting large-scale industrial projects (as indicated by the growing list of industrial supporters). The second principal objective of Papyrus was to provide an open and highly customizable tool for defining domain-specific languages and corresponding tools via the UML profile mechanism. This facilitates interchange with other tools supporting the UML standard and also takes advantage of the widespread acceptance of UML 2. Consequently, Papyrus provides an efficient and effective alternative to custom and proprietary DSL tools, without losing the benefits of an international standard.

At present, a significantly revised version of Papyrus is under development with a major delivery milestone set for mid-year 2009. This version will support all 13 diagram types of UML2 and will also provide the fullest support for profiles of any UML tool currently available, whether open-source, or commercial. The next major deliverable is set for mid-year 2010, when Papyrus will be integrated into the principal Eclipse release schedule whereby key components are packaged and released jointly each year.

Finally,it is intended that Papyrus serve as an experimental platform for researchers who need to construct proof of concept prototypes. Built on top of Eclipse as an open-source project, Papyrus is an ideal candidate for this purpose.

Reference

[1] OMG: UML Version v2.1.2, http://www.omg.org/spec/UML/2.1.2/

20 The Model-Integrated Computing Tool Suite

Janos Sztipanovits, Gabor Karsai, Sandeep Neema, and Ted Bapty

ISIS - Vanderbilt University, Nashville TN 37203, USA

Abstract. Embedded system software development is challenging, owing to a tight integration of the software and its physical environment, profoundly impacting the software technology that can be applied for constructing embedded systems. Modeling and model-based design are central to capture all essential aspects of embedded systems. Vanderbilt University's Model Integrated Computing tool suite, driven by the recognition of the need for integrated systems and software modeling, provides a reusable infrastructure for model-based design of embedded systems. The suite includes metaprogrammable model-builder (GME), model-transformation engine (UDM/GReAT), tool-integration framework (OTIF), and design space exploration tool (DESERT). The application of the MIC tool suite in constructing a tool chain for Automotive Embedded System (VCP) is presented.

20.1 Introduction

The tight integration of the software and its physical environment has profound impact on the nature of the software technology to be applied for constructing embedded systems. The reason can be best explained by Brook's argument, which remains valid nearly 20 years after its publication: the essential complexity of large-scale software systems is in their *conceptual construct* [1]. In embedded systems, the conceptual construct of the software is combined with the conceptual construct of its physical environment; therefore, the methods and tools developed for managing complexity must include both the physical and computational sides. The common "denominator" for representing, relating and analyzing all essential aspects of embedded systems is *modeling* and *model-based design*.

The significance of modeling in software engineering has been recognized from the early nineties (see. e.g. Harel [2]). The recognition of the need for integrated system and software modeling led us to pursue the construction of a reusable infrastructure for model-based design. We have built several generations of tool environments since the late eighties directed to a wide range of application domains starting with signal processing [3]. Important milestones in the development of the Model-Integrated Computing (MIC) tool suite have been the following: (a) introduction of multiple-view modeling, programmable model builder and model-based integration of distributed applications in chemical process industry [4], (b) use of complex, multiple-view graphical modeling tool with object-oriented database backend for the diagnosability analysis of the International Space Station design [5], (c) generation of high-performance parallel signal processing applications from models [6], and (d) use of embedded models for the structural adaptation of signal processing systems [7].

The technology advancements related to different applications led to the formulation of the MIC architecture concept [8] with domain-specific modeling languages (DSML) and modeling at its center. The main challenge has been the dichotomy between domain-specificity and reusability. The construction of the meta-level tool architecture [9] [10] was followed by the first appearance of the metaprogrammable model builder [11] that ultimately led to the subsequent development of the metaprogrammable MIC tool suite [12].

20.2 Components of the MIC Tool Suite

Domain-specific modeling is at the center of the MIC development approach: domain-specific models are created in the design process, they are analyzed via formal and simulation-based analysis techniques, and they are used to construct and generate the implementation of applications, i.e. the actual software code that performs, for instance, control functions. The domain-specific models in an MIC engineering process are constructed in domain-specific modeling languages (DSML) whose syntax and semantics are precisely defined using metamodels and model transformations. However, MIC cannot exist without tooling: tools that assist the designer and developer in modeling, analysis, generation, evolution, and the maintenance of systems. This section provides a summary of the tools of MIC: the Generic Modeling Environment (GME), the Universal Data Model (UDM) package, the Graph Rewriting and Transformation language (GReAT), the Design Space Exploration Tool (DESERT), and the Open Tool Integration Framework (OTIF). These tools form a metaprogrammable tool suite that are connected by shared meta models as shown in Fig. 20.1. The tools can be extended by 'best of breed' analysis, simulation and verification tools connected via model transformation to build domain-specific toolchains.

20.2.1 The Generic Modeling Environment (GME)

GME is the core MIC tool [13] that is used for both meta-modeling and modeling. GME is metaprogrammable: it can load metaprograms generated from metamodels and "morph" itself into a domain-specific modeling environment. GME is primarily a visual modeling tool (although textual model elements are also supported). GME is equipped with a metaprogram that configures it to behave as the metamodeling tool: it understands a UML-like notation (the metamodeling language), and an associated translator program can generate the metaprogram from the metamodel. The GME metamodeling approach is based on the use of stylized UML class diagrams and Object Constraint Language (OCL) constraints [14]. These metamodels capture the abstract syntax and well-formedness rules of the modeling language. Abstract syntax defines the set of

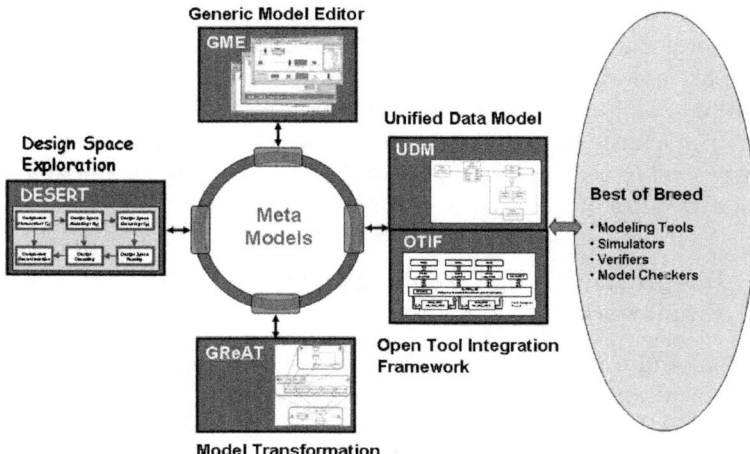

Fig. 20.1. The MIC Tool Suite

concepts, their attributes, and relationships one can use for building models in the language. For example, in a control system design language that supports event-driven control, the abstract syntax includes concepts like "states", "events", and "finite state machine", etc., relationships like "transitions", and attributes like "guard expression", "initial state", etc. The well-formedness rules of a language formally describe the constraints that the models need to satisfy in order to be syntactically correct.

20.2.2 Transforming the Models: UDM and GReAT

GME is a general purpose modeling environment, and it provides a set of Application Programming Interfaces (API-s) to access models. These low-level programmatic interfaces allow building software tools using traditional languages that access and possibly manipulate models. As a higher-level, more formal alternative to the API-s we have created tools that allow structured access to models on one hand, and allow the transformation of those models into other kinds of objects on the other hand. The first step is facilitated by the tool called "Universal Data Model" (UDM), and the second is done using the "Graph Rewriting and Transformation" (GReAT) language.

UDM is a metaprogrammable software tool [15] that generates domain-specific classes and API-s to access the models within GME, in XML form, in ODBC-based data-bases. The advantage of using UDM is that tools that access models could be developed using the concepts from the domain-specific modeling language (e.g. "Assembly", and "TimeTrigger"), instead of the generic concepts of GME. GReAT is a (graphical) modeling language for describing transformations on models [16]. The transformation specification is built upon metamodels of the input and the target of the transformations, and is expressed with the help of sequenced rewriting (or transformation) rules. The key point here is that both the input and the target must have a defined metamodel (i.e. abstract syntax with well-formedness rules). Often target models use some lower level modeling language, like a modeling language of simple finite transition systems. Note that in the ultimate, a target metamodel may represent the instruction set of a (real or virtual) machine. In practice, target metamodels often consist of concepts that correspond to code patterns (e.g. while-loop) that are instantiated with the values of attributes of the concept instances.

20.2.3 Integrating Design Tools: The Open Tool Integration Framework

The MIC toolsuite is often used to build not only a single tool, but tool chains consisting of various modeling, analysis, and generation tools, where many tools could be non-MIC tools. In this case one faces a tool integration problem: namely, how to construct integrated tool chains from tools that were not designed to work together. The MIC toolsuite includes a framework, called Open Tool Integration Framework (OTIF) that supports the construction of such integrated tool chains [17]. The tool integration problem that OTIF provides a tool for is as follows. In an engineering workflow various design tools are used, and the data ("models") need to be exchanged between the design tools. Each design tool has its own format (i.e. DSML) for storing models. The workflow implicitly specifies an ordering among the tools, and the direction of "model flow" defines the producer/consumer relationships between specific pairs of tools. We also assume that models are available in a packaged, "batch" form. OTIF provides a skeleton architecture for building tool chains that follow this model. OTIF has been implemented as a set of components, some of which are metaprogrammable, and it relies on the UDM and GReAT tools. It has been used to construct a number of tool chains consisting of MIC and other tools.

20.2.4 Design Space Exploration

When large-scale systems are constructed, in the early design phases it is often unclear what implementation choices could be used in order to achieve the required performance. In embedded systems, frequently multiple implementations are available for components (e.g. software on a general purpose processor, software on a DSP, FPGA, or an ASIC), and it is not obvious how to make a choice, if the number of components is large. Another meta-programmable MIC tool can assist in this process. This tool is called DESERT (for Design Space Exploration Tool) [18]. DESERT expects that the DSML allows the expression of alternatives for components in a complex model. A model, with hierarchically layers alternatives, defines a design space. Once a design spaces is modeled, one can attach applicability conditions to the design alternatives. These conditions are symbolic logical expressions that express when a particular alternative is to be chosen. Conditions could also link alternatives in different components via implication. One example for this feature is: "if alternative A is chosen in component C1, then alternative X must be chosen in component C2". During the design process, engineers want to evaluate alternative designs, which are constrained by high-level design parameters like latency, jitter, power consumption, etc. DESERT provides an environment in which the design space can be rapidly pruned by applying the constraints, thereby restricting the applicable aletrnatives.

20.3 Application Example: Vehicle Control Platform

The MIC tools have been used in numerous projects, and various tool chains have been constructed using it. Application experience includes large systems engineering projects such as the International Space Station diagnosability analysis [5], automotive manufacturing execution systems deployed in major plants [19] and prototypes for integrated design environments [12]. In this section we describe a tool chain that illustrates how the metaprogrammable tools have been used to solve the construction and integration of non-trivial tool architecture for embedded control applications. This is called the Vehicle Control Platform (VCP) tool chain [20] and it was built for constructing vehicle control software. The design flow starts with specifying controller components in the form of behavioral models, using Simulink/Stateflow.

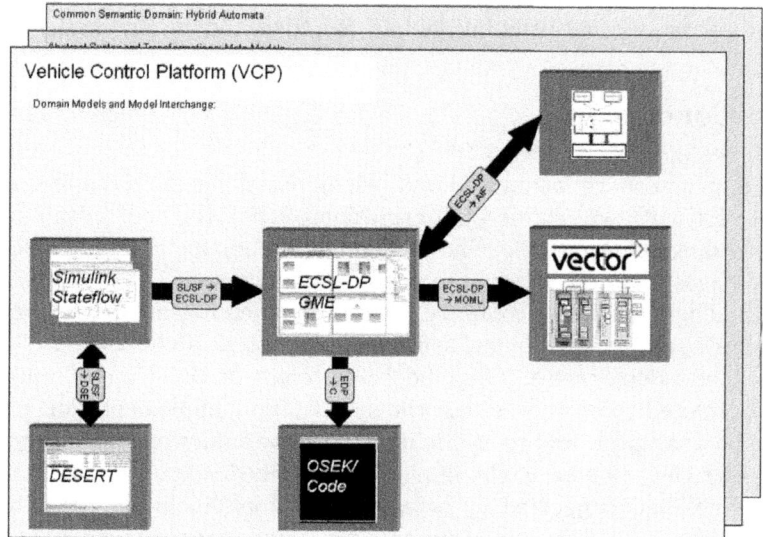

Fig. 20.2. The VCP Tool Suite

This step primarily consists of building up a library of controller blocks. The next step is design space modeling, which happens in a DSML called ECSL-DP, and which is supported by GME. During the construction of the design space models, the designer constructs hierarchical designs for the controllers, with possible alternative implementations on various levels of the hierarchy. The designer specifies component structures and component interactions. Note that the elementary components are from the behavioral models built in the first step. Once the design space modeling is finished, the designer can explore alternative designs with the help of DESERT. This stage will result in specific point design(s) that satisfy all design parameters. The specific designs are also captured in ECSL-DP. ECSL-DP has provisions for mapping designs into distributed electronic control units (ECU-s) and buses in the vehicle, and this mapping is specified in the models. Once the design models are finished a number of analysis steps can take place. Here we mention two: one can perform a schedulability analysis using a tool called AIRES [21], or one can perform a behavioral simulation using tools from the Vector toolsuite [22]. Note that this behavioral simulation on the ECSL-DP models is an alternative to the simulation of Simulink/Stateflow models and it can potentially be more accurate because of the finer details ECSL-DP captures. The result of these analyses (e.g. end-to-end latency in the system) can be annotated back into the ECSL-DP models. Finally, executable code (in C) is generated that runs on the real-time operating system OSEK. Figure 20.2 shows the high-level architecture and workflow in the tool chain. For the tool chain we have built the ECSL-DP modeling tool using GME, created various tool adaptors, and built a number of model transformation tools using GReAT. The five model translators contain, on average, 50 transformation rules, and process practical models with acceptable speed: 1-2 minutes, maximum. These model translators are automatically invoked as and when they are needed by OTIF.

20.4 Conclusion

In this paper we have introduced and briefly described the metaprogrammable toolsuite for MIC. We showed the evolution of the MIC tool suite and its use in a wide range of domain-specific tool chains supporting complex design flows. In each stages of a design flow, the actual state of the design is expressed using a DSML. These languages comprise the required heterogeneous abstractions for expressing controller dynamics, software and system architecture, component behavior, and deployment. The models expressed in these DSMLs need to be precisely related to each other via the specification/implementation interfaces, need to be analyzable and their fidelity need to be sufficiently precise to accurately predict the behavior of the implemented embedded controller. In addition, the design flow is supported by heterogeneous tools including modeling tools, formal verification tools, simulators, test generators, language design tools, code generators, debuggers, and performance analysis tools must all cooperate to assist developers and engineers struggling to construct the required systems. If the DSMLs are only informally specified then mismatched tool semantics may introduce mismatched interpretations of requirements, models and analysis results. This is particularly problematic in the safety critical real-time and embedded systems domain, where semantic ambiguities may produce conflicting results across different tools. Our current efforts focus on the formal, transformational specification of structural [23] and behavioral [24] semantics for DSMLs.

References

[1] Brooks Jr., F.P.: No silver bullet: Essence and accidents of software engineering. IEEE Computer Magazine, 10–19 (April 1987)
[2] Harel, D.: Biting the silver bullet. IEEE Computer Magazine, 8–19 (January 1992)
[3] Sztipanovits, J., Karsai, G., Biegl, C.: Graph model based approach to the representation, interpretation and execution of real time signal processing systems. International Journal of Intelligent Systems 3(3), 269–280 (1988)
[4] Karsai, G., Sztipanovits, J., Franke, H., Padalkar, S., DeCaria, F.: Model-embedded on-line problem solving environment for chemical engineering. In: Proceedings of the International Conference on Engineering of Complex Computer Systems, Ft. Lauderdale, FL, pp. 361–368 (November 1995)
[5] Misra, A., Sztipanovits, J., Underbrik, A., Carnes, R., Purves, B.: Diagnosability of dynamical systems. In: Third Intenational Workshop on Principles of Diagnosis, Rosario, Orcas Island, WA, pp. 239–244 (May 1992)
[6] Abbott, B., Bapty, T., Biegl, C., Karsai, G., Sztipanovits, J.: Model-based software synthesis. IEEE Software, 42–53 (May 1993)
[7] Sztipanovits, J., Wilkes, D., Karsai, G., Lynd, L.: The multigraph and structural adaptivity. IEEE Transaction on Signal Processing 41(8), 2695–2716 (1993)
[8] Sztipanovits, J., Karsai, G., Biegl, C., Bapty, T., Ledeczi, A., Malloy, D.: Multigraph: An architecture for model-integrated computing. In: Proceedings of the International Conference on Engineering of Complex Computer Systems, Ft. Lauderdale, FL, pp. 361–368 (November 1995)
[9] Sztipanovits, J., Karsai, G.: Model-integrated computing. IEEE Computer 22(5), 110–112 (1997)
[10] Nordstrom, G., Sztipanovits, J., Karsai, G.: Meta-level extension of the multigraph architecture. In: Engineering of Computer-Based Systems Conference, Jerusalem, Israel, pp. 61–68 (May 1998)
[11] Nordstrom, G., Sztipanovits, J., Karsai, G., Ledeczi, A.: Metamodeling - rapid design and evolution of domain-specific modeling environments. In: Proceedings of the IEEE ECBS 1999 Conference, Nashville, TN, pp. 68–74 (April 1999)
[12] ISIS: Mic tool distribution
[13] Ledeczi, A., Bakay, A., Maroti, M., Volgyesi, P., Nordstrom, G., Sprinkle, J.: Composing domain-specific design environments. IEEE Computer Magazine, 44–51 (November 1997)
[14] Object Management Group: UML 2.0 OCL Specification (2003)
[15] Bakay, A.: The udm framework
[16] Karsai, G., Agrawal, A., Shi, F.: On the use of graph transformations for the formal specification of model interpreters. Journal of Universal Computer Science 9(11), 1296–1321 (2003)
[17] Karsai, G., Lang, A., Neema, S.: Design patterns for open tool integration. Journal of Software and System Modeling 4(1) (2004)
[18] Neema, S., Sztipanovits, J., Karsai, G., Butts, K.: Constraint-based design space exploration and model synthesis. In: Alur, R., Lee, I. (eds.) EMSOFT 2003. LNCS, vol. 2855, pp. 290–305. Springer, Heidelberg (2003)
[19] Earl, L., Amit, M., Janos, S.: Increasing productivity at saturn. IEEE Computer Magazine, 35–44 (August 1998)

[20] Porter, J., Karsai, G., Volgyesi, P., Nine, H., Humke, P., Hemingway, G., Thibodeaux, R., Sztipanovits, J.: Towards model-based integration of tool and techniques for embedded control system design, verification, and implementation. In: Chaudron, M.R.V. (ed.) Models in Software Engineering. LNCS, vol. 5421, pp. 20–34. Springer, Heidelberg (2009)

[21] Zonghua, G., Wang, S., Kodase, S., Shin, G.K.: An end-to-end tool chain for multi-view modeling and analysis of avionics mission computing software. In: 24th IEEE International Real-Time Systems Symposium (RTSS 2003), Cancun, Mexico (September 2003)

[22] Vector Informatik Group: The vector tools

[23] Jackson, E., Sztipanovits, J.: Formalizing the structural semantics of domain-specific modeling languages. Journal of Software and Systems Modeling (2009) (to appear)

[24] Chen, K., Sztipanovits, J., Neema, S.: Compositional specification of behavioral semantics. In: Design, Automation, and Test in Europe: The Most Influential Papers of 10 Years DATE, pp. 253–256 (April 2008)

21 Application of Quality Standards to Multiple Artifacts with a Universal Compliance Solution

Tibor Farkas[1], Torsten Klein[2], and Harald Röbig[2]

[1] Fraunhofer Institute FOKUS, Kaiserin-Augusta-Allee 31, 10589 Berlin, Germany
tibor.farkas@fokus.fraunhofer.com
[2] Carmeq GmbH, Carnotstrasse 4, 10587 Berlin, Germany
{torsten.klein,harald.roebig}@carmeq.com

Abstract. For standards compliance achievement in model-based engineering of embedded real-time systems, model analyzers and code checkers are constituted in early development phases to lower error rates and to eliminate time-consuming quality reviews. However, solutions available today only address a single modeling language and examinations are localized to specific development environments. Fulfilling more advanced traceability examinations required by procedural and technical quality standards, compliance checking has to be applicable across different modeling tools and development workflows. Furthermore it should cover correlation analyses that include model-to-model, model-to-file and model-to-database comparisons on multiple artifacts. This chapter introduces a novel compliance solution Assessment Studio that supports universal guideline and traceability checking with automated analyses in multi-domain modeling environments. MESA, our meta-modeling approach for guideline checking, was enhanced to support multiple meta-models with associated artifacts. Therefore, we use a XML-based transformation and proof mechanism by automatically executing rules written in LINQ, adding auto-correction and metrics measurement capabilities. Several case studies demonstrate the feasibility of this approach at Volkswagen.

Keywords: Automated Review, Compliance, Traceability, Guideline Checking, Assessment, Metrics, LINQ, XML, CMMI, MISRA, AUTOSAR.

21.1 Introduction

In today's automotive global market, the pressures facing industrial and technology standards compliance is more important than ever. In recent years many automotive manufacturers have passed laws and regulations that require embedded systems engineers to adopt certain new standards and policies into their model-based product development for more transparency, better interoperability and standards compliance [1]. Vehicle functions are designed in workflows across interdisciplinary development groups whereas engineering is mostly based on a vast number of artifacts: As depicted in [2], rich vehicle function requirements are managed with DOORS, conceptual designs are modeled in Unified Modeling Language (UML) or Systems Modeling Language (SysML), algorithmic control is modeled for simulation and code generation with MATLAB/Simulink/Stateflow (ML/SL/SF) or on a physical plane with ASCET-MD and finally, huge test specifications are determined with tools like Microsoft Excel or Classification Tree Editor XL (CTE). In model-based engineering of embedded real-time systems compliance to local and global quality requirements will become more important for car manufacturers (OEM) and its suppliers [3]. At the global level, there are two primary types of compliance: Procedural and technical [4]. Procedural guidelines generically describe 'what' has to be done (e.g. in ISO/IEC 61508 [5] or CMMI [6]) versus technical related guidelines such as MAAB [7], MISRA [1] or AUTOSAR [8] style guides that give instructions 'how' the quality requirements have to be realized. Not fulfilling such compliance standards could have bad impacts on traceability in workflows.

Compliance targets different process levels in the V-Model. Nevertheless, being able to achieve compliance to procedural and technical standards on different development stages and tools is very challenging. Guidelines and policies are written mostly in an individual textual representation with a different kind of graphical illustrations included [7, 1, 8]. Paper documents have considerable disadvantages: Complex models with more than thousands of model elements – a model size that is easily reached today – cannot be feasibly checked by a human. Today in a model-based development only insufficient automatic verifiability is possible, because multiple models, files, glossaries and test specifications are involved. Especially a formal notation is missing that could be used for automated examinations. Automated quality review tools available today are dedicated to one specific development environment or target only a unique modeling language [9, 10]. However, for traceability checks comparisons on multiple artifacts like model-to-model, model-to-file or model-to-database are needed. Using a different checker for each tool environment or development stage makes compliance achievements ineffective and opaque.

21.2 Idea: Meta-modeling for Constraint Definition

Our idea was to have a tool independent compliance achievement that includes multiple artifacts in correlation with traceability analysis and furthermore allows comparisons across workflows. Therefore, our approach was to formalize narrative rule descriptions to computable expressions. This was introduced in MESA [2] with the approach to adopt concepts of the Model-Driven Architecture (MDA). MDA uses open standards for meta-modeling like Meta-Object Facility (MOF) and the Object Constraint Language (OCL) for constraint definition by executable expressions. A common tool-chain at Volkswagen (VW) was selected (DOORS, ML/SL/SF, CTE) to exemplify the feasibility. Practical use of automated guideline examination and constraint violation analysis was demonstrated through a software prototype (Automotive Software Development

Rule Checker) to achieve VW modeling guidelines based on Mathworks Automotive Advisory Board style guide (MAAB). Artifact specific meta-models were developed for DOORS, ML/SL/SF, CTE and a central meta-model repository was build to show, how conformance checking with OCL could be constituted on a multitude of different development artifacts stored to that repository as described in [2, 4]. This chapter introduces Assessment Studio [11], an enhanced approach for automated compliance achievement and model assessment based on the general idea of MESA (Fig. 1 shows an example). In difference to our former work, Assessment Studio supports multiple-check languages and file-based examinations, if a global artifact repository is not present. Therefore, we investigated for distributed environments a direct artifact meta-model transformation to a common file-basis using Extensible Markup Language (XML). Furthermore, an utilization of different languages for checking and auto-correction (e.g. M-script, DXL-script, VBScript) for various artifact and model examinations is targeted. Implied is a new duality concept that supports universal constraint checking with LINQ (Language INtegrated Query [12]) to handle multiple artifact assessments. Via integrated tool-adapters dedicated scripting languages are also supported in Assessment Studio.

21.3 Approach: Universal Compliance Achievement

A first idea to automate rule checking is in tool programming of check code using a scripting language supported by the development environment itself. This could be M programming with ML/SL/SF or DXL scripting with DOORS, for instance. Scripts would be then executed with eventually available batch programs, such as [9, 10]. Problematic is the case when common tool environments do not offer such an integrated programming environment – like ASCET-MD, Excel or CTE or if traceability is needed between tools (e.g. references between requirements to a model). Redundant and ineffective scripts would be possible if multiple environments should be addressed in an script-based examination. Solutions that focus on the application of a single modeling language are depended and limited to their environments. Affected and related artifacts are not included into checks. Therefore, this delimited scripting approach seems feasible for some reasons, but is not sustainable for continuous workflows in model-driven development. In our approach independence of a vendor specific tool environment is necessary. This is provided by using artifact meta-models that are transformed to XML as open file format. Each developed artifact has an in-memory meta-model representation of its information structure, when working with it. Technically, this is a tree structured schema – an abstract collection of meta-data – consisting of a set of components: Element and attribute declarations and complex and simple type definitions. These components are usually created by processing and managing documents, code or models, which contain the source language definitions of these components. XML itself has achieved tremendous adoption as a basis for structuring data, whether with XML Metadata Interchange (XMI), AUTOSAR ECU/SWC definitions, in numerous ASAM standards or simply as data-dumps exported from databases. Some tools already use a structured XML file format like UML/SysML tools, AUTOSAR development environments or CTE. Other tools support exporting their in-memory model data to XML structured files like ASCET-MD or Microsoft Office applications. For environments that originally do not support XML persistence, mostly third party converter programs are available on the market like SimEx [13] for ML/SL/SF or EXERPT [14] for DOORS.

We make an assertion, that an artifact is an instance of the tool specific meta-model in general, containing specific data that could be stored into a XML file. Besides, data in this file contains also tool specific information (e.g. settings for simulation or compilation). Having all requirements data, model data and test specification data as a XML representation, we have to choose a global and tool independent programming language for the definition of executable rules on a high abstraction level. OCL was used in MESA to formalize guidelines as executable expressions (rules) with OSLO [15] and to show, how textual guidelines could be transformed into a more formal specification (descriptive check algorithm). This was feasible for value checks on attributes or uniformity of properties (e.g. *Object.BackgroundColor = white*). Due more complex rules we investigated that large OCL constructs are hard to read by a human. Furthermore, the capability of the OCL interpreter OSLO was limited to basic equations and operations of first order logic. More advanced checks required enhanced functionality that is specialized for querying and modifying, grouping, joining and managing object collections. In addition, we needed regular expression evaluation for fast pattern matching, compact control statements (e.g. *while, for each, if-then-else*) for traversing large object trees and rule packaging (library of rules) capabilities.

The following guideline for ASCET-MD prescribes that all basic operators such as addition, subtraction, multiplication and division should only have two ingoing operands. The code excerpt introduces an algorithm written in LINQ. It was constituted for the correct and coherent modeling of algorithms.

Example: LINQ expression of an executable guideline.

```
string[] arithOperators = { "+", "-", "*", "/" };

var Operators =
from Obj in Artefact["ASCET-Model"].Descendants("Operator")
 where arithOperators.Contains(Obj.Attribute("type").Value) &&
 Obj.Descendants("ReturnPort").First().Elements().Count() > 2
select Obj;

if(Operators.Count() > 0) Result = "FAIL"; else Result = "PASS";
```

(Excerpt taken from instructions in MISRA modeling guide for ASCET-MD [11].)

First a set of basic operands are defined in an array *arithOperators*. Next the *Artefact*-statement references the corresponding model file (ASCET-Model.xml) and loads an instance into memory. According to the ASCET meta-model the *from*-clause selects all nested child elements from operator descendants and defines a general query named *Operators*. In the **where**-clause multiple conditions could be defined for our guideline. In this guideline, first of all only the basic operands are filtered with a comparison of two quantities from all operands inside the model. The second condition selects all operands where the amount of ingoing ports (named *ReturnPort* in ASCET meta-model) exceeds two. Finally the **select**-clause gets all those defected elements that have more than two operands and store them into our *Operators* collection. Therefore, if the *Operators* contains defected elements, the guideline check result is set to *fail* else to *pass*. To allow guideline checks including multiple artifacts, in one expression multiple queries could be defined on different artifacts. After gathering all required information from different artifacts with multiple queries, we could join them, compare them, group them etc. for different examinations. Because those larger examples exceed our space in this chapter, we only show an example for one artifact. LINQ to XML, a sub-component of the LINQ project [12], aims to address our demands. LINQ to XML is a modernized in-memory XML programming API designed to take advantage of the latest .NET Framework 3.5 language. It provides both DOM and XQuery/XPath like functionality in a consistent programming language (C# syntax) across the different data access technologies. Therefore, we used LINQ instead of OCL in our new approach to achieve enhanced capabilities for examination. Similar to Structured Query Language (SQL), LINQ defines a set of query operators for objects that can be used to query, group and filter object oriented data stored in arrays, collections or class hierarchies. It operates on object trees, XML files and relational databases.

21.4 Case Studies: Compliance with Modeling Standards

With Assessment Studio, we developed an extension mechanism that implements a specialized and dual query processing engine that executes LINQ queries on multiple transformed (XML files) artifacts for traceability checks and comparisons on the one side and executes dedicated script queries (M-, DXL-, VB-Scripts) over tool API's for auto-correction on the other side. Furthermore, we defined a methodology that translates check results to a defined achievement of failure categories and prepare them in a pie chart for later assessments.

In our cases we implemented MISRA, MAAB and AUTOSAR basic compliance achievements in model-based engineering of embedded software. In addition, most companies use enhanced or customized guidelines on top of basics guidelines. We used those standards as our basis to transform depicted guidelines into executable rule queries with LINQ. For automated compliance adherence, query operations have to be executed by a query processing engine. Such an engine was developed with Assessment Studio for the descriptive usage of LINQ. Furthermore, it contains integrated modules for reporting, analysis, rule management (authoring and versioning) and electronic documentation.

In addition to automotive standards compliance checks we defined key performance indicators (metrics) written as LINQ expression that define a set of values used to measure artifacts, like efficiency measurements of model-based

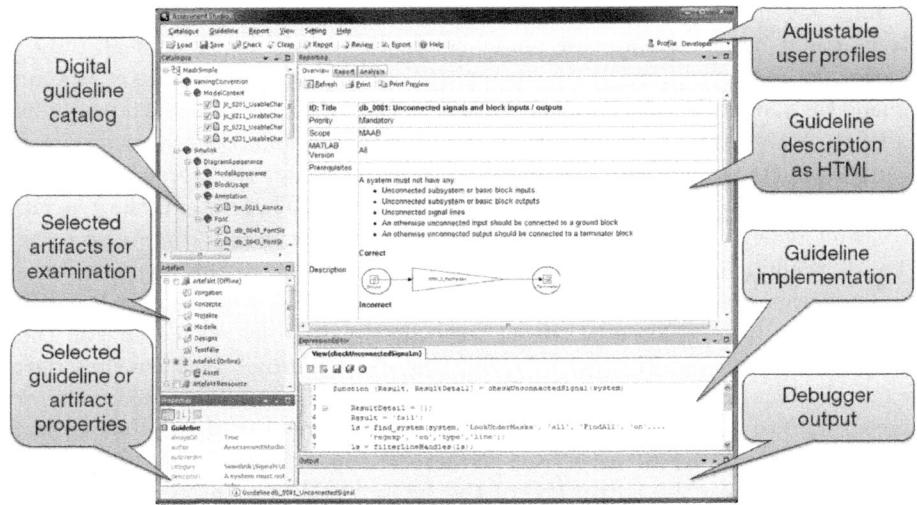

Fig. 21.1. Assessment Studio Overview

designs. The measurement is done by the query processing engine that summarizes information, such as: (a) Quantitative indicators which can be presented as a number, (b) Practical indicators that interface with existing company processes, (c) Directional indicators specifying whether an organization is getting better or not, (d) Actionable indicators are sufficiently in an organization's control to effect change. In the end comparisons were carried out for relevant characteristics [16] in a quality management system: Process assessment, prospective development, conciseness, clarity, correctness, consistency, completeness and predictions. In several case studies Assessment Studio (Fig. 1) enabled developers to use a more open and generic proof mechanism for standards compliance achievement on model-level by checking rules, using auto-correction features (in ML/SL/SF with M-script execution) and measuring metrics for efficiency to monitor and document quality goals within several development stages. We have developed conformance checks for MISRA, MAAB, CTE and AUTOSAR compliance in model-based engineering in conjunction with requirements management with DOORS (based on RIF), with models such as UML/SysML, ML/SL/SF and ASCET-MD and for test cases specifications made with Microsoft Excel spreadsheets or classification trees with CTE.

21.5 Conclusion

Organizations implement internal procedural and technical standards so they are able to operate more efficiently. The question of adhering to standards conformance (technical compliance) is becoming increasingly challenging for automotive businesses. Compliance adherence will increase further over time because of the adoption of common standards (e.g. AUTOSAR). Typically, very specialized quality assurance solutions in model-based engineering of embedded real-time systems are used by companies that are seeking certification with standards bodies such as the International Standards Organization (ISO), or that need to comply with requirements for special issues, like safety issues in software code (e.g. MISRA). Code checking is done today, but is not sufficient due to an increased use of model-based development and mutual relations to process related quality standards. Moving more and more from code towards a software-model in embedded real-time systems engineering – beginning at requirements specification, over functional design and simulation down to the code development – the overall quality compliance at model-level is gaining more importance.

Ensuring standards compliance in model-based engineering is a difficult task – especially, because besides a model other related files are involved, such as specification documents, data files, glossaries and test data. They all have to be considered in an examination or for traceability checks. Another tangible reason for early quality compliance achievement is that a model serves as specification and as source for auto-code generation. Of course, not every conformance check could be automated. However, reducing most of time consuming manual artifact reviews for standards conformance saves immediately costs and enhances product quality at the end.

In model-based engineering quality assurance and standards compliance software helps organizations with code checkers and model analyzers adhere to existing and emerging safety and security regulations. Unfortunately, automated quality review solutions available today are dedicated to one specific development environment or targeting even just one unique modeling language. Using different kind of checker tools for each development environment or development stage makes compliance achievements ineffective, reports ambiguous and opaque.

Today universal automatisms for ensuring the standards conformance of multiple artifacts by guideline checking and especially executable and tool independent rules for quality assurance are missing. Therefore the intention of this chapter was to present an overall guideline checker solution Assessment Studio for embedded software and systems engineering based on former research work MESA. Assessment Studio offers development groups a universal mechanism for quality assurance and standards compliance by automatically checking rules on XML-basis, using auto-correction via tool-adapters and measuring key performance indicators to monitor and document quality goals in all development stages.

With the presented approach errors and defects could be avoided during the development process. Further it is possible to formalize executable rules in LINQ that concern more than one tool and check them automatically in correlation and for comparisons. Additionally it was stated, that it also could be used for the evaluation of quality metrics and measurements to achieve estimations. In this way, automotive manufacturers and its suppliers can automate reviews globally, better track and report information to regulatory bodies, passing quality audits and reducing the risk of non-compliance.

References

[1] The Motor Industry Software Reliability Association: MISRA-C: Guidelines for the Use of the C Language in Vehicle Based Systems. MISRA-C (1998)

[2] Farkas, T., Röbig. H.: Automatisierte, werkzeugübergreifende richtlinienprüfung zur unterstützung des automotive-entwicklungsprozesses. In: Conrad, M., Giese, H., Rumpe, B., Schätz, B. (eds.) Proceedings of Model Based Engineering of Embedded Systems III (MBEES III), Dagstuhl, Germany, TU Braunschweig Report TUBS-SSE 2007-01 (2007)

[3] Mack, M.: Ascet autocode ensures misra conformance, etas, http://www.etas.com

[4] Rech, J., Bunse, C.: Quality Improvement in Automotive Software Engineering using a Model-Based Approach. In: Model-Driven Software Development Integrating Quality Assurance (Premier Reference Source). Idea Group Publishing (2008)

[5] ISO 61508, I.O.f.S..I.E.C.: IEC-61508 Functional safety of electrical/electronic/programmable electronic safety-related system. IEC (1998)

[6] Kneuper, R.: CMMI: Verbesserung von Softwareprozessen mit Capability Maturity Model Integration. dPunkt Verlag, Heidelberg (2006)

[7] MAAB The MathWorks Automotive Advisory Board: Control Algorithm Modeling Guidelines Using MATLAB, Simulink, and Stateflow, Version 2.0, http://www.mathworks.com/industries/auto/maab.html

[8] AUTOSAR: Automotive open system architecture specification, deliverables: Applying simulink to autosar; applying ascet to autosar, http://www.autosar.org

[9] The MathWorks, Simulink Verification and Validation Toolbox, http://www.mathworks.de/products/simverification

[10] Stürmer, I., Kreuz, I., Schäfer, W., Schürr, A.: The mate approach: Enhanced simulink and statfelow model transformation. In: Proceedings of MathWorks Automotive Conference, Dearborn (MI), USA (June 2007)

[11] Match Technologies, Assessment Studio, http://www.match-technologies.com

[12] Microsoft Developer Network: The LINQ Project, http://msdn.microsoft.com/en-us/netframework/aa904594.aspx

[13] IT-Power Consultants, SimEx - Konvertierung von Simulink/Stateflow Modelldaten in das XML-Format, http://www.itpower.de

[14] EXTESSY, EXCERPT - Extessy Engineering Requirements Platform, http://www.extessy.com

[15] OSLO Open Source Library for OCL, http://oslo-project.berlios.de

[16] International Organization for Standardization: ISO 9126: Software Engineering - Product quality, Part 1-4. International Organization for Standardization (2007)

Author Index

Bapty, Ted   369
Buckl, Christian   271

Chen, DeJiu   345
Cuenot, Philippe   297

Derler, Patricia   107
Dumoulin, Cédric   361

Espinoza, Huascar   129

Farcas, Claudiu   155
Farcas, Emilia   155
Farkas, Tibor   377
Feilkas, Martin   317
Feng, Lei   345
Freund, Ulrich   355
Frey, Patrick   297

Geburzi, Anne   329
Gérard, Sébastien   129, 361
Giese, Holger   3, 17

Henkler, Stefan   309
Hirsch, Martin   309
Hölzl, Florian   317
Huhn, Michaela   201
Hungar, Hardi   201

Johansson, Rolf   297

Karsai, Gabor   57, 369
Klar, Felix   337
Klein, Torsten   377
Knoll, Alois   271
Krüger, Ingolf   155

Legros, Elodie   323
Levendovszky, Tihamer   241
Lönn, Henrik   297
Lund, Mass Soldal   77

Menarini, Massimiliano   155

Naderlinger, Andreas   107
Neema, Sandeep   369
Neumann, Stefan   17
Niggemann, Oliver   17, 329

Papadopoulos, Yiannis   297
Persson, Magnus   345
Pree, Wolfgang   107
Priesterjahn, Claudia   309

Qureshi, Tahir Naseer   345

Refsdal, Atle   77
Reiser, Mark-Oliver   297
Resmerita, Stefan   107
Röbig, Harald   377
Rumpe, Bernhard   57, 241

Sandberg, Anders   297
Schäfer, Wilhelm   309, 323
Schätz, Bernhard   3, 17, 241
Schieferdecker, Ina   271
Schürr, Andy   323, 337
Selic, Bran   129, 361
Servat, David   297
Sprinkle, Jonathan   57, 241
Stølen, Ketil   77
Stroop, Joachim   329
Stürmer, Ingo   323
Sztipanovits, Janos   369

Tavakoli Kolagari, Ramin   297
Terrier, François   129
Tessier, Patrick   361
Tichy, Matthias   309
Törngren, Martin   297, 345

Vangheluwe, Hans   57

Weber, Matthias   297
Weisemöller, Ingo   337

Zander, Justyna   271

GPSR Compliance

The European Union's (EU) General Product Safety Regulation (GPSR) is a set of rules that requires consumer products to be safe and our obligations to ensure this.

If you have any concerns about our products, you can contact us on ProductSafety@springernature.com

In case Publisher is established outside the EU, the EU authorized representative is:

Springer Nature Customer Service Center GmbH
Europaplatz 3
69115 Heidelberg, Germany

Batch number: 09478804

Printed by Printforce, the Netherlands